INSTRUCTOR'S SOLUTIONS MANUAL

A First Course in Abstract Algebra

Seventh Edition

JOHN B. FRALEIGH
University of Rhode Island

Addison
Wesley

Boston San Francisco New York
London Toronto Sydney Tokyo Singapore Madrid
Mexico City Munich Paris Cape Town Hong Kong Montreal

ISBN   0-321-17340-6

1 2 3 4 5 6 7 8 9 10 EP 05 04 03 02

CONTENTS

VI. Extension Fields

VII. Advanced Group Theory

VIII. Groups in Topology

IX. Factorization

X. Automorphisms and Galois Theory

Preface

This manual contains solutions to all exercises in the text, except those odd-numbered exercises for which fairly lengthy complete solutions are given in the answers at the back of the text. Then reference is simply given to the text answers to save typing.

I prepared these solutions myself. While I tried to be accurate, there are sure to be the inevitable mistakes and typos. An author reading proof rends to see what he or she wants to see. However, the instructor should find this manual adequate for the purpose for which it is intended.

Morgan, Vermont   J.B.F
July, 2002

0. Sets and Relations

1. $\{\sqrt{3}, -\sqrt{3}\}$ **2.** The set is empty.

3. $\{1, -1, 2, -2, 3, -3, 4, -4, 5, -5, 6, -6, 10, -10, 12, -12, 15, -15, 20, -20, 30, -30,$
$60, -60\}$

4. $\{-10, -9, -8, -7, -6, -5, -4, -3, -2, -1, 0, 1, 2, 3, 4, 5, 6, 7, 8, 9, 10, 11\}$

5. It is not a well-defined set. (Some may argue that no element of $\mathbb{Z}^+$ is large, because every element exceeds only a finite number of other elements but is exceeded by an infinite number of other elements. Such people might claim the answer should be $\varnothing$.)

6. $\varnothing$ **7.** The set is $\varnothing$ because $3^3 = 27$ and $4^3 = 64$.

8. It is not a well-defined set.   **9.** $\mathbb{Q}$

10. The set containing all numbers that are (positive, negative, or zero) integer multiples of 1, 1/2, or 1/3.

11. $\{(a, 1), (a, 2), (a, c), (b, 1), (b, 2), (b, c), (c, 1), (c, 2), (c, c)\}$

12. a. It is a function. It is not one-to-one since there are two pairs with second member 4. It is not onto B because there is no pair with second member 2.

 b. (Same answer as Part(**a**).)

 c. It is not a function because there are two pairs with first member 1.

 d. It is a function. It is one-to-one. It is onto B because every element of B appears as second member of some pair.

 e. It is a function. It is not one-to-one because there are two pairs with second member 6. It is not onto B because there is no pair with second member 2.

 f. It is not a function because there are two pairs with first member 2.

13. Draw the line through P and x, and let y be its point of intersection with the line segment CD.

14. a. $\phi : [0, 1] \to [0, 2]$ where $\phi(x) = 2x$ **b.** $\phi : [1, 3] \to [5, 25]$ where $\phi(x) = 5 + 10(x - 1)$

 c. $\phi : [a, b] \to [c, d]$ where $\phi(x) = c + \frac{d-c}{b-a}(x - a)$

15. Let $\phi : S \to \mathbb{R}$ be defined by $\phi(x) = \tan(\pi(x - \frac{1}{2}))$.

16. a. $\varnothing$; cardinality 1   **b.** $\varnothing, \{a\}$; cardinality 2   **c.** $\varnothing, \{a\}, \{b\}, \{a, b\}$; cardinality 4

 d. $\varnothing, \{a\}, \{b\}, \{c\}, \{a, b\}, \{a, c\}, \{b, c\}, \{a, b, c\}$; cardinality 8

17. Conjecture: $|\mathcal{P}(A)| = 2^s = 2^{|A|}$.

 Proof The number of subsets of a set A depends only on the cardinality of A, not on what the elements of A actually are. Suppose $B = \{1, 2, 3, \cdots, s - 1\}$ and $A = \{1, 2, 3, \cdots, s\}$. Then A has all the elements of B plus the one additional element s. All subsets of B are also subsets of A; these are precisely the subsets of A that do not contain s, so the number of subsets of A not containing s is $|\mathcal{P}(B)|$. Any other subset of A must contain s, and removal of the s would produce a subset of B. Thus the number of subsets of A containing s is also $|\mathcal{P}(B)|$. Because every subset of A either contains s or does not contain s (but not both), we see that the number of subsets of A is $2|\mathcal{P}(B)|$.

We have shown that if A has one more element that B, then $|\mathcal{P}(A)| = 2|\mathcal{P}(B)|$. Now $|\mathcal{P}(\varnothing)| = 1$, so if $|A| = s$, then $|\mathcal{P}(A)| = 2^s$.

18. We define a one-to-one map ϕ of B^A onto $\mathcal{P}(A)$. Let $f \in B^A$, and let $\phi(f) = \{x \in A \mid f(x) = 1\}$. Suppose $\phi(f) = \phi(g)$. Then $f(x) = 1$ if and only if $g(x) = 1$. Because the only possible values for $f(x)$ and $g(x)$ are 0 and 1, we see that $f(x) = 0$ if and only if $g(x) = 0$. Consequently $f(x) = g(x)$ for all $x \in A$ so $f = g$ and ϕ is one to one. To show that ϕ is onto $\mathcal{P}(A)$, let $S \subseteq A$, and let $h : A \to \{0, 1\}$ be defined by $h(x) = 1$ if $x \in S$ and $h(x) = 0$ otherwise. Clearly $\phi(h) = S$, showing that ϕ is indeed onto $\mathcal{P}(A)$.

19. Picking up from the hint, let $Z = \{x \in A \mid x \notin \phi(x)\}$. We claim that for any $a \in A, \phi(a) \neq Z$. Either $a \in \phi(a)$, in which case $a \notin Z$, or $a \notin \phi(a)$, in which case $a \in Z$. Thus Z and $\phi(a)$ are certainly different subsets of A; one of them contains a and the other one does not.

Based on what we just showed, we feel that the power set of A has cardinality greater than $|A|$. Proceeding naively, we can start with the infinite set $\mathbb{Z}$, form its power set, then form the power set of that, and continue this process indefinitely. If there were only a finite number of infinite cardinal numbers, this process would have to terminate after a fixed finite number of steps. Since it doesn't, it appears that there must be an infinite number of different infinite cardinal numbers.

The set of everything is not logically acceptable, because the set of all subsets of the set of everything would be larger than the set of everything, which is a fallacy.

20. a. The set containing precisely the two elements of A and the three (different) elements of B is $C = \{1, 2, 3, 4, 5\}$ which has 5 elements.

i) Let $A = \{-2, -1, 0\}$ and $B = \{1, 2, 3, \cdots\} = \mathbb{Z}^+$. Then $|A| = 3$ and $|B| = \aleph_0$, and A and B have no elements in common. The set C containing all elements in either A or B is $C = \{-2, -1, 0, 1, 2, 3, \cdots\}$. The map $\phi : C \to B$ defined by $\phi(x) = x + 3$ is one to one and onto B, so $|C| = |B| = \aleph_0$. Thus we consider $3 + \aleph_0 = \aleph_0$.

ii) Let $A = \{1, 2, 3, \cdots\}$ and $B = \{1/2, 3/2, 5/2, \cdots\}$. Then $|A| = |B| = \aleph_0$ and A and B have no elements in common. The set C containing all elements in either A of B is $C = \{1/2, 1, 3/2, 2, 5/2, 3, \cdots\}$. The map $\phi : C \to A$ defined by $\phi(x) = 2x$ is one to one and onto A, so $|C| = |A| = \aleph_0$. Thus we consider $\aleph_0 + \aleph_0 = \aleph_0$.

b. We leave the plotting of the points in $A \times B$ to you. Figure 0.14 in the text, where there are $\aleph_0$ rows each having $\aleph_0$ entries, illustrates that we would consider that $\aleph_0 \cdot \aleph_0 = \aleph_0$.

21. There are $10^2 = 100$ numbers (.00 through .99) of the form .##, and $10^5 = 100,000$ numbers (.00000 through .99999) of the form .#####. Thus for .##### $\cdots$, we expect $10^{\aleph_0}$ sequences representing all numbers $x \in \mathbb{R}$ such that $0 \leq x \leq 1$, but a sequence trailing off in 0's may represent the same $x \in \mathbb{R}$ as a sequence trailing of in 9's. At any rate, we should have $10^{\aleph_0} \geq |[0, 1]| = |\mathbb{R}|$; see Exercise 15. On the other hand, we can represent numbers in $\mathbb{R}$ using any integer base $n > 1$, and these same $10^{\aleph_0}$ sequences using digits from 0 to 9 in base $n = 12$ would not represent all $x \in [0, 1]$, so we have $10^{\aleph_0} \leq |\mathbb{R}|$. Thus we consider the value of $10^{\aleph_0}$ to be $|\mathbb{R}|$. We could make the same argument using any other integer base $n > 1$, and thus consider $n^{\aleph_0} = |\mathbb{R}|$ for $n \in \mathbb{Z}^+, n > 1$. In particular, $12^{\aleph_0} = 2^{\aleph_0} = |\mathbb{R}|$.

22. $\aleph_0, |\mathbb{R}|, 2^{|\mathbb{R}|}, 2^{(2^{|\mathbb{R}|})}, 2^{(2^{(2^{|\mathbb{R}|})})}$ **23. 1.** There is only one partition $\{\{a\}\}$ of a one-element set $\{a\}$.

24. There are two partitions of $\{a, b\}$, namely $\{\{a, b\}\}$ and $\{\{a\}, \{b\}\}$.

25. There are five partitions of $\{a, b, c\}$, namely $\{\{a, b, c\}\}$, $\{\{a\}, \{b, c\}\}$, $\{\{b\}, \{a, c\}\}$, $\{\{c\}, \{a, b\}\}$, and $\{\{a\}, \{b\}, \{c\}\}$.

26. 15. The set $\{a, b, c, d\}$ has 1 partition into one cell, 7 partitions into two cells (four with a 1,3 split and three with a 2,2 split), 6 partitions into three cells, and 1 partition into four cells for a total of 15 partitions.

27. 52. The set $\{a, b, c, d, e\}$ has 1 partition into one cell, 15 into two cells, 25 into three cells, 10 into four cells, and 1 into five cells for a total of 52. (Do a combinatorics count for each possible case, such as a 1,2,2 split where there are 15 possible partitions.)

28. *Reflexive:* In order for $x \, \mathcal{R} \, x$ to be true, x must be in the same cell of the partition as the cell that contains x. This is certainly true.

Transitive: Suppose that $x \, \mathcal{R} \, y$ and $y \, \mathcal{R} \, z$. Then x is in the same cell as y so $\overline{x} = \overline{y}$, and y is in the same cell as z so that $\overline{y} = \overline{z}$. By the transitivity of the set equality relation on the collection of cells in the partition, we see that $\overline{x} = \overline{z}$ so that x is in the same cell as z. Consequently, $x \, \mathcal{R} \, z$.

29. Not an equivalence relation; 0 is not related to 0, so it is not reflexive.

30. Not an equivalence relation; $3 \geq 2$ but $2 \not\geq 3$, so it is not symmetric.

31. It is an equivalence relation; $\overline{0} = \{0\}$ and $\overline{a} = \{a, -a\}$ for $a \in \mathbb{R}, a \neq 0$.

32. It is not an equivalence relation; $1 \, \mathcal{R} \, 3$ and $3 \, \mathcal{R} \, 5$ but we do not have $1 \, \mathcal{R} \, 5$ because $|1 - 5| = 4 > 3$.

33. (See the answer in the text.)

34. It is an equivalence relation;

$$\overline{1} = \{1, 11, 21, 31, \cdots\}, \quad \overline{2} = \{2, 12, 22, 32, \cdots\}, \quad \cdots, \quad \overline{10} = \{10, 20, 30, 40, \cdots\}.$$

35. (See the answer in the text.)

36. **a.** Let h, k, and m be positive integers. We check the three criteria.
Reflexive: $h - h = n0$ so $h \sim h$.
Symmetric: If $h \sim k$ so that $h - k = ns$ for some $s \in \mathbb{Z}$, then $k - h = n(-s)$ so $k \sim h$.
Transitive: If $h \sim k$ and $k \sim m$, then for some $s, t \in \mathbb{Z}$, we have $h - k = ns$ and $k - m = nt$. Then $h - m = (h - k) + (k - m) = ns + nt = n(s + t)$, so $h \sim m$.

b. Let $h, k \in \mathbb{Z}^+$. In the sense of this exercise, $h \sim k$ if and only if $h - k = nq$ for some $q \in \mathbb{Z}$. In the sense of Example 0.19, $h \equiv k \pmod{n}$ if and only if h and k have the same remainder when divided by n. Write $h = nq_1 + r_1$ and $k = nq_2 + r_2$ where $0 \leq r_1 < n$ and $0 \leq r_2 < n$. Then

$$h - k = n(q_1 - q_2) + (r_1 - r_2)$$

and we see that $h - k$ is a multiple of n if and only if $r_1 = r_2$. Thus the conditions are the same.

c. **a.** $\overline{0} = \{\cdots, -2, 0, 2, \cdots\}$, $\overline{1} = \{\cdots, -3, -1, 1, 3, \cdots\}$

b. $\overline{0} = \{\cdots, -3, 0, 3, \cdots\}$, $\overline{1} = \{\cdots, -5, -2, 1, 4, \cdots\}$, $\overline{2} = \{\cdots, -1, 2, 5, \cdots\}$

c. $\overline{0} = \{\cdots, -5, 0, 5, \cdots\}$, $\overline{1} = \{\cdots, -9, -4, 1, 6, \cdots\}$, $\overline{2} = \{\cdots, -3, 2, 7, \cdots\}$, $\overline{3} = \{\cdots, -7, -2, 3, 8, \cdots\}$, $\overline{4} = \{\cdots, -1, 4, 9, \cdots\}$

37. The name *two-to-two function* suggests that such a function f should carry every pair of distinct points into two distinct points. Such a function is one-to-one in the conventional sense. (If the domain has only one element, the function cannot fail to be two-to-two, because the only way it can fail to be two-to-two is to carry two points into one point, and the set does not have two points.) Conversely, every function that is one-to-one in the conventional sense carries each pair of distinct points into two distinct points. Thus the functions conventionally called one-to-one are precisely those that carry two points into two points, which is a much more intuitive unidirectional way of regarding them. Also, the standard way of trying to show that a function is one-to-one is precisely to show that it does not fail to be two-to-two. That is, proving that a function is one-to-one becomes more natural in the two-to-two terminology.

1. Introduction and Examples

1. $i^3 = i^2 \cdot i = -1 \cdot i = -i$ **2.** $i^4 = (i^2)^2 = (-1)^2 = 1$ **3.** $i^{23} = (i^2)^{11} \cdot i = (-1)^{11} \cdot i = (-1)i = -i$

4. $(-i)^{35} = (i^2)^{17}(-i) = (-1)^{17}(-i) = (-1)(-i) = i$

5. $(4 - i)(5 + 3i) = 20 + 12i - 5i - 3i^2 = 20 + 7i + 3 = 23 + 7i$

6. $(8 + 2i)(3 - i) = 24 - 8i + 6i - 2i^2 = 24 - 2i - 2(-1) = 26 - 2i$

7. $(2 - 3i)(4 + i) + (6 - 5i) = 8 + 2i - 12i - 3i^2 + 6 - 5i = 14 - 15i - 3(-1) = 17 - 15i$

8. $(1 + i)^3 = (1 + i)^2(1 + i) = (1 + 2i - 1)(1 + i) = 2i(1 + i) = 2i^2 + 2i = -2 + 2i$

9. $(1 - i)^5 = 1^5 + \frac{5}{1}1^4(-i) + \frac{5 \cdot 4}{2 \cdot 1}1^3(-i)^2 + \frac{5 \cdot 4}{2 \cdot 1}1^2(-i)^3 + \frac{5}{1}1^1(-i)^4 + (-i)^5 = 1 - 5i + 10i^2 - 10i^3 + 5i^4 - i^5 = 1 - 5i - 10 + 10i + 5 - i = -4 + 4i$

10. $|3 - 4i| = \sqrt{3^2 + (-4)^2} = \sqrt{9 + 16} = \sqrt{25} = 5$ **11.** $|6 + 4i| = \sqrt{6^2 + 4^2} = \sqrt{36 + 16} = \sqrt{52} = 2\sqrt{13}$

12. $|3 - 4i| = \sqrt{3^2 + (-4)^2} = \sqrt{25} = 5$ and $3 - 4i = 5(\frac{3}{5} - \frac{4}{5}i)$

13. $|-1 + i| = \sqrt{(-1)^2 + 1^2} = \sqrt{2}$ and $-1 + i = \sqrt{2}(-\frac{1}{\sqrt{2}} + \frac{1}{\sqrt{2}}i)$

14. $|12 + 5i| = \sqrt{12^2 + 5^2} = \sqrt{169}$ and $12 + 5i = 13(\frac{12}{13} + \frac{5}{13}i)$

15. $|-3 + 5i| = \sqrt{(-3)^2 + 5^2} = \sqrt{34}$ and $-3 + 5i = \sqrt{34}(-\frac{3}{\sqrt{34}} + \frac{5}{\sqrt{34}}i)$

16. $|z|^4(\cos 4\theta + i \sin 4\theta) = 1(1 + 0i)$ so $|z| = 1$ and $\cos 4\theta = 1$ and $\sin 4\theta = 0$. Thus $4\theta = 0 + n(2\pi)$ so $\theta = n\frac{\pi}{2}$ which yields values $0, \frac{\pi}{2}, \pi$, and $\frac{3\pi}{2}$ less than 2π. The solutions are

$$z_1 = \cos 0 + i \sin 0 = 1, \quad z_2 = \cos \frac{\pi}{2} + i \sin \frac{\pi}{2} = i,$$

$$z_3 = \cos \pi + i \sin \pi = -1, \quad \text{and} \quad z_4 = \cos \frac{3\pi}{2} + i \sin \frac{3\pi}{2} = -i.$$

17. $|z|^4(\cos 4\theta + i \sin 4\theta) = 1(-1 + 0i)$ so $|z| = 1$ and $\cos 4\theta = -1$ and $\sin 4\theta = 0$. Thus $4\theta = \pi + n(2\pi)$ so $\theta = \frac{\pi}{4} + n\frac{\pi}{2}$ which yields values $\frac{\pi}{4}, \frac{3\pi}{4}, \frac{5\pi}{4}$, and $\frac{7\pi}{4}$ less than 2π. The solutions are

$$z_1 = \cos \frac{\pi}{4} + i \sin \frac{\pi}{4} = \frac{1}{\sqrt{2}} + \frac{1}{\sqrt{2}}i, \quad z_2 = \cos \frac{3\pi}{4} + i \sin \frac{3\pi}{4} = -\frac{1}{\sqrt{2}} + \frac{1}{\sqrt{2}}i,$$

$$z_3 = \cos \frac{5\pi}{4} + i \sin \frac{5\pi}{4} = -\frac{1}{\sqrt{2}} - \frac{1}{\sqrt{2}}i, \quad \text{and} \quad z_4 = \cos \frac{7\pi}{4} + i \sin \frac{7\pi}{4} = \frac{1}{\sqrt{2}} - \frac{1}{\sqrt{2}}i.$$

18. $|z|^3(\cos 3\theta + i\sin 3\theta) = 8(-1 + 0i)$ so $|z| = 2$ and $\cos 3\theta = -1$ and $\sin 3\theta = 0$. Thus $3\theta = \pi + n(2\pi)$ so $\theta = \frac{\pi}{3} + n\frac{2\pi}{3}$ which yields values $\frac{\pi}{3}, \pi$, and $\frac{5\pi}{3}$ less than 2π. The solutions are

$$z_1 = 2(\cos \frac{\pi}{3} + i\sin \frac{\pi}{3}) = 2(\frac{1}{2} + \frac{\sqrt{3}}{2}i) = 1 + \sqrt{3}i, \qquad z_2 = 2(\cos \pi + i\sin \pi) = 2(-1 + 0i) = -2,$$

and

$$z_3 = 2(\cos \frac{5\pi}{3} + i\sin \frac{5\pi}{3}) = 2(\frac{1}{2} - \frac{\sqrt{3}}{2}i) = 1 - \sqrt{3}i.$$

19. $|z|^3(\cos 3\theta + i\sin 3\theta) = 27(0 - i)$ so $|z| = 3$ and $\cos 3\theta = 0$ and $\sin 3\theta = -1$. Thus $3\theta = 3\pi/2 + n(2\pi)$ so $\theta = \frac{\pi}{2} + n\frac{2\pi}{3}$ which yields values $\frac{\pi}{2}, \frac{7\pi}{6}$, and $\frac{11\pi}{6}$ less than 2π. The solutions are

$$z_1 = 3(\cos \frac{\pi}{2} + i\sin \frac{\pi}{2}) = 3(0 + i) = 3i, \qquad z_2 = 3(\cos \frac{7\pi}{6} + i\sin \frac{7\pi}{6}) = 3(-\frac{\sqrt{3}}{2} - \frac{1}{2}i) = -\frac{3\sqrt{3}}{2} - \frac{3}{2}i$$

and

$$z_3 = 3(\cos \frac{11\pi}{6} + i\sin \frac{11\pi}{6}) = 3(\frac{\sqrt{3}}{2} - \frac{1}{2}i) = \frac{3\sqrt{3}}{2} - \frac{3}{2}i.$$

20. $|z|^6(\cos 6\theta + i\sin 6\theta) = 1 + 0i$ so $|z| = 1$ and $\cos 6\theta = 1$ and $\sin 6\theta = 0$. Thus $6\theta = 0 + n(2\pi)$ so $\theta = 0 + n\frac{2\pi}{6}$ which yields values $0, \frac{\pi}{3}, \frac{2\pi}{3}, \pi, \frac{4\pi}{3}$, and $\frac{5\pi}{3}$ less than 2π. The solutions are

$$z_1 = 1(\cos 0 + i\sin 0) = 1 + 0i = 1, \qquad z_2 = 1(\cos \frac{\pi}{3} + i\sin \frac{\pi}{3}) = \frac{1}{2} + \frac{\sqrt{3}}{2}i,$$

$$z_3 = 1(\cos \frac{2\pi}{3} + i\sin \frac{2\pi}{3}) = -\frac{1}{2} + \frac{\sqrt{3}}{2}i, \qquad z_4 = 1(\cos \pi + i\sin \pi) = -1 + 0i = -1,$$

$$z_5 = 1(\cos \frac{4\pi}{3} + i\sin \frac{4\pi}{3}) = -\frac{1}{2} - \frac{\sqrt{3}}{2}i, \qquad z_6 = 1(\cos \frac{5\pi}{3} + i\sin \frac{5\pi}{3}) = \frac{1}{2} - \frac{\sqrt{3}}{2}i.$$

21. $|z|^6(\cos 6\theta + i\sin 6\theta) = 64(-1 + 0i)$ so $|z| = 2$ and $\cos 6\theta = -1$ and $\sin 6\theta = 0$. Thus $6\theta = \pi + n(2\pi)$ so $\theta = \frac{\pi}{6} + n\frac{2\pi}{6}$ which yields values $\frac{\pi}{6}, \frac{\pi}{2}, \frac{5\pi}{6}, \frac{7\pi}{6}, \frac{3\pi}{2}$ and $\frac{11\pi}{6}$ less than 2π. The solutions are

$$z_1 = 2(\cos \frac{\pi}{6} + i\sin \frac{\pi}{6}) = 2(\frac{\sqrt{3}}{2} + \frac{1}{2}i) = \sqrt{3} + i,$$

$$z_2 = 2(\cos \frac{\pi}{2} + i\sin \frac{\pi}{2}) = 2(0 + i) = 2i,$$

$$z_3 = 2(\cos \frac{5\pi}{6} + i\sin \frac{5\pi}{6}) = 2(-\frac{\sqrt{3}}{2} + \frac{1}{2}i) = -\sqrt{3} + i,$$

$$z_4 = 2(\cos \frac{7\pi}{6} + i\sin \frac{7\pi}{6}) = 2(-\frac{\sqrt{3}}{2} - \frac{1}{2}i) = -\sqrt{3} - i,$$

$$z_5 = 2(\cos \frac{3\pi}{2} + i\sin \frac{3\pi}{2}) = 2(0 - i) = -2i,$$

$$z_6 = 2(\cos \frac{11\pi}{6} + i\sin \frac{11\pi}{6}) = 2(\frac{\sqrt{3}}{2} - \frac{1}{2}i) = \sqrt{3} - i.$$

22. $10 + 16 = 26 > 17$, so $10 +_{17} 16 = 26 - 17 = 9$.   **23.** $8 + 6 = 14 > 10$, so $8 +_{10} 6 = 14 - 10 = 4$.

24. $20.5 + 19.3 = 39.8 > 25$, so $20.5 +_{25} 19.3 = 39.8 - 25 = 14.8$.

25. $\frac{1}{2} + \frac{7}{8} = \frac{11}{8} > 1$, so $\frac{1}{2} +_1 \frac{7}{8} = \frac{11}{8} - 1 = \frac{3}{8}$.   **26.** $\frac{3\pi}{4} + \frac{3\pi}{2} = \frac{9\pi}{4} > 2\pi$, so $\frac{3\pi}{4} +_{2\pi} \frac{3\pi}{2} = \frac{9\pi}{4} - 2\pi = \frac{\pi}{4}$.

27. $2\sqrt{2} + 3\sqrt{2} = 5\sqrt{2} > \sqrt{32} = 4\sqrt{2}$, so $2\sqrt{2} +_{\sqrt{32}} 3\sqrt{2} = 5\sqrt{2} - 4\sqrt{2} = \sqrt{2}$.

28. 8 is not in $\mathbb{R}_6$ because $8 > 6$, and we have only defined $a +_6 b$ for $a, b \in \mathbb{R}_6$.

29. We need to have $x + 7 = 15 + 3$, so $x = 11$ will work. It is easily checked that there is no other solution.

30. We need to have $x + \frac{3\pi}{2} = 2\pi + \frac{3\pi}{4} = \frac{11\pi}{4}$, so $x = \frac{5\pi}{4}$ will work. It is easy to see there is no other solution.

31. We need to have $x + x = 7 + 3 = 10$, so $x = 5$ will work. It is easy to see that there is no other solution.

32. We need to have $x + x + x = 7 + 5$, so $x = 4$ will work. Checking the other possibilities 0, 1, 2, 3, 5, and 6, we see that this is the only solution.

33. An obvious solution is $x = 1$. Otherwise, we need to have $x + x = 12 + 2$, so $x = 7$ will work also. Checking the other ten elements, in $\mathbb{Z}_{12}$, we see that these are the only solutions.

34. Checking the elements $0, 1, 2, 3 \in \mathbb{Z}_4$, we find that they are all solutions. For example, $3+_4 3+_4 3+_4 3 = (3 +_4 3) +_4 (3 +_4 3) = 2 +_4 2 = 0$.

35. $\zeta^0 \leftrightarrow 0, \qquad \zeta^3 = \zeta^2\zeta \leftrightarrow 2 +_8 5 = 7, \qquad \zeta^4 = \zeta^2\zeta^2 \leftrightarrow 2 +_8 2 = 4, \qquad \zeta^5 = \zeta^4\zeta \leftrightarrow 4 +_8 5 = 1,$
$\zeta^6 = \zeta^3\zeta^3 \leftrightarrow 7 +_8 7 = 6, \qquad \zeta^7 = \zeta^3\zeta^4 \leftrightarrow 7 +_8 4 = 3$

36. $\zeta^0 \leftrightarrow 0, \qquad \zeta^2 = \zeta\zeta \leftrightarrow 4 +_7 4 = 1, \qquad \zeta^3 = \zeta^2\zeta \leftrightarrow 1 +_7 4 = 5, \qquad \zeta^4 = \zeta^2\zeta^2 \leftrightarrow 1 +_7 1 = 2,$
$\zeta^5 = \zeta^3\zeta^2 \leftrightarrow 5 +_7 1 = 6, \qquad \zeta^6 = \zeta^3\zeta^3 \leftrightarrow 5 +_7 5 = 3$

37. If there were an isomorphism such that $\zeta \leftrightarrow 4$, then we would have $\zeta^2 \leftrightarrow 4 +_6 4 = 2$ and $\zeta^4 = \zeta^2\zeta^2 \leftrightarrow 2 +_6 2 = 4$ again, contradicting the fact that an isomorphism $\leftrightarrow$ must give a *one-to-one correpondence*.

38. By Euler's fomula, $e^{ia}e^{ib} = e^{i(a+b)} = \cos(a + b) + i\sin(a + b)$. Also by Euler's formula,

$$
\begin{aligned}
e^{ia}e^{ib} &= (\cos a + i\sin a)(\cos b + i\sin b) \\
&= (\cos a\cos b - \sin a\sin b) + i(\sin a\cos b + \cos a\sin b).
\end{aligned}
$$

The desired formulas follow at once.

39. (See the text answer.)

40. a. We have $e^{3\theta} = \cos 3\theta + i\sin 3\theta$. On the other hand,

$$
\begin{aligned}
e^{3\theta} &= (e^\theta)^3 = (\cos\theta + i\sin\theta)^3 \\
&= \cos^3\theta + 3i\cos^2\theta\sin\theta - 3\cos\theta\sin^2\theta - i\sin^3\theta \\
&= (\cos^3\theta - 3\cos\theta\sin^2\theta) + i(3\cos^2\theta\sin\theta - \sin^3\theta).
\end{aligned}
$$

Comparing these two expressions, we see that

$$
\cos 3\theta = \cos^3\theta - 3\cos\theta\sin^2\theta.
$$

b. From Part(a), we obtain

$$
\cos 3\theta = \cos^3\theta - 3(\cos\theta)(1 - \cos^2\theta) = 4\cos^3\theta - 3\cos\theta.
$$

2. Binary Operations

1. $b*d=e, \quad c*c=b, \quad [(a*c)*e]*a=[c*e]*a=a*a=a$

2. $(a*b)*c=b*c=a$ and $a*(b*c)=a*a=a$, so the operation might be associative, but we can't tell without checking all other triple products.

3. $(b*d)*c=e*c=a$ and $b*(d*c)=b*b=c$, so the operation is not associative.

4. It is not commutative because $b*e=c$ but $e*b=b$.

5. Now $d*a=d$ so fill in d for $a*d$. Also, $c*b=a$ so fill in a for $b*c$. Now $b*d=c$ so fill in c for $d*b$. Finally, $c*d=b$ so fill in b for $d*c$.

6. $d*a=(c*b)*a=c*(b*a)=c*b=d$. In a similar fashion, substituting $c*b$ for d and using the associative property, we find that $d*b=c, d*c=c$, and $d*d=d$.

7. It is not commutative because $1-2 \neq 2-1$. It is not associative because $2=1-(2-3) \neq (1-2)-3 = -4$.

8. It is commutative because $ab+1=ba+1$ for all $a,b \in \mathbb{Q}$. It is not associative because $(a*b)*c = (ab+1)*c = abc+c+1$ but $a*(b*c)=a*(bc+1)=abc+a+1$, and we need not have $a=c$.

9. It is commutative because $ab/2=ba/2$ for all $a,b \in \mathbb{Q}$. It is associative because $a*(b*c)=a*(bc/2)= [a(bc/2)]/2=abc/4$, and $(a*b)*c=(ab/2)*c=[(ab/2)c]/2=abc/4$ also.

10. It is commutative because $2^{ab}=2^{ba}$ for all $a,b \in \mathbb{Z}^+$. It is not associative because $(a*b)*c=2^{ab}*c= 2^{(2^{ab})c}$, but $a*(b*c)=a*2^{bc}=2^{a(2^{bc})}$.

11. It is not commutative because $2*3 = 2^3 = 8 \neq 9 = 3^2 = 3*2$. It is not associative because $a*(b*c)=a*b^c=a^{(b^c)}$, but $(a*b)*c=a^b*c=(a^b)^c=a^{bc}$, and $bc \neq b^c$ for some $b,c \in \mathbb{Z}^+$.

12. If S has just one element, there is only one possible binary operation on S; the table must be filled in with that single element. If S has two elements, there are 16 possible operations, for there are four places to fill in a table, and each may be filled in two ways, and $2 \cdot 2 \cdot 2 \cdot 2 = 16$. There are 19,683 operations on a set S with three elements, for there are nine places to fill in a table, and $3^9 = 19,683$. With n elements, there are n^2 places to fill in a table, each of which can be done in n ways, so there are $n^{(n^2)}$ possible tables.

13. A commutative binary operation on a set with n elements is completely determined by the elements on or above the *main diagonal* in its table, which runs from the upper left corner to the lower right corner. The number of such places to fill in is

$$n + \frac{n^2 - n}{2} = \frac{n^2 + n}{2}.$$

Thus there are $n^{(n^2+n)/2}$ possible commutative binary operations on an n-element set. For $n=2$, we obtain $2^3=8$, and for $n=3$ we obtain $3^6=729$.

14. It is incorrect. Mention should be made of the underlying set for $*$ and the universal quantifier, *for all*, should appear.

A binary operation $*$ on a set S is **commutative** if and only if $a*b=b*a$ for all $a,b \in S$.

15. The definition is correct.

16. It is incorrect. Replace the final S by H.

17. It is not a binary operation. Condition 2 is violated, for $1 * 1 = 0$ and $0 \notin \mathbb{Z}^+$.

18. This does define a binary operation.

19. This does define a binary operation.

20. This does define a binary operation.

21. It is not a binary operation. Condition 1 is violated, for $2 * 3$ might be any integer greater than 9.

22. It is not a binary operation. Condition 2 is violated, for $1 * 1 = 0$ and $0 \notin \mathbb{Z}^+$.

23. a. Yes. $\begin{bmatrix} a & -b \\ b & a \end{bmatrix} + \begin{bmatrix} c & -d \\ d & c \end{bmatrix} = \begin{bmatrix} a+c & -(b+d) \\ b+d & a+c \end{bmatrix}$.

    b. Yes. $\begin{bmatrix} a & -b \\ b & a \end{bmatrix}\begin{bmatrix} c & -d \\ d & c \end{bmatrix} = \begin{bmatrix} ac-bd & -(ad+bc) \\ ad+bc & ac-bd \end{bmatrix}$.

24. F T F F F T T T F   **25.** (See the answer in the text.)

26. We have $(a*b)*(c*d) = (c*d)*(a*b) = (d*c)*(a*b) = [(d*c)*a]*b$, where we used commutativity for the first two steps and associativity for the last.

27. The statement is true. Commutativity and associativity assert the equality of certain computations. For a binary operation on a set with just one element, that element is the result of every computation involving the operation, so the operation must be commutative and associative.

28.
$*$	a	b
a	b	a
b	a	a

The statement is false. Consider the operation on $\{a, b\}$ defined by the table. Then $(a*a)*b = b*b = a$ but $a*(a*b) = a*a = b$.

29. It is associative.
    Proof: $[(f+g)+h](x) = (f+g)(x) + h(x) = [f(x)+g(x)] + h(x) = f(x) + [g(x)+h(x)] = f(x) + [(g+h)(x)] = [f+(g+h)](x)$ because addition in $\mathbb{R}$ is associative.

30. It is not commutative. Let $f(x) = 2x$ and $g(x) = 5x$. Then $(f-g)(x) = f(x) - g(x) = 2x - 5x = -3x$ while $(g-f)(x) = g(x) - f(x) = 5x - 2x = 3x$.

31. It is not associative. Let $f(x) = 2x, g(x) = 5x$, and $h(x) = 8x$. Then $[f - (g-h)](x) = f(x) - (g-h)(x) = f(x) - [g(x) - h(x)] = f(x) - g(x) + h(x) = 2x - 5x + 8x = 5x$, but $[(f-g) - h](x) = (f-g)(x) - h(x) = f(x) - g(x) - h(x) = 2x - 5x - 8x = -11x$.

32. It is commutative.
    Proof: $(f \cdot g)(x) = f(x) \cdot g(x) = g(x) \cdot f(x) = (g \cdot f)(x)$ because multiplication in $\mathbb{R}$ is commutative.

33. It is associative.
    Proof: $[(f \cdot g) \cdot h](x) = (f \cdot g)(x) \cdot h(x) = [f(x) \cdot g(x)] \cdot h(x) = f(x) \cdot [g(x) \cdot h(x)] = [f \cdot (g \cdot h)](x)$ because multiplication in $\mathbb{R}$ is associative.

34. It is not commutative. Let $f(x) = x^2$ and $g(x) = x + 1$. Then $(f \circ g)(3) = f(g(3)) = f(4) = 16$ but $(g \circ f)(3) = g(f(3)) = g(9) = 10$.

35. It is not true. Let $*$ be $+$ and let $*'$ be $\cdot$ and let $S = \mathbb{Z}$. Then $2 + (3 \cdot 5) = 17$ but $(2 + 3) \cdot (2 + 5) = 35$.

36. Let $a, b \in H$. By definition of H, we have $a * x = x * a$ and $b * x = x * b$ for all $x \in S$. Using the fact that $*$ is associative, we then obtain, for all $x \in S$,

$$(a * b) * x = a * (b * x) = a * (x * b) = (a * x) * b = (x * a) * b = x * (a * b).$$

This shows that $a * b$ satisfies the defining criterion for an element of H, so $(a * b) \in H$.

37. Let $a, b \in H$. By definition of H, we have $a * a = a$ and $b * b = b$. Using, one step at a time, the fact that $*$ is associative and commutative, we obtain

$$
\begin{aligned}
(a * b) * (a * b) &= [(a * b) * a] * b = [a * (b * a)] * b = [a * (a * b)] * b \\
&= [(a * a) * b] * b = (a * b) * b = a * (b * b) = a * b.
\end{aligned}
$$

This show that $a * b$ satisfies the defining criterion for an element of H, so $(a * b) \in H$.

3. Isomorphic Binary Structures

1. i) ϕ must be one to one.   ii) $\phi[S]$ must be all of S'.   iii) $\phi(a * b) = \phi(a) *' \phi(b)$ for all $a, b \in S$.

2. It is an isomorphism; ϕ is one to one, onto, and $\phi(n + m) = -(n + m) = (-n) + (-m) = \phi(n) + \phi(m)$ for all $m, n \in \mathbb{Z}$.

3. It is not an isomorphism; ϕ does not map $\mathbb{Z}$ onto $\mathbb{Z}$. For example, $\phi(n) \neq 1$ for all $n \in \mathbb{Z}$.

4. It is not an isomorphism because $\phi(m + n) = m + n + 1$ while $\phi(m) + \phi(n) = m + 1 + n + 1 = m + n + 2$.

5. It is an isomorphism; ϕ is one to one, onto, and $\phi(a + b) = \frac{a+b}{2} = \frac{a}{2} + \frac{b}{2} = \phi(a) + \phi(b)$.

6. It is not an isomorphism because ϕ does not map $\mathbb{Q}$ onto $\mathbb{Q}$. $\phi(a) \neq -1$ for all $a \in \mathbb{Q}$.

7. It is an isomorphism because ϕ is one to one, onto, and $\phi(xy) = (xy)^3 = x^3 y^3 = \phi(x)\phi(y)$.

8. It is not an isomorphism because ϕ is not one to one. All the 2×2 matrices where the entries in the second row are double the entries above them in the first row are mapped into 0 by ϕ.

9. It is an isomorphism because for 1×1 matrices, $[a][b] = [ab]$, and $\phi([a]) = a$ so ϕ just removes the brackets.

10. It is an isomorphism. For any base $a \neq 1$, the exponential function $f(x) = a^x$ maps $\mathbb{R}$ one to one onto $\mathbb{R}^+$, and ϕ is the exponential map with $a = 0.5$. We have $\phi(r + s) = 0.5^{(r+s)} = (0.5^r)(0.5^s) = \phi(r)\phi(s)$.

11. It is not an isomorphism because ϕ is not one to one; $\phi(x^2) = 2x$ and $\phi(x^2 + 1) = 2x$.

12. It is not an isomorphism because ϕ is not one to one: $\phi(\sin x) = \cos 0 = 1$ and $\phi(x) = 1$.

13. No, because ϕ does not map F onto F. For all $f \in F$, we see that $\phi(f)(0) = 0$ so, for example, no function is mapped by ϕ into $x + 1$.

14. It is an isomorphism. By calculus, $\phi(f) = f$, so ϕ is the identity map which is always an isomorphism of a binary structure with itself.

15. It is not an isomorphism because ϕ does not map F onto F. Note that $\phi(f)(0) = 0 \cdot f(0) = 0$. Thus there is no element of F that is mapped by ϕ into the constant function 1.

16. a. For ϕ to be an isomorphism, we must have

$$m * n = \phi(m-1) * \phi(n-1) = \phi((m-1) + (n-1)) = \phi(m+n-2) = m+n-1.$$

The identity element is $\phi(0) = 1$.

b. Using the fact that ϕ^{-1} must also be an isomorphism, we must have

$$m * n = \phi^{-1}(m+1) * \phi^{-1}(n+1) = \phi^{-1}((m+1) + (n+1)) = \phi^{-1}(m+n+2) = m+n+1.$$

The identity element is $\phi^{-1}(0) = -1$.

17. a. For ϕ to be an isomorphism, we must have

$$m * n = \phi(m-1) * \phi(n-1) = \phi((m-1) \cdot (n-1)) = \phi(mn - m - n + 1) = mn - m - n + 2.$$

The identity element is $\phi(1) = 2$.

b. Using the fact that ϕ^{-1} must also be an isomorphism, we must have

$$m * n = \phi^{-1}(m+1) * \phi^{-1}(n+1) = \phi^{-1}((m+1) \cdot (n+1)) = \phi^{-1}(mn + m + n + 1) = mn + m + n.$$

The identity element is $\phi^{-1}(1) = 0$.

18. a. For ϕ to be an isomorphism, we must have

$$a * b = \phi\left(\frac{a+1}{3}\right) * \phi\left(\frac{b+1}{3}\right) = \phi\left(\frac{a+1}{3} + \frac{b+1}{3}\right) = \phi\left(\frac{a+b+2}{3}\right) = a+b+1.$$

The identity element is $\phi(0) = -1$.

b. Using the fact that ϕ^{-1} must also be an isomorphism, we must have

$$a * b = \phi^{-1}(3a-1) * \phi^{-1}(3b-1) = \phi^{-1}((3a-1) + (3b-1)) = \phi^{-1}(3a + 3b - 2) = a + b - \frac{1}{3}.$$

The identity element is $\phi^{-1}(0) = 1/3$.

19. a. For ϕ to be an isomorphism, we must have

$$a * b = \phi\left(\frac{a+1}{3}\right) * \phi\left(\frac{b+1}{3}\right) = \phi\left(\frac{a+1}{3} \cdot \frac{b+1}{3}\right) = \phi\left(\frac{ab + a + b + 1}{9}\right) = \frac{ab + a + b - 2}{3}.$$

The identity element is $\phi(1) = 2$.

b. Using the fact that ϕ^{-1} must also be an isomorphism, we must have

$$a * b = \phi^{-1}(3a-1) \cdot \phi^{-1}(3b-1) = \phi^{-1}((3a-1) \cdot (3b-1)) = \phi^{-1}(9ab - 3a - 3b + 1) = 3ab - a - b + \frac{2}{3}.$$

The identity element is $\phi^{-1}(1) = 2/3$.

20. Computing $\phi(x * y)$ is done by first executing the binary operation $*$, and then performing the map ϕ. Computing $\phi(x) *' \phi(y)$ is done by first performing the map ϕ, and then executing the binary operation $*'$. Thus, reading in left to right order of peformance, the isomorphism property is

$$\text{(binary operation)(map)} = \text{(map)(binary operation)}$$

which has the formal appearance of commutativity.

21. The definition is incorrect. It should be stated that $\langle S, *\rangle$ and $\langle S', *'\rangle$ are binary structures, ϕ must be one to one and onto S', and the universal quantifier "for all $a, b \in S$" should appear in an appropriate place.

Let $\langle S, *\rangle$ and $\langle S', *'\rangle$ be binary structures. A map $\phi : S \to S'$ is an **isomorphism** if and only if ϕ is one to one and onto S', and $\phi(a * b) = \phi(a) *' \phi(b)$ for all $a, b \in S$.

22. It is badly worded. The "for all $s \in S$" applies to the equation and not to the "is an identity for $*$".

Let $*$ be a binary operation on a set S. An element e of S is an **identity element** for $*$ if and only if $s * e = e * s = s$ for all $s \in S$.

23. Suppose that e and $\bar{e}$ are two identity elements and, viewing each in turn as an identity element, compute $e * \bar{e}$ in two ways.

24. a. Let $*$ be a binary operation on a set S. An element e_L of S is a **left identity element** for $*$ if and only if $e_L * s = s$ for all $s \in S$.

b. Let $*$ be a binary operation on a set S. An element e_R of S is a **right identity element** for $*$ if and only if $s * e_R = s$ for all $s \in S$.

A one-sided identity element is not unique. Let $*$ be defined on S by $a * b = a$ for all $a, b \in S$. Then every $b \in S$ is a right identity. Similarly, a left identity is not unique. If in the proof of Theorem 3.13, we replace e by e_L and $\bar{e}$ by $\bar{e}_L$ everywhere, and replace the word "identity" by "left identity", the first incorrect statement would be, "However, regarding $\bar{e}_L$ as left identity element, we must have $e_L * \bar{e}_L = e_L$."

25. No, if $\langle S* \rangle$ has a left identity element e_L and a right identity element e_R, then $e_L = e_R$.

Proof Because e_L is a left identity element we have $e_L * e_R = e_R$, but viewing e_R as right identity element, $e_L * e_R = e_L$. Thus $e_L = e_R$.

26. *One-to-one:* Suppose that $\phi^{-1}(a') = \phi^{-1}(b')$ for $a', b' \in S'$. Then $a' = \phi(\phi^{-1}(a')) = \phi(\phi^{-1}(b')) = b'$, so ϕ^{-1} is one to one.

Onto: Let $a \in S$. Then $\phi^{-1}(\phi(a)) = a$, so ϕ^{-1} maps S' onto S.

Homomorphism property: Let $a', b' \in S'$. Now

$$\phi(\phi^{-1}(a' *' b')) = a' *' b'.$$

Because ϕ is an isomorphism,

$$\phi(\phi^{-1}(a') * \phi^{-1}(b\,')) = \phi(\phi^{-1}(a')) *' \phi(\phi^{-1}(b')) = a' *' b'$$

also. Because ϕ is one to one, we conclude that

$$\phi^{-1}(a' * b') = \phi^{-1}(a') *' \phi^{-1}(b').$$

27. *One-to-one:* Let $a, b \in S$ and suppose $(\psi \circ \phi)(a) = (\psi \circ \phi)(b)$. Then $\psi(\phi(a)) = \psi(\phi(b))$. Because ψ is one to one, we conclude that $\phi(a) = \phi(b)$. Because ϕ is one to one, we must have $a = b$.

Onto: Let $a'' \in S''$. Because ψ maps S' onto S'', there exists $a' \in S'$ such that $\psi(a') = a''$. Because ϕ maps S onto S', there exists $a \in S$ such that $\phi(a) = a'$. Then $(\psi \circ \phi)(a) = \psi(\phi(a)) = \psi(a') = a''$, so $\psi \circ \phi$ maps S onto S''.

Homomorphism property: Let $a, b \in S$. Since ϕ and ψ are isomorphisms, $(\psi \circ \phi)(a * b) = \psi(\phi(a * b)) = \psi(\phi(a) *' \phi(b)) = \psi(\phi(a)) *'' \psi(\phi(b)) = (\psi \circ \phi)(a) *'' (\psi \circ \phi)(b)$.

28. Let $\langle S, * \rangle, \langle S', *' \rangle$ and $\langle S'', *'' \rangle$ be binary structures.

    Reflexive: Let $\iota : S \to S$ be the identity map. Then ι maps S one to one onto S and for $a, b \in S$, we have $\iota(a * b) = a * b = \iota(a) * \iota(b)$, so ι is an isomorphism of S with itself, that is $S \simeq S$.

    Symmetric: If $S \simeq S'$ and $\phi : S \to S'$ is an isomorphism, then by Exercise 26, $\phi^{-1} : S' \to S$ is an isomorphism, so $S' \simeq S$.

    Transitive: Suppose that $S \simeq S'$ and $S' \simeq S''$, and that $\phi : S \to S'$ and $\psi : S' \to S''$ are isomorphisms. By Exercise 27, we know that $\psi \circ \phi : S \to S''$ is an isomorphism, so $S \simeq S''$.

29. Let $\langle S, * \rangle$ and $\langle S', *' \rangle$ be isomorphic binary structures and let $\phi : S \to S'$ be an isomorphism. Suppose that $*$ is commutative. Let $a', b' \in S'$ and let $a, b \in S$ be such that $\phi(a) = a'$ and $\phi(b) = b'$. Then $a' *' b' = \phi(a) *' \phi(b) = \phi(a * b) = \phi(b * a) = \phi(b) *' \phi(a) = b' *' a'$, showing that $*'$ is commutative.

30. Let $\langle S, * \rangle$ and $\langle S', *' \rangle$ be isomorphic binary structures and let $\phi : S \to S'$ be an isomorphism. Suppose that $*$ is associative. Let $a', b', c' \in S'$ and let $a, b, c \in S$ be such that $\phi(a) = a', \phi(b) = b'$ and $\phi(c) = c'$. Then

$$\begin{aligned}(a' *' b') *' c' &= (\phi(a) *' \phi(b)) *' \phi(c) = \phi(a * b) *' \phi(c) = \phi((a * b) * c)) \\ &= \phi(a * (b * c)) = \phi(a) *' \phi(b * c) = \phi(a) *' (\phi(b) *' \phi(c)) = a' *' (b' *' c'),\end{aligned}$$

showing that $*'$ is associative.

31. Let $\langle S, * \rangle$ and $\langle S', *' \rangle$ be isomorphic binary structures and let $\phi : S \to S'$ be an isomorphism. Suppose that S has the property that for each $c \in S$ there exists $x \in S$ such that $x * x = c$. Let $c' \in S'$, and let $c \in S$ such that $\phi(c) = c'$. Find $x \in S$ such that $x * x = c$. Then $\phi(x * x) = \phi(c) = c'$, so $\phi(x) *' \phi(x) = c'$. If we denote $\phi(x)$ by x', then we see that $x' *' x' = c'$, so S' has the analagous property.

32. Let $\langle S, * \rangle$ and $\langle S', *' \rangle$ be isomorphic binary structures and let $\phi : S \to S'$ be an isomorphism. Suppose that S has the property that there exists $b \in S$ such that $b * b = b$. Let $b' = \phi(b)$. Then $b' *' b' = \phi(b) *' \phi(b) = \phi(b * b) = \phi(b) = b'$, so S' has the analogous property.

33. Let $\phi : \mathbb{C} \to H$ be defined by $\phi(a + bi) = \begin{bmatrix} a & -b \\ b & a \end{bmatrix}$ for $a, b \in \mathbb{R}$. Clearly ϕ is one to one and onto H.

a. We have $\phi((a + bi) + (c + di)) = \phi((a + c) + (b + d)i) = \begin{bmatrix} a + c & -(b + d) \\ b + d & a + c \end{bmatrix} = \begin{bmatrix} a & -b \\ b & a \end{bmatrix} +$

$\begin{bmatrix} c & -d \\ d & c \end{bmatrix} = \phi(a + bi) + \phi(c + di)$.

b. We have $\phi((a + bi) \cdot (c + di)) = \phi((ac - bd) + (ad + bc)i) = \begin{bmatrix} ac - bd & -(ad + bc) \\ ad + bc & ac - bd \end{bmatrix} =$

$\begin{bmatrix} a & -b \\ b & a \end{bmatrix} \cdot \begin{bmatrix} c & -d \\ d & c \end{bmatrix} = \phi(a + bi) \cdot \phi(c + di)$.

34. Let the set be $\{a, b\}$. We need to decide whether interchanging the names of the letters everywhere in the table and then writing the table again in the order a first and b second gives the same table or a different table. The same table is obtained if and only if in the body of the table, diagonally opposite entries are different. Four such tables exist, since there are four possible choices for the first row; Namely, the tables

*	a	b
a	a	a
b	b	b

*	a	b
a	a	b
b	a	b

*	a	b
a	b	a
b	b	a

and

*	a	b
a	b	b
b	a	a

.

The other 12 tables can be paired off into tables giving the same algebraic structure. One table of each pair is listed below. The number of different algebraic structures is therefore $4 + 12/2 = 10$.

*	a	b
a	a	a
b	a	a

*	a	b
a	a	a
b	a	b

*	a	b
a	a	a
b	b	a

*	a	b
a	a	b
b	a	a

*	a	b
a	b	a
b	a	a

*	a	b
a	a	b
b	b	a

4. Groups

1. No. G_3 fails. **2.** Yes **3.** No. G_1 fails. **4.** No. G_3 fails. **5.** No. G_1 fails.

6. No. G_2 fails.

7. The group $\langle U_{1000}, \cdot \rangle$ of solutions of $z^{1000} = 1$ in $\mathbb{C}$ under multiplication has 1000 elements and is abelian.

8.

$\cdot_8$	1	3	5	7
1	1	3	5	7
3	3	1	7	5
5	5	7	1	3
7	7	5	3	1

9. Denoting the operation in each of the three groups by $*$ and the identity element by e for the moment, the equation $x * x * x * x = e$ has four solutions in $\langle U, \cdot \rangle$, one solution in $\langle \mathbb{R}, + \rangle$, and two solutions in $\langle \mathbb{R}^*, \cdot \rangle$.

10. a. *Closure:* Let nr and ns be two elements of $n\mathbb{Z}$. Now $nr + ns = n(r + s) \in n\mathbb{Z}$ so $n\mathbb{Z}$ is closed under addition.

Associative: We know that addition of integers is associative.

Identity: $0 = n0 \in n\mathbb{Z}$, and 0 is the additive identity element.

Inverses: For each $nm \in n\mathbb{Z}$, we also have $n(-m) \in n\mathbb{Z}$ and $nm + n(-m) = n(m - m) = n0 = 0$.

b. Let $\phi : \mathbb{Z} \to n\mathbb{Z}$ be defined by $\phi(m) = nm$ for $m \in \mathbb{Z}$. Clearly ϕ is one to one and maps $\mathbb{Z}$ onto $n\mathbb{Z}$. For $r, s \in \mathbb{Z}$, we have $\phi(r + s) = n(r + s) = nr + ns = \phi(r) + \phi(s)$. Thus ϕ is an isomorphism of $\langle \mathbb{Z}, + \rangle$ with $\langle n\mathbb{Z}, + \rangle$.

11. Yes, it is a group. Addition of diagonal matrices amounts to adding in $\mathbb{R}$ entries in corresponding positions on the diagonals, and that addition is associative. The matrix with all entries 0 is the additive identity, and changing the sign of the entries in a matrix yields the additive inverse of the matrix.

12. No, it is not a group. Multiplication of diagonal matrices amounts to muliplying in $\mathbb{R}$ entries in corresponding positions on the diagonals. The matrix with 1 at all places on the diagonal is the identity element, but a matrix having a diagonal entry 0 has no inverse.

13. Yes, it is a group. See the answer to Exercise 12.

14. Yes, it is a group. See the answer to Exercise 12.

15. No. The matrix with all entries 0 is upper triangular, but has no inverse.

16. Yes, it is a group. The sum of upper-triangular matrices is again upper triangular, and addition amounts to just adding entries in $\mathbb{R}$ in corresponding positions.

17. Yes, it is a group.

    Closure: Let A and B be upper triangular with determinant 1. Then entry c_{ij} in row i and column j in $C = AB$ is 0 if $i > j$, because for each product $a_{ik}b_{kj}$ where $i > j$ appearing in the computation of c_{ij}, either $k < i$ so that $a_{ik} = 0$ or $k \geq i > j$ so that $b_{kj} = 0$. Thus the product of two upper-triangular matrices is again upper triangular. The equation $\det(AB) = \det(A) \cdot \det(B)$, shows that the product of two matrices of determinant 1 again has determinant 1.

    Associative: We know that matrix multiplication is associative.

    Identity: The $n \times n$ identity matrix I_n has determinant 1 and is upper triangular.

    Inverse: The product property $1 = \det(I_n) = \det(A^{-1}A) = \det(A^{-1}) \cdot \det(A)$ shows that if $\det(A) = 1$, then $\det(A^{-1}) = 1$ also.

18. Yes, it is a group. The relation $\det(AB) = \det(A) \cdot \det(B)$ show that the set of $n \times n$ matrices with determinant ± 1 is closed under multiplication. We know matrix multiplication is associative, and $\det(I_n) = 1$. As in the preceding solution, we see that $\det(A) = \pm 1$ implies that $\det(A^{-1}) = \pm 1$, so we have a group.

19. **a.** We must show that S is closed under $*$, that is, that $a + b + ab \neq -1$ for $a, b \in S$. Now $a + b + ab = -1$ if and only if $0 = ab + a + b + 1 = (a + 1)(b + 1)$. This is the case if and only if either $a = -1$ or $b = -1$, which is not the case for $a, b \in S$.

    b. *Associative:* We have

    $$a * (b * c) = a * (b + c + bc) = a + (b + c + bc) + a(b + c + bc) = a + b + c + ab + ac + bc + abc$$

    and

    $$(a * b) * c = (a + b + ab) * c = (a + b + ab) + c + (a + b + ab)c = a + b + c + ab + ac + bc + abc.$$

Identity: 0 acts as identity elemenr for $*$, for $0 * a = a * 0 = a$.

Inverses: $\frac{-a}{a+1}$ acts as inverse of a, for

$$a * \frac{-a}{a+1} = a + \frac{-a}{a+1} + a\frac{-a}{a+1} = \frac{a(a+1) - a - a^2}{a+1} = \frac{0}{a+1} = 0.$$

c. Because the operation is commutative, $2 * x * 3 = 2 * 3 * x = 11 * x$. Now the inverse of 11 is $-11/12$ by Part(**b**). From $11 * x = 7$, we obtain

$$x = \frac{-11}{12} * 7 = \frac{-11}{12} + 7 + \frac{-11}{12}7 = \frac{-11 + 84 - 77}{12} = \frac{-4}{12} = -\frac{1}{3}.$$

20.

	e	a	b	c
c	c	a	b	c
a	a	e	c	b
b	b	c	e	a
c	c	b	a	e

Table I

	e	a	b	c
e	e	a	b	c
a	a	e	c	b
b	b	c	a	e
c	c	b	e	a

Table II

	e	a	b	c
e	e	a	b	c
a	a	b	c	e
b	b	c	e	a
c	c	e	a	b

Table III

Table I is structurally different from the others because every element is its own inverse. Table II can be made to look just like Table III by interchanging the names a and b everywhere to obtain

	e	b	a	c
e	e	b	a	c
b	b	e	c	a
a	a	c	b	e
c	c	a	e	b

and rewriting this table in the order e, a, b, c.

a. The symmetry of each table in its main diagonal shows that all groups of order 4 are commutative.

b. Table III gives the group U_4, upon replacing e by 1, a by i, b by -1, and c by $-i$.

c. Take $n = 2$. There are four 2×2 diagonal matrices with entries ± 1, namely

$$E = \begin{bmatrix} 1 & 0 \\ 0 & 1 \end{bmatrix}, A = \begin{bmatrix} -1 & 0 \\ 0 & 1 \end{bmatrix}, B = \begin{bmatrix} 1 & 0 \\ 0 & -1 \end{bmatrix}, \text{ and } C = \begin{bmatrix} -1 & 0 \\ 0 & -1 \end{bmatrix}.$$

If we write the table for this group using the letters E, A, B, C in that order, we obtain Table I with the letters capitalized.

21. A binary operation on a set $\{x, y\}$ of two elements that produces a group is completely determined by the choice of x or y to serve as identity element, so just 2 of the 16 possible tables give groups. For a set $\{x, y, z\}$ of three elements, a group binary operation is again determined by the choice x, y, or z to serve as identity element, so there are just 3 of the 19,683 binary operations that give groups. (Recall that there is only one way to fill out a group table for $\{e, a\}$ and for $\{e, a, b\}$ if you require e to be the identity element.)

22. The orders $G_1 G_3 G_2, G_3 G_1 G_2$, and $G_3 G_2 G_1$ are not acceptable. The identity element e occurs in the statement of G_3, which must not come before e is defined in G_2.

23. Ignoring spelling, punctuation and grammar, here are some of the mathematical errors.

a. The statement "$x =$ identity" is wrong.

b. The identity element should be e, not (e). It would also be nice to give the properties satisfied by the identity element and by inverse elements.

c. Associativity is missing. Logically, the identity element should be mentioned before inverses. The statement "an inverse exists" is not quantified correctly: for each element of the set, an inverse exists. Again, it would be nice to give the properties satisfied by the identity element and by inverse elements.

d. Replace " such that for all $a, b \in G$" by " if for all $a \in G$". Delete "under addition" in line 2. The element should be e, not $\{e\}$. Replace "$= e$" by "$= a$" in line 3.

24. We need only make a table that has e as an identity element and has an e in each row and each column of the body of the table to satisfy axioms G_2 and G_3. Then we make some row or column contain some element twice, and it can't be a group, so G_1 must fail.

$*$	e	a	b
e	e	a	b
a	a	e	b
b	b	a	e

25. F T T F F T T T F T

26. Multiply both sides of the equation $a * b = a * c$ on the left by the inverse of a, and simplify, using the axioms for a group.

27. Show that $x = a' * b$ is a solution of $a * x = b$ by substitution and the axioms for a group. Then show that it is the only solution by multiplying both sides of the equation $a * x = b$ on the left by a' and simplifying, using the axioms for a group.

28. Let $\phi : G \to G'$ be a group isomorphism of $\langle G, * \rangle$ onto $\langle G', *' \rangle$, and let $a, a' \in G$ such that $a * a' = e$. Then $\phi(e) = \phi(a * a') = \phi(a) *' \phi(a')$. Now $\phi(e)$ is the identity element of G' by Theorem 3.14. Thus the equation $\phi(a) *' \phi(a') = \phi(e)$ shows that $\phi(a)$ and $\phi(a')$ are inverse pairs in G', which was to be shown.

29. Let $S = \{x \in G \mid x' \neq x\}$. Then S has an even number of elements, because its elements can be grouped in pairs x, x'. Because G has an even number of elements, the number of elements in G but not in S (the set $G - S$) must be even. The set $G - S$ is nonempty because it contains e. Thus there is at least one element of $G - S$ other than e, that is, at least one element other than e that is its own inverse.

30. a. We have $(a * b) * c = (|a| b) * c \ |(|a|b)|c = |ab|c$. We also have $a *(b * c) = a * (|b|c) = |a||b|c = |ab|c$, so $*$ is associative.

b. We have $1 * a = |1| a = a$ for all $a \in \mathbb{R}^*$ so 1 is a left identity element. For $a \in \mathbb{R}^*, 1/|a|$ is a right inverse.

c. It is not a group because both $1/2$ and $-1/2$ are right inverse of 2.

d. The one-sided definition of a group, mentioned just before the exercises, must be all left sided or all right sided. We must not mix them.

31. Let $\langle G, * \rangle$ be a group and let $x \in G$ such that $x * x = x$. Then $x * x = x * e$, and by left cancellation, $x = e$, so e is the only idempotent element in a group.

32. We have $e = (a*b)*(a*b)$, and $(a*a)*(b*b) = e*e = e$ also. Thus $a*b*a*b = a*a*b*b$. Using left and right cancellation, we have $b*a = a*b$.

33. Let $P(n) = (a*b)^n = a^n * b^n$. Since $(a*b)^1 = a*b = a^1 * b^1$, we see $P(1)$ is true. Suppose $P(k)$ is true. Then $(a*b)^{k+1} = (a*b)^k * (a*b) = (a^k * b^k) * (a*b) = [a^k * (b^k * a)] * b = [a^k * (a*b^k)] * b = [(a^k * a) * b^k] * b = (a^{k+1} * b^k) * b = a^{k+1} * (b^k * b) = a^{k+1} * b^{k+1}$. This completes the induction argument.

34. The elements $e, a, a^2, a^3, \cdots, a^m$ aren't all different since G has only m elements. If one of $a, a^2, a^3, \cdots,$ a^m is e, then we are done. If not, then we must have $a^i = a^j$ where $i < j$. Repeated left cancellation of a yields $e = a^{j-i}$.

35. We have $(a*b)*(a*b) = (a*a)*(b*b)$, so $a*[b*(a*b)] = a*[a*(b*b)]$ and left cancellation yields $b*(a*b) = a*(b*b)$. Then $(b*a)*b = (a*b)*b$ and right cancellation yields $b*a = a*b$.

36. Let $a*b = b*a$. Then $(a*b)' = (b*a)' = a'*b'$ by Corollary 4.17. Conversely, if $(a*b)' = a'*b'$, then $b'*a' = a'*b'$. Then $(b'*a')' = (a'*b')'$ so $(a')'*(b')' = (b')'*(a')'$ and $a*b = b*a$.

37. We have $a*b*c = a*(b*c) = e$, which implies that $b*c$ is the inverse of a. Therefore $(b*c)*a = b*c*a = e$ also.

38. We need to show that a left identity element is a right identity element and that a left inverse is a right inverse. Note that $e*e = e$. Then $(x'*x)*e = x'*x$ so $(x')'*(x'*x)*e = (x')'*(x'*x)$. Using associativity, $[(x')'*x']*x*e = [(x')'*x']*x$. Thus $(e*x)*e = e*x$ so $x*e = x$ and e is a right identity element also. If $a'*a = e$, then $(a'*a)*a' = e*a' = a'$. Multiplication of $a'*a*a' = a'$ on the left by $(a')'$ and associativity yield $a*a' = e$, so a' is also a right inverse of a.

39. Using the hint, we show there is a left identity element and that each element has a left inverse. Let $a \in G$; we are given that G is nonempty. Let e be a solution of $y*a = a$. We show at $e*b = b$ for any $b \in G$. Let c be a solution of the equation $a*x = b$. Then $e*b = e*(a*c) = (e*a)*c = a*c = b$. Thus e is a left identity. Now for each $a \in G$, let a' be a solution of $y*a = e$. Then a' is a left inverse of a. By Exercise 38, G is a group.

40. It is easy to see that $\langle G, * \rangle$ is a group, because the order of multiplication in G is simply reversed: $(a*b)*c = a*(b*c)$ follows at once from $c \cdot (b \cdot a) = (c \cdot b) \cdot a$, the element e continues to act as identity element, and the inverse of each element is unchanged.

Let $\phi(a) = a'$ for $a \in G$, where a' is the inverse of a in the group $\langle G, \cdot \rangle$. Uniqueness of inverses and the fact that $(a')' = a$ show at once that ϕ is one to one and onto G. Also, $\phi(a \cdot b) = (a \cdot b)' = b' \cdot a' = a' * b' = \phi(a) * \phi(b)$, showing that ϕ is an isomorphism of $\langle G, \cdot \rangle$ onto $\langle G, * \rangle$.

41. Let $a, b \in G$. If $g*a*g' = g*b*g'$, then $a = b$ by group cancellation, so i_g is a one-to-one map. Because $i_g(g'*a*g) = g*g'*a*g*g' = a$, we see that i_g maps G onto G. We have $i_g(a*b) = g*a*b*g' = g*a*(g'*g)*b*g' = (g*a*g')*(g*b*g') = i_g(a)*i_g(b)$, so i_g satisfies the homomorphism property also, and is thus an isomorphism.

5. Subgroups

1. Yes   **2.** No, there is no identity element.   **3.** Yes   **4.** Yes   **5.** Yes

6. No, the set is not closed under addition.   **7.** $\mathbb{Q}^+$ and $\{\pi^n \mid n \in \mathbb{Z}\}$

8. No. If $\det(A) = \det(B) = 2$, then $\det(AB) = \det(A)\det(B) = 4$. The set is not closed under multiplication.

9. Yes   **10.** Yes, see Exercise 17 of Section 4.

11. No. If $\det(A) = \det(B) = -1$, then $\det(AB) = \det(A)\det(B) = 1$. The set is not closed under multiplication.

12. Yes, see Exercise 17 of Section 4.

13. Yes. Suppose that $(A^T)A = I_n$ and $(B^T)B = I_n$. Then we have $(AB)^T AB = B^T(A^T A)B = B^T I_n B = B^T B = I_n$, so the set of these matrices is closed under multiplication. Since $I_n^T = I_n$ and $I_n I_n = I_n$, the set contains the identity. For each A in the set, the equation $(A^T)A = I_n$ shows that A has an inverse A^T. The equation $(A^T)^T A^T = AA^T = I_n$ shows that A^T is in the given set. Thus we have a subgroup.

14. a) No, $\tilde{F}$ is not closed under addition.   b) Yes

15. a) Yes   b) No, it is not even a subset of $\tilde{F}$.

16. a) No, it is not closed under addition.   b) Yes

17. a) No, it is not closed under addition.   b) Yes

18. a) No, it is not closed under addition.   b) No, it is not closed under multiplication.

19. a) Yes   b) No, the zero constant function is not in $\tilde{F}$.

20. $G_1 \leq G_1,\ G_1 < G_4$ $G_2 < G_1,\ G_2 \leq G_2,\ G_2 < G_4,\ G_2 < G_7,\ G_2 < G_8$ $G_3 \leq G_3,\ G_3 < G_5$

$G_4 \leq G_4$ $G_5 \leq G_5$ $G_6 \leq G_5,\ G_6 \leq G_6$ $G_7 < G_1,\ G_7 < G_4,\ G_7 \leq G_7$

$G_8 < G_1,\ G_8 < G_4,\ G_8 < G_7,\ G_8 \leq G_8$ $G_9 < G_3,\ G_9 < G_5,\ G_9 \leq G_9$

21. a. -50, -25, 0, 25, 50   **b.** $4, 2, 1, \frac{1}{2}, \frac{1}{4}$ **c.** $1, \pi, \pi^2, \frac{1}{\pi}, \frac{1}{\pi^2}$ **22.** $\begin{bmatrix} 0 & -1 \\ -1 & 0 \end{bmatrix}, \begin{bmatrix} 1 & 0 \\ 0 & 1 \end{bmatrix}$

23. All the matrices $\begin{bmatrix} 1 & n \\ 0 & 1 \end{bmatrix}$ for $n \in \mathbb{Z}$.   **24.** All the matrices $\begin{bmatrix} 3^n & 0 \\ 0 & 2^n \end{bmatrix}$ for $n \in \mathbb{Z}$.

25. All matrices of the form $\begin{bmatrix} 4^n & 0 \\ 0 & 4^n \end{bmatrix}$ or $\begin{bmatrix} 0 & -2^{2n+1} \\ -2^{2n+1} & 0 \end{bmatrix}$ for $n \in \mathbb{Z}$.

26. G_1 is cyclic with generators 1 and -1. G_2 is not cyclic. G_3 is not cyclic.
G_4 is cyclic with generators 6 and -6. G_5 is cyclic with generators 6 and $\frac{1}{6}$. G_6 is not cyclic.

   To get the answers for Exercises 27 - 35, the student computes the given element to succesive powers (or summands). The first power (number of summands) that gives the identity element is the order of the cyclic subgroup. After students have studied Section 9, you might want to come back here and show them the easy way to handle the row permutations of the identity matrix in Exercises 33 - 35 by writing the permutation as a product of disjoint cycles. For example, in Exercise 35, row 1 is in row 4 place, row 4 is in row 2 place, and row 2 is in row 1 place, corresponding to the cycle (1,4,2). Row 3 is left fixed.

27. 4   **28.** 2   **29.** 3   **30.** 5   **31.** 4   **32.** 8   **33.** 2   **34.** 4   **35.** 3

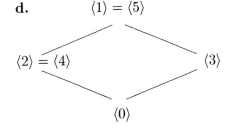

36. a.

$+_6$	0	1	2	3	4	5
0	0	1	2	3	4	5
1	1	2	3	4	5	0
2	2	3	4	5	0	1
3	3	4	5	0	1	2
4	4	5	0	1	2	3
5	5	0	1	2	3	4

b. $\langle 0 \rangle = \{0\}$
$\langle 2 \rangle = \langle 4 \rangle = \{0, 2, 4\}$
$\langle 3 \rangle = \{0, 3\}$
$\langle 1 \rangle = \langle 5 \rangle = \mathbb{Z}_6$

c. 1 and 5

d.
$$\langle 1 \rangle = \langle 5 \rangle$$
$$\langle 2 \rangle = \langle 4 \rangle \qquad \langle 3 \rangle$$
$$\langle 0 \rangle$$

37. Incorrect, the closure condition must be stated.

A **subgroup** of a group G is a subset H of G that is closed under the induced binary operation from G, contains the identity element e of G, and contains the inverse h^{-1} of each $h \in H$.

38. The definition is correct. **39.** T F T F F F F F T F

40. In the Klein 4-group, the equation $x^2 = e$ has all four elements of the group as solutions.

41. *Closure:* Let $a, b \in H$ so that $\phi(a), \phi(b) \in \phi[H]$. Now $(a * b) \in H$ because $H \leq G$. Since ϕ is an isomorphism, $\phi(a) *' \phi(b) = \phi(a * b) \in \phi[H]$, so $\phi[H]$ is closed under $*'$.

Identity: By Theorem 3.14, $e' = \phi(e) \in \phi[H]$.

Inverses: Let $a \in H$ so that $\phi(a) \in \phi[H]$. Then $a^{-1} \in H$ because H is a subgroup of G. We have $e' = \phi(e) = \phi(a^{-1} * a) = \phi(a^{-1}) *' \phi(a)$, so $\phi(a)^{-1} = \phi(a^{-1}) \in \phi[H]$.

42. Let a be a generator of G. We claim $\phi(a)$ is a generator of G'. Let $b' \in G'$. Because ϕ maps G onto G', there exists $b \in G$ such that $\phi(b) = b'$. Because a generates G, there exists $n \in \mathbb{Z}$ such that $b = a^n$. Because ϕ is an isomorphism, $b' = \phi(b) = \phi(a^n) = \phi(a)^n$. Thus G' is cyclic.

43. *Closure:* Let $S = \{hk \mid h \in H, k \in K\}$ and let $x, y \in S$. Then $x = hk$ and $y = h'k'$ for some $h, h' \in H$ and $k, k' \in K$. Because G is abelian, we have $xy = hkh'k' = (hh')(kk')$. Because H and K are subgroups, we have $hh' \in H$ and $kk' \in K$, so $xy \in S$ and S is closed under the induced operation.

Identity: Because H and K are subgroups, $e \in H$ and $e \in K$ so $e = ee \in S$.

Inverses: For $x = hk$, we have $h^{-1} \in H$ and $k^{-1} \in K$ because H and K are subgroups. Then $h^{-1}k^{-1} \in S$ and because G is abelian, $h^{-1}k^{-1} = k^{-1}h^{-1} = (hk)^{-1} = x^{-1}$, so the inverse of x is in S. Hence S is a subgroup.

44. If H is empty, then there is no $a \in H$.

45. Let H be a subgroup of G. Then for $a, b \in H$, we have $b^{-1} \in H$ and ab^{-1} H because H must be closed under the induced operation.

Conversely, suppose that H is nonempty and $ab^{-1} \in H$ for all $a, b \in H$. Let $a \in H$. Then taking $b = a$, we see that $aa^{-1} = e$ is in H. Taking $a = e$, and $b = a$, we see that $ea^{-1} = a^{-1} \in H$. Thus H contains the identity element and the inverse of each element. For closure, note that for $a, b \in H$, we also have $a, b^{-1} \in H$ and thus $a(b^{-1})^{-1} = ab \in H$.

46. Let $B = \{e, a, a^2, a^3, \cdots, a^{n-1}\}$ be a cyclic group of n elements. Then $a^{-1} = a^{n-1}$ also generates G, because $(a^{-1})^i = (a^i)^{-1} = a^{n-i}$ for $i = 1, 2, \cdots, n-1$. Thus if G has only one generator, we must have $n - 1 = 1$ and $n = 2$. Of course, $G = \{e\}$ is also cyclic with one generator.

47. *Closure:* Let $a, b \in H$. Because G is abelian, $(ab)^2 = a^2 b^2 = ee = e$ so $ab \in H$ and H is closed under the induced operation.

Identity: Because $ee = e$, we see $e \in H$.

Inverses: Because $aa = e$, we see that each element of H is its own inverse. Thus H is a subgroup.

48. *Closure:* Let $a, b \in H$. Because G is abelian, $(ab)^n = a^n b^n = ee = e$ so $ab \in H$ and H is closed under the induced operation.

Identity: Because $e^n = e$, we see that $e \in H$.

Inverses: Let $a \in H$. Because $a^n = e$, we see that the inverse of a is a^{n-1} which is in H because H is closed under the induced operation. Thus H is a subgroup of G.

49. Let G have m elements. Then the elements $a, a^2, a^3, \cdots, a^{m+1}$ cannot all be different, so $a^i = a^j$ for some $i < j$. Then multiplication by a^{-i} shows that $e = a^{j-i}$, and we can take $j - i$ as the desired n.

50. Let $a \in H$ and let H have n elements. Then the elements $a, a^2, a^3, \cdots, a^{n+1}$ are all in H (because H is closed under the operation) and cannot all be different, so $a^i = a^j$ for some $i < j$. Then multiplication by a^{-i} shows that $e = a^{j-i}$ so $e \in H$. Also, $a^{-1} \in H$ because $a^{-1} = a^{j-i-1}$. This shows that H is a subgroup of G.

51. *Closure:* Let $x, y \in H_a$. Then $xa = ax$ and $ya = ay$. We then have $(xy)a = x(ya) = x(ay) = (xa)y = (ax)y = a(xy)$, so $xy \in H_a$ and H_a is closed under the operation.

Identity: Because $ea = ae = a$, we see that $e \in H_a$.

Inverses: From $xa = ax$, we obtain $xax^{-1} = a$ and then $ax^{-1} = x^{-1}a$, showing that $x^{-1} \in H_a$, which is thus a subgroup.

52. a. *Closure:* Let $x, y \in H_S$. Then $xs = sx$ and $ys = sy$ for all $s \in S$. We then have $(xy)s = x(ys) = x(sy) = (xs)y = (sx)y = s(xy)$ for all $s \in S$, so $xy \in H_S$ and H_S is closed under the operation.

Identity: Because $es = se = s$ for all $s \in S$, we see that $e \in H_S$.

Inverses: From $xs = sx$ for all $s \in S$, we obtain $xsx^{-1} = s$ and then $sx^{-1} = x^{-1}s$ for all $s \in S$, showing that $x^{-1} \in H_S$, which is thus a subgroup.

b. Let $a \in H_G$. Then $ag = ga$ for all $g \in G$; in particular, $ab = ba$ for all $b \in H_G$ because H_G is a subset of G. This shows that H_G is abelian.

53. *Reflexive:* Let $a \in G$. Then $aa^{-1} = e$ and $e \in H$ since H is a subgroup. Thus $a \sim a$.

Symmetric: Let $a, b \in G$ and $a \sim b$, so that $ab^{-1} \in H$. Since H is a subgroup, we have $(ab^{-1})^{-1} = ba^{-1} \in H$, so $b \sim a$.

Transitive: Let $a, b, c \in G$ and $a \sim b$ and $b \sim c$. Then $ab^{-1} \in H$ and $bc^{-1} \in H$ so $(ab^{-1})(bc^{-1}) = ac^{-1} \in H$, and $a \sim c$.

54. *Closure:* Let $a, b \in H \cap K$. Then $a, b \in H$ and $a, b \in K$. Because H and K are subgroups, we have $ab \in H$ and $ab \in K$, so $ab \in H \cap K$.

Identity: Because H and K are subgroups, we have $e \in H$ and $e \in K$ so $e \in H \cap K$.

Inverses: Let $a \in H \cap K$ so $a \in H$ and $a \in K$. Because H and K are subgroups, we have $a^{-1} \in H$ and $a^{-1} \in K$, so $a^{-1} \in H \cap K$.

55. Let G be cyclic and let a be a generator for G. For $x, y \in G$, there exist $m, n \in \mathbb{Z}$ such that $x = a^m$ and $y = b^n$. Then $xy = a^m b^n = a^{m+n} = a^{n+m} = a^n a^m = yx$, so G is abelian.

56. We can show it if G is abelian. Let $a, b \in G$ so that $a^n, b^n \in G_n$. Then $a^n b^n = (ab)^n$ because G is abelian, so G_n is closed under the induced operation. Also $e = e^n \in G_n$. Finally $(a^n)^{-1} = (a^{-1})^n \in G_n$, so G_n is indeed a subgroup of G.

57. Let G be a group with no proper nontrivial subgroups. If $G = \{e\}$, then G is of course cyclic. If $G \neq \{e\}$, then let $a \in G, a \neq e$. We know that $\langle a \rangle$ is a subgroup of G and $\langle a \rangle \neq \{e\}$. Because G has no proper nontrivial subgroups, we must have $\langle a \rangle = G$, so G is indeed cyclic.

6. Cyclic Groups

1. $42 = 9 \cdot 4 + 6, q = 4, r = 6$ **2.** $-42 = 9(-5) + 3, q = -5, r = 3$ **3.** $-50 = 8(-7) + 6, q = -7, r = 6$

4. $50 = 8 \cdot 6 + 2, q = 6, r = 2$ **5.** 8   **6.** 8   **7.** 60

8. 1, 2, 3, and 4 are relative prime to 5 so the answer is 4.

9. 1, 3, 5, and 7 are relatively prime to 8 so the answer is 4.

10. 1, 5, 7, and 11 are relatively prime to 12 so the answer is 4.

11. 1, 7, 11, 13, 17, 19, 23, 29, 31, 37, 41, 43, 47, 49, 53, and 59 are relatively prime to 60 so the answer is 16.

12. There is one automorphism; 1 must be carried into the only generator which is 1.

13. There are 2 automorphisms; 1 can be carried into either of the generators 1 or 5

14. There are 4 automorphisms; 1 can be carried into any of the generators 1, 3, 5, or 7.

15. There are 2 automorphisms; 1 can be carried into either of the generators 1 or -1.

16. There are 4 automorphisms; 1 can be carried into any of the generators 1, 5, 7, or 11.

17. $\gcd(25, 30) - 5$ and $30/5 - 6$ so $\langle 25 \rangle$ has 6 elements

18. $\gcd(30, 42) = 6$ and $42/6 = 7$ so $\langle 30 \rangle$ has 7 elements.

19. The polar angle for i is $\pi/2$, so it generates a subgroup of 4 elements.

20. The polar angle for $(1+i)/\sqrt{2}$ is $\pi/4$, so it generates a subgroup of 8 elements.

21. The absolute value of $1 + i$ is $\sqrt{2}$, so it generates an infinite subgroup of $\aleph_0$ elements.

22. Subgroup diagram:

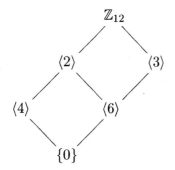

23. (See the answer in the text.)

24. Subgroup diagram:

25. 1, 2, 3, 6   **26.** 1, 2, 4, 8   **27.** 1, 2, 3, 4, 6, 12   **28.** 1, 2, 4, 5, 10, 20   **29.** 1, 17

30. Incorrect; n must be minimal in $\mathbb{Z}^+$ with that property.

An element a of a group G has **order** $n \in \mathbb{Z}^+$ if $a^n = e$ and $a^m \neq e$ for $m \in \mathbb{Z}^+$ where $m < n$.

31. The definition is correct.

32. T F F F T F F F T T   f) The Klein 4-group is an example. g) 9 generates $\mathbb{Z}_{20}$.

33. The Klein 4-group   **34.** $\langle \mathbb{R}, + \rangle$ **35.** $\mathbb{Z}_2$

36. No such example exists. Every infinite cyclic group is isomorphic to $\langle \mathbb{Z}, + \rangle$ which has just two generators, 1 and -1.

37. $\mathbb{Z}_8$ has generators 1, 3, 5, and 7.   **38.** i and $-i$

39. Corresponding to polar angles $n(2\pi/6)$ for $n = 1$ and 5, we have $\frac{1}{2}(1 \pm i\sqrt{3})$.

40. Corresponding to polar angles $n(2\pi/8)$ for $n = 1,7,3$, and 5, we have $\frac{1}{\sqrt{2}}(1 \pm i)$ and $\frac{1}{\sqrt{2}}(-1 \pm i)$.

41. Corresponding to polar angles $n(2\pi/12)$ for $n = 1,11,5$, and 7, we have $\frac{1}{2}(\sqrt{3} \pm i)$ and $\frac{1}{2}(-\sqrt{3} \pm i)$.

42. Expressing two elements of the group as powers of the same generator, their product is the generator raised to the sum of the powers, and addition of integers is commutative.

43. Asuming the subgroup isn't just $\{e\}$, let a be a generator of the cyclic group, and let n be the smallest positive integer power of a that is in the subgroup. For a^m in the subgroup, use the division algorithm for n divided by m and the choice of n to argue that $n = qm$ for some integer q, so that $a^m = (a^n)^q$.

44. By the homomorphism property $\phi(ab) = \phi(a)\phi(b)$ extended by induction, we have $\phi(a^n) = (\phi(a))^n$ for all $n \in \mathbb{Z}+$. By Theorem 3.14, we know that $\phi(a^0) = \phi(e) = e'$. The equation $e' = \phi(e) = \phi(aa^{-1}) = \phi(a)\phi(a^{-1})$ shows that $\phi(a^{-1}) = (\phi(a))^{-1}$. Extending this last equation by induction, we see that $\phi(a^{-n}) = (\phi(a))^{-n}$ for all negative integers $-n$. Because G is cyclic with generator a, this means that for all $g = a^n \in G, \phi(g) = \phi(a^n) = [\phi(a)]^n$ is completely determined by the value $\phi(a)$.

45. The equation $(n_1r + m_1s) + (n_2r + m_2s) = (n_1 + n_2)r + (m_1 + m_2)s$ shows that the set is closed under addition. Because $0r + 0s = 0$, we see that 0 is in the set. Because $[(-m)r + (-n)s] + (mr + ns) = 0$, we see that the set contains the inverse of each element. Thus it is a subgroup of $\mathbb{Z}$.

46. Let n be the order of ab so that $(ab)^n = e$. Multiplying this equation on the left by b and on the right by a, we find that $(ba)^{n+1} = bea = (ba)e$. Cancellation of the first factor ba from both sides shows that $(ba)^n = e$, so the order of ba is $\leq n$. If the order of ba were less than n, a symmetric argument would show that the order of ab is less than n, contrary to our choice of n. Thus ba has order n also.

47. a. As a subgroup of the cyclic group $\langle \mathbb{Z}, + \rangle$, the subgroup $G = r\mathbb{Z} \cap s\mathbb{Z}$ is cyclic. The positive generator of G is the **least common multiple** of r and s.

b. The least common multiple of r and s is rs if and only if r and s are relative prime, so that they have no common prime factor.

c. Let $d = ir + js$ be the gcd of r and s, and let $m = kr = qs$ be the least common multiple of r and s. Then $md = mir + mjs = qsir + krjs = (qi + kj)rs$, so rs is a divisor of md. Now let $r = ud$ and let $s = vd$. Then $rs = uvdd = (uvd)d$, and $uvd = rv = su$ is a multiple of r and s, and hence $uvd = mt$. Thus $rs = mtd = (md)t$, so md is divisor of rs. Hence $md = rs$.

48. Note that every group is the union of its cyclic subgroups, because every element of the group generates a cyclic subgroup that contains the element. Let G have only a finite number of subgroups, and hence only a finite number of cyclic subgroups. Now none of these cyclic subgroups can be infinite, for every infinite cyclic group is isomorphic to $\mathbb{Z}$ which has an infinite number of subgroups, namely $\mathbb{Z}, 2\mathbb{Z}, 3\mathbb{Z}, \cdots$. Such subgroups of an infinite cyclic subgroup of G would of course give an infinite number of subgroups of G, contrary to hypothesis. Thus G has only finite cyclic subgroups, and only a finite number of those. We see that the set G can be written as a finite union of finite sets, so G is itself a finite set.

49. The Klein 4-group V is a counterexample.

50. Note that $xax^{-1} \neq e$ because $xax^{-1} = e$ would imply that $xa = x$ and $a = e$, and we are given that a has order 2. We have $(xax^{-1})^2 = xax^{-1}xax^{-1} = xex^{-1} = xx^{-1} = e$. Because a is given to be the *unique* element in G of order 2, we see that $xax^{-1} = a$, and upon multiplication on the right by x, we obtain $xa = ax$ for all $x \in G$.

51. The positive integers less that pq and relatively prime to pq are those that are not multiples of p and are not multiples of q. There are $p - 1$ multiples of q and $q - 1$ multiples of p that are less than pq. Thus there are $(pq - 1) - (p - 1) - (q - 1) = pq - p - q + 1 = (p - 1)(q - 1)$ positive integers less than pq and relatively prime to pq.

52. The positive integers less than p^r and relatively prime to p^r are those that are not multiples of p. There are $p^{r-1} - 1$ multiples of p less than p^r. Thus we see that there are $(p^r - 1) - (p^{r-1} - 1) = p^r - p^{r-1} = p^{r-1}(p-1)$ positive integers less than p^r and relatively prime to p^r.

53. It is no loss of generality to supppose that $G = \mathbb{Z}_n$ and that we are considering the equation $mx = 0$ for a positive integer m dividing n. Clearly $0, n/m, 2n/m, \cdots, (m-1)n/m$ are m solutions of $mx = 0$. If r is any solution in $\mathbb{Z}_n$ of $mx = 0$, then n is a divisor of mr, so that $mr = qn$. But then $r = q(n/m) < n$, so that q must be one of $0, 1, 2, \cdots, m-1$, and we see that the solutions exhibited above are indeed all the solutions.

54. There are exactly d solutions, where d is the gcd of m and n. Working in $\mathbb{Z}_n$ again, we see that $0, n/d, 2n/d, \cdots, (d-1)n/d$ are solutions of $mx = 0$. If r is any solution, then n divides mr so that $mr = nq$ and $r = nq/m$. Write $m = m_1 d$ and $n = n_1 d$ so that the gcd of m_1 and n_1 is 1. Then $r = nq/m$ can be written as $r = n_1 dq/m_1 d = n_1 q/m_1$. Since m_1 and n_1 are relatively prime, we conclude that m_1 divides q; let $q = m_1 s$. Then $r = n_1 q/m_1 = n_1 m_1 s/m_1 = n_1 s = (n/d)s$. Since $r < n$, we have $n_1 s < n = n_1 d$ so $s < d$. consequently, s must be one of the numbers $0, 1, 2, \cdots, d-1$ and we see that the solutions exhibited above are indeed all the solutions.

55. All positive integers less than p are relatively prime to p because p is prime, and hence they all generate $\mathbb{Z}_p$. Thus $\mathbb{Z}_p$ has no proper cyclic subgroups, and thus no proper subgroups, because as a cyclic group, $\mathbb{Z}_p$ has only cyclic subgroups.

56. a. Let a be a generator of H and let b be a generator of K. Because G is abelian, we have $(ab)^{rs} = (a^r)^s (b^s)^r = e^r e^s = e$. We claim that no lower power of ab is equal to e, for suppose that $(ab)^n = a^n b^n = e$. Then $a^n = b^{-n} = c$ must be an element of both H and K, and thus generates a subgroup of H of order dividing r which must also be a subgroup of K of order dividing s. Because r and s are relatively prime, we see that we must have $c = e$, so $a^n = b^n = e$. But then n is divisible by both r and s, and because r and s are relatively prime, we have $n = rs$. Thus ab generates the desired cyclic subgroup of G of order rs.

b. Let d be the gcd of r and s, and let $s = dq$ so that q and r are relatively prime and $rq = rs/d$ is the least common multiple of r and s (see Exercise 47c). Let a and b be generators of H and K respectively. Then $|\langle a \rangle| = r$ and $|\langle b^d \rangle| = q$ where r and q are relatively prime. Part(**a**) shows that the element ab^d generates a cyclic subgroup of order rq which is the least common multiple of r and s.

7. Generators and Cayley Digraphs

1. 0, 1, 2, 3, 4, 5, 6, 7, 8, 9, 10, 11   **2.** 0, 2, 4, 6, 8, 10   **3.** 0, 2, 4, 6, 8, 10, 12, 14, 16

4. 0, 6, 12, 18, 24, 30   **5.** $\cdots, -24, -18, -12, -6, 0, 6, 12, 18, 24, \cdots$

6. $\cdots, -15, -12, -9, -6, -3, 0, 3, 6, 9, 12, 15, \cdots$

7. a: Starting at the vertex $a^2 b$, we travel three solid lines in the direction of the arrow, arriving at $a^3 b$.

b. Starting at the vertex ab, we travel three solid lines in the direction of the arrow and then one dashed line, arriving at a^2.

c. Starting at the vertex b, we travel two solid lines in the direction of the arrow and then one dashed line, arriving at a^2.

8.

	e	a	b	c
e	e	a	b	c
a	a	e	c	b
b	b	c	e	a
c	c	b	a	e

9. (See the answer in the text.)

10.

	e	a	b	c	d	f
e	e	a	b	c	d	f
a	a	c	f	e	b	d
b	b	d	e	f	a	c
c	c	e	d	a	f	b
d	d	f	c	b	e	a
f	f	b	a	d	c	e

11. Choose a pair of generating directed arcs, call them *arc1* and *arc2*. Start at any vertex of the digraph, and see if the sequences *arc1, arc2* and *arc2, arc1* lead to the same vertex. (This corresponds to asking if the two corresponding group generators commute.) The group is commutative if and only if these two sequences lead to the same vertex for every pair of generating directed arcs.

12. It is not commutative, for a followed by b leads to ab, while b followed by a leads to a^3b.

13. If more than one element of the cyclic group is used to generate the Cayley digraph, it may not be obvious from the digraph that the group is cyclic. See, for example, Figure 7.9, where 5 actually generates the group $\mathbb{Z}_6$ having these digraphs generated by 2 and 3.

14. No, it does not contain the identity 0.

15. (See the answer in the text.)

16. Here is a Cayley digraph.

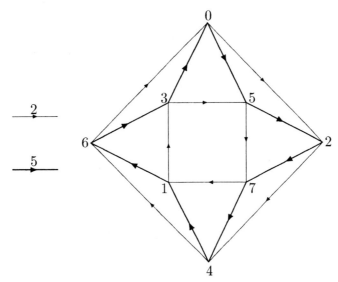

17. **a.** Starting from the vertex representing the identity, every path though the graph that terminates at that same vertex represents a product of generators or their inverses that is equal to the identity and thus gives a relation.

b. $a^4 = e, b^2 = e, (ab)^2 = e$.

18. The diagram in Figure 7.13a which represents the Klein 4-group, and a square with solid clockwise arrows edges which represents $\mathbb{Z}_4$.

19. Generalizing Figure 7.13b, form a regular $2n$-gon with alternately solid and dashed edges, without arrows. The four properties listed after Example 7.10 in the text are satisfied and the digraph represent a nonabelian group of order $2n$ for $n \geq 3$.

8. Groups of Permutations

1. $\begin{pmatrix} 1 & 2 & 3 & 4 & 5 & 6 \\ 1 & 2 & 3 & 6 & 5 & 4 \end{pmatrix}$ **2.** $\begin{pmatrix} 1 & 2 & 3 & 4 & 5 & 6 \\ 2 & 4 & 1 & 5 & 6 & 3 \end{pmatrix}$ **3.** $\begin{pmatrix} 1 & 2 & 3 & 4 & 5 & 6 \\ 3 & 4 & 1 & 6 & 2 & 5 \end{pmatrix}$

4. $\begin{pmatrix} 1 & 2 & 3 & 4 & 5 & 6 \\ 5 & 1 & 6 & 2 & 4 & 3 \end{pmatrix}$ **5.** $\begin{pmatrix} 1 & 2 & 3 & 4 & 5 & 6 \\ 2 & 6 & 1 & 5 & 4 & 3 \end{pmatrix}$

6. Starting with 1 and applying σ repeatedly, we see that σ takes 1 to 3 to 4 to 5 to 6 to 2 to 1, so σ^6 is the smallest possible power of σ that is the identity permutation. It is easily checked that σ^6 carries 2, 3, 4, 5 and 6 to themselves also, so σ^6 is indeed the identity and $|\langle\sigma\rangle| = 6$.

7. $\tau^2 = \begin{pmatrix} 1 & 2 & 3 & 4 & 5 & 6 \\ 4 & 3 & 2 & 1 & 5 & 6 \end{pmatrix}$ and it is clear that $(\tau^2)^2$ is the identity. Thus we have $|\langle\tau^2\rangle| = 2$.

8. Because σ^6 is the identity permutation (see Exercise 6), we have

$$\sigma^{100} = (\sigma^6)^{16}\sigma^4 = \sigma^4 = \begin{pmatrix} 1 & 2 & 3 & 4 & 5 & 6 \\ 6 & 5 & 2 & 1 & 3 & 4 \end{pmatrix}.$$

9. We find that μ^2 is the identity permutation, so $\mu^{100} = (\mu^2)^{50}$ is also the identity permutation.

10. $\{\mathbb{Z}, 17\mathbb{Z}, 3\mathbb{Z}, \langle\pi\rangle\}$ is a subcollection of isomorphic groups, as are $\{\mathbb{Z}_6, G\}, \{\mathbb{Z}_2, S_2\}, \{S_6\}, \{\mathbb{Q}\}, \{\mathbb{R}, \mathbb{R}^+\}$, $\{\mathbb{R}^*\}, \{\mathbb{Q}^*\}$, and $\{\mathbb{C}^*\}$.

11. $\{1, 2, 3, 4, 5, 6\}$ **12.** $\{1, 2, 3, 4\}$ **13.** $\{1, 5\}$

14. We see that ϵ, ρ, and ρ^2 give the three positions of the triangle in Fig. 8.9 obtained by rotations. The permutations $\phi, \rho\phi$, and $\rho^2\phi$ amount geometrically to turning the triangle over (ϕ) and then rotating it to obtain the other three positions.

15. A similar labeling for D_4 is $\epsilon, \rho, \rho^2, \rho^3, \phi, \rho\phi, \rho^2\phi, \rho^3\phi$ where their ϕ is our μ_1. They correspond to our elements in the order $\rho_0, \rho_1, \rho_2, \rho_3, \mu_1, \delta_1, \mu_2, \delta_2$.

16. σ may have the action of any of the six possible permutations of the set $\{1, 2, 4\}$, so there are six possibilities for σ.

17. There are 4 possibilities for $\sigma(1)$, then 3 possibilities for $\sigma(3)$, then 2 possibilities for $\sigma(4)$, and then 1 possibility for $\sigma(5)$, for $4 \cdot 3 \cdot 2 \cdot 1 = 24$ possibilities in all.

18. a. $\langle\rho_1\rangle = \langle\rho_2\rangle = \{\rho_0, \rho_1, \rho_2\}$ and $\langle\mu_1\rangle = \{\rho_0, \mu_1\}$.

b.

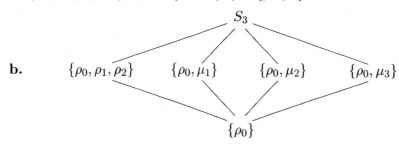

19. (See the answer in the text.)

20. This group is not isomorphic to S_3 because it is abelian and S_3 is nonabelian. It is isomorphic to $\mathbb{Z}_6$.

	ρ^0	ρ	ρ^2	ρ^3	ρ^4	ρ^5
ρ^0	ρ^0	ρ	ρ^2	ρ^3	ρ^4	ρ^5
ρ	ρ	ρ^2	ρ^3	ρ^4	ρ^5	ρ^0
ρ^2	ρ^2	ρ^3	ρ^4	ρ^5	ρ^0	ρ
ρ^3	ρ^3	ρ^4	ρ^5	ρ^0	ρ	ρ^2
ρ^4	ρ^4	ρ^5	ρ^0	ρ	ρ^2	ρ^3
ρ^5	ρ^5	ρ^0	ρ	ρ^2	ρ^3	ρ^4

21. (See the answer in the text.)

22. We list matrices in order corresponding to the permutations $\rho_0, \rho_1, \rho_2, \rho_3, \mu_1, \mu_2, \delta_1, \delta_2$ of D_4. Thus the fifth matrix listed, which corresponds to $\mu_1 = \begin{pmatrix} 1 & 2 & 3 & 4 \\ 2 & 1 & 4 & 3 \end{pmatrix}$ is the matrix obtained from the identity by interchanging row 1 with row 2 and row 3 with row 4.

$$\begin{bmatrix} 1 & 0 & 0 & 0 \\ 0 & 1 & 0 & 0 \\ 0 & 0 & 1 & 0 \\ 0 & 0 & 0 & 1 \end{bmatrix} \quad \begin{bmatrix} 0 & 0 & 0 & 1 \\ 1 & 0 & 0 & 0 \\ 0 & 1 & 0 & 0 \\ 0 & 0 & 1 & 0 \end{bmatrix} \quad \begin{bmatrix} 0 & 0 & 1 & 0 \\ 0 & 0 & 0 & 1 \\ 1 & 0 & 0 & 0 \\ 0 & 1 & 0 & 0 \end{bmatrix} \quad \begin{bmatrix} 0 & 1 & 0 & 0 \\ 0 & 0 & 1 & 0 \\ 0 & 0 & 0 & 1 \\ 1 & 0 & 0 & 0 \end{bmatrix}$$

$$\begin{bmatrix} 0 & 1 & 0 & 0 \\ 1 & 0 & 0 & 0 \\ 0 & 0 & 0 & 1 \\ 0 & 0 & 1 & 0 \end{bmatrix} \quad \begin{bmatrix} 0 & 0 & 0 & 1 \\ 0 & 0 & 1 & 0 \\ 0 & 1 & 0 & 0 \\ 1 & 0 & 0 & 0 \end{bmatrix} \quad \begin{bmatrix} 0 & 0 & 1 & 0 \\ 0 & 1 & 0 & 0 \\ 1 & 0 & 0 & 0 \\ 0 & 0 & 0 & 1 \end{bmatrix} \quad \begin{bmatrix} 1 & 0 & 0 & 0 \\ 0 & 0 & 0 & 1 \\ 0 & 0 & 1 & 0 \\ 0 & 1 & 0 & 0 \end{bmatrix}$$

23. The identity and flipping over on the vertical axis that falls on the vertical line segment of the figure give the only symmetries. The symmetry group is isomorphic to $\mathbb{Z}_2$.

24. As symmetries other than the identity, the figure admits a rotation through 180°, a flip in the vertical line shown, and a flip in the analogous horizontal line (not shown). This group of four elements is isomorphic to the Klein 4-group.

25. If we join endpoints of the line segments, we have a square with the given lines as its diagonals. The symmetries of that square produce all the symmetries of the given figure, so the group of symmetries is isomorphic to D_4.

26. The only symmetries are those obtained by sliding the figure to the left or to the right. We consider the vertical line segments to be one unit apart. For each integer n, we can slide the figure n units to the right if $n > 0$ and $|n|$ units to the left if $n < 0$, leaving the figure alone if $n = 0$. A moment of thought shows that performing the symmetry corresponding to an integer n and then the one corresponding to an integer m yields the symmetry corresponding to $n + m$. We see that the symmetry group is isomorphic to $\mathbb{Z}$.

27. (See the answer in the text.)

28. Replace the final "to" by "onto".

A **permutation** of a set S is a one-to-one map of S onto S.

29. The definition is correct. **30.** This one-to-one map of $\mathbb{R}$ onto $\mathbb{R}$ is a permutation.

31. This is not a permutation; it is neither one to one nor onto. Note that $f_2(3) = f_2(-3) = 9$ and $f_2(x) = -1$ has no solution.

32. This one-to-one map of $\mathbb{R}$ onto $\mathbb{R}$ is a permutation.

33. This is not a permutation, it is not a map *onto* $\mathbb{R}$. Note that $f_4(x) \neq -1$ for any $x \in \mathbb{R}$.

34. This is not a permutation. Note that $f_5(2) = f_5(-1) = 0$, so f_5 is not one to one.

35. T F T T T T F F F T

36. Every proper subgroup of S_3 is abelian, for such a subgroup has order either 1, 2, or 3 by Exercise 18b.

37. Function composition is associative and there is an identity element, so we have a *monoid*.

38. Let ρ denote the rotation through $2\pi/n$ radians and let ϕ denote the reflection (flip) an axis through a vertex that bisects the vertex angle there. The diagram below shows the top part of a Cayley digraph consisting of two concentric n-gons whose $2n$ vertices correspond to the elements of D_n. We let ϵ denote the identity element.

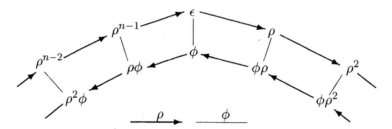

39. If x is a fixed element of G, then mapping each g in G into xg gives a permutation λ_x of G. The map ϕ of G into S_G that carries each x in G into λ_x is then an isomorphism of G with a subgroup of the group S_G.

40. Yes, it is a subgroup.

Closure: If $\sigma(b) = b$ and $\mu(b) = b$, then $(\sigma\mu)(b) = \sigma(\mu(b)) = \sigma(b) = b$.

Identity: The identity carries every element into itself, and hence carries b into b.

Inverses: If $\sigma(b) = b$, then $\sigma^{-1}(b) = b$.

41. No, the set need not be closed under the operation if B has more than one element. Suppose that σ and μ are in the given set, that $b, c \in B$ and $\sigma(b) = c$ but that $\mu(c) \notin B$. Then $(\mu\sigma)(b) = \mu(\sigma(b)) = \mu(c) \notin B$, so $\mu\sigma$ is not in the given set.

42. No, an inverse need not exist. Suppose $A = \mathbb{Z}$ and $B = \mathbb{Z}^+$, and let $\sigma : A \to A$ be defined by $\sigma(n) = n + 1$. Then σ is in the given set, but σ^{-1} is not because $\sigma^{-1}(1) = 0 \notin \mathbb{Z}^+$.

43. Yes, it is a subgroup. Use the proof in Exercise 40, but replace b by B and ( ) by [ ] everywhere.

44. The order of D_n is $2n$ because the regular n-gon can be rotated to n possible positions, and then turned over and rotated to give another n positions. The rotations of the n-gon, without turning it over, clearly form a cyclic subgroup of order n.

45. The group has 24 elements, for any one of the 6 faces can be on top, and for each such face on top, the cube can be rotated in four different positions leaving that face on top. The four such rotations, leaving the top face on top and the bottom face on the bottom, form a cyclic subgroup of order 4. There are two more such rotation groups of order 4, one formed by the rotations leaving the front and back faces in those positions, and one formed by the rotations leaving the side faces in those positions. One exhibits a subgroup of order three by taking hold of a pair of diagonally opposite vertices and rotating through the three possible positions, corresponding to the three edges emanating from each vertex. There are four such diagonally opposite pairs of vertices, giving the desired four groups of order three.

46. Let $n \geq 3$, and let $\rho \in S_n$ be defined by $\rho(1) = 2, \rho(2) = 3, \rho(3) = 1$, and $\rho(m) = m$ for $3 < m \leq n$. Let $\mu \in S_n$ be defined by $\mu(1) = 1, \mu(2) = 3, \mu(3) = 2$, and $\mu(m) = m$ for $3 < m \leq n$. Then $\rho\mu \neq \mu\rho$ so S_n is not commutative. (Note that if $n = 3$, then ρ is our element ρ_1 and μ is our element μ_1 in S_3.)

47. Suppose $\sigma(i) = m \neq i$. Find $\gamma \in S_n$ such that $\gamma(i) = i$ and $\gamma(m) = r$ where $r \neq m$. (Note this is possible because $n \geq 3$.) Then $(\sigma\gamma)(i) = \sigma(\gamma(i)) = \sigma(i) = m$ while $(\gamma\sigma)(i) = \gamma(\sigma(i)) = \gamma(m) = r$, so $\sigma\gamma \neq \gamma\sigma$. Thus $\sigma\gamma = \gamma\sigma$ for all $\gamma \in S_n$ only if σ is the identity permutation.

48. Let c be an element in both $\mathcal{O}_{a,\sigma}$ and $\mathcal{O}_{b,\sigma}$. Then there exist integers r and s such that $\sigma^r(a) = c$ and $\sigma^s(b) = c$. Then $\sigma^{r-s}(a) = \sigma^{-s}(\sigma^r(a)) = \sigma^{-s}(c) = b$. Therefore, for each integer $n \in \mathbb{Z}$, we see that $\sigma^n(b) = \sigma^{n+r-s}(a)$. Hence $\{\sigma^n(b) \mid n \in \mathbb{Z}\} = \{\sigma^n(a) \mid n \in \mathbb{Z}\}$.

49. Let $A = \{a_1, a_2, \cdots a_n\}$. Let $\sigma \in S_A$ be defined by $\sigma(a_i) = a_{i+1}$ for $1 \leq i < n$ and $\sigma(a_n) = a_1$. (Note that σ essentially performs a rotation if the elements of A are spaced evenly about a circle.) It is clear that σ^n is the identity permutation and $|\langle\sigma\rangle| = n = |A|$. We let $H = \langle\sigma\rangle$. Let a_i and a_j be given; suppose $i < j$. Then $\sigma^{j-i}(a_i) = a_j$ and $\sigma^{i-j}(a_j) = a_i$, so H is transitive on A.

50. Let $\langle\sigma\rangle$ be transitive on A and let $a \in A$. Then $\{\sigma^n(a) \mid n \in \mathbb{Z}\}$ must include all elements of A, that is, $\mathcal{O}_{a,\sigma} = A$.

Conversely, suppose that $\mathcal{O}_{a,\sigma} = A$ for some $a \in A$. Then $\{\sigma^n(a) \mid n \in \mathbb{Z}\} = A$. Let $b, c \in A$ and let $b = \sigma^r(a)$ and $c = \sigma^s(a)$. Then $\sigma^{s-r}(b) = \sigma^s(\sigma^{-r}(b)) = \sigma^s(a) = c$, showing that $\langle\sigma\rangle$ is transitive on A.

51. a. The person would see all possible products $a *' b$ and all instances of the associative property for $*'$ in G'.

b. *Associativity:* Let $a, b, c \in G'$. Then $a *' (b *' c) = a *' (c * b) = (c * b) * a = c * (b * a) = (b * a) *' c = (a *' b) *' c$ where we used the fact that G is a group and the definition of $*'$.

Identity: We have $e *' a = a * e = a$ and $a *' e = e * a = a$ for all $a \in G'$.

Inverses: Let $a \in G'$ and let a^{-1} be the inverse of a in G. Then $a^{-1} *' a = a * a^{-1} = e = a^{-1} * a = a *' a^{-1}$, so a^{-1} is also the inverse of a in G'.

52. To start, we show that ρ_a is a permutation of G. If $\rho_a(x) = \rho_a(y)$, then $xa = ya$ and $x = y$ by group cancellation, so ρ_a is one to one. Because $\rho_a(xa^{-1}) = xa^{-1}a = x$, we see that ρ_a maps G onto G. Thus ρ_a is a permutation of the set G. Let $G'' = \{\rho_a \mid a \in G\}$.

For $a, b \in G$, we have $(\rho_a\rho_b)(x) = \rho_a(\rho_b(x)) = \rho_a(xb) = xba = \rho_{ba}(x)$, showing that G'' is closed under permutation multipliction. Because ρ_e is the identity permutation and because $\rho_{a^{-1}}\rho_a = \rho_e$, we see that G'' is a subgroup of the group S_G of all permutations of G.

Let $\phi : G \to G''$ be defined by $\phi(a) = \rho_{a^{-1}}$. Clearly ϕ is one to one and maps G onto G''. From the equation $\rho_a \rho_b = \rho_{ba}$ derived above, we have $\phi(ab) = \rho_{(ab)^{-1}} = \rho_{b^{-1}a^{-1}} = \rho_{a^{-1}}\rho_{b^{-1}} = \phi(a)\phi(b)$, which is the homomorphism property for ϕ. Therefore ϕ is an isomorphism of G onto G''.

53. a. Let us show that that the $n \times n$ permutation matrices form a subgroup of the group $GL(n, \mathbb{R})$ of all invertible $n \times n$ matrices under matrix multiplication. If P_1 and P_2 are two of these permutation matrices, then the exercise stated that $P_1 P_2$ is the matrix that produces the same reordering of the rows of P_2 as the reordering of the rows of I_n that produced P_1. Thus $P_1 P_2$ can again be obtained from the identity matrix I_n by reordering its rows, so it is a permutation matrix. The matrix I_n is the identity permutation matrix. If P is obtained from I_n by a reordering the rows that puts row i in the position j, then P^{-1} is the matrix obtained from I_n by putting row j in position i. Thus the $n \times n$ permutation matrices do form a group under permutation multiplication.

Let us number the elements of G from 1 to n, and number the rows of I_n from 1 to n, say from top to the bottom in the matrix. Theorem 8.16, says we can associate with each $g \in G$ a permutation (reordering) of the elements of G, which we can now think of as a reordering of the numbers from 1 to n, which we can in turn think of as a reordering of the rows of the matrix I_n, which is in turn produced by multiplying I_n on the left by a permutation matrix P. The effect of left multiplication of a matrix by a permutation matrix, explained in the exercise, shows that this association of g with P is an isomorpism of G with a subgroup of the group of all permutation matrices.

b. Proceeding as in the second paragraph of Part(a), we number the elements $e, a, b,$ and c of the Klein 4-group in Table 5.11 with the numbers 1, 2, 3, and 4 respectively. Looking at Table 5.11, we see that left multiplication of each of e, a, b, c by a produces the sequence a, e, c, b. Applying the same reordering to the numbers 1, 2, 3, 4 produces the reordering 2, 1, 4, 3. Thus we associate with a the matrix obtained from the I_4 by interchanging rows 1 and 2 and interchanging rows 3 and 4. Proceeding in this fashion with the other three elements, we obtain these pairings requested in the exercise.

$$
e \leftrightarrow \begin{bmatrix} 1 & 0 & 0 & 0 \\ 0 & 1 & 0 & 0 \\ 0 & 0 & 1 & 0 \\ 0 & 0 & 0 & 1 \end{bmatrix} \quad
a \leftrightarrow \begin{bmatrix} 0 & 1 & 0 & 0 \\ 1 & 0 & 0 & 0 \\ 0 & 0 & 0 & 1 \\ 0 & 0 & 1 & 0 \end{bmatrix} \quad
b \leftrightarrow \begin{bmatrix} 0 & 0 & 1 & 0 \\ 0 & 0 & 0 & 1 \\ 1 & 0 & 0 & 0 \\ 0 & 1 & 0 & 0 \end{bmatrix} \quad
c \leftrightarrow \begin{bmatrix} 0 & 0 & 0 & 1 \\ 0 & 0 & 1 & 0 \\ 0 & 1 & 0 & 0 \\ 1 & 0 & 0 & 0 \end{bmatrix}
$$

9. Orbits, Cycles, and the Alternating Groups

1. $\{1, 2, 5\}, \{3\}, \{4, 6\}$ **2.** $\{1, 5, 7, 8\}, \{2, 3, 6\}, \{4\}$ **3.** $\{1, 2, 3, 4, 5\}, \{6\}, \{7, 8\}$ **4.** $\mathbb{Z}$

5. $\{2n \mid n \in \mathbb{Z}\}, \{2n + 1 \mid n \in \mathbb{Z}\}$ **6.** $\{3n \mid n \in \mathbb{Z}\}, \{3n + 1 \mid n \in \mathbb{Z}\}, \{3n + 2 \mid n \in \mathbb{Z}\}$

7. $\begin{pmatrix} 1 & 2 & 3 & 4 & 5 & 6 & 7 & 8 \\ 4 & 1 & 3 & 5 & 8 & 6 & 2 & 7 \end{pmatrix}$ **8.** $\begin{pmatrix} 1 & 2 & 3 & 4 & 5 & 6 & 7 & 8 \\ 3 & 7 & 2 & 8 & 5 & 4 & 1 & 6 \end{pmatrix}$ **9.** $\begin{pmatrix} 1 & 2 & 3 & 4 & 5 & 6 & 7 & 8 \\ 5 & 4 & 3 & 7 & 8 & 6 & 2 & 1 \end{pmatrix}$

10. $(1, 8), (3, 6, 4)(5, 7)$ and $(1, 8)(3, 4)(3, 6)(5, 7)$

11. $(1, 3, 4)(2, 6)(5, 8, 7)$ and $(1, 4)(1, 3)(2, 6)(5, 7)(5, 8)$

12. $(1, 3, 4, 7, 8, 6, 5, 2)$ and $(1, 2)(1, 5)(1, 6)(1, 8)(1, 7)(1, 4)(1, 3)$

13. (See the answer in the text.)

14. The greatest order is 6 and comes from a product of disjoint cycles of lengths 2 and 3.

15. The greatest order is 6 and comes from a cycle of length 6.

16. The greatest order is 12, coming from a product of disjoint cycles of lengths 4 and 3.

17. The greatest order is 30 and comes from a product of disjoint cycles of lengths 2, 3, and 5.

18. The greatest order is 105 and comes from a product of disjoint cycles of lengths 3, 5, and 7.

19. (See the text answer.)   **20.** The definition is correct.

21. The definition is incorrect; $(1,4,5)$ is a cycle in S_5, but it has three orbits, $\{1,4,5\}, \{2\}$, and $\{3\}$.

A permutation σ of a finite set is a **cycle** if and only if σ has at most one orbit of cardinality greater than 1.

22. The definition is incorrect; it must be specified as a subgroup of some S_n.

The **alternating group** A_n is the subgroup of S_n consisting of the even permutations in S_n.

23. F T F F F F T T T F

24. The even permutations in S_3 are $\rho_0 = (12)(12), \rho_1 = (1,2,3) = (1,3)(1,2)$, and $\rho_2 = (1,3,2) = (1,2)(1,3)$.

	ρ_0	ρ_1	ρ_2
ρ_0	ρ_0	ρ_1	ρ_2
ρ_1	ρ_1	ρ_2	ρ_0
ρ_2	ρ_2	ρ_0	ρ_1

25. Viewing a permutation σ in S_n as permuting the rows of the identity matrix I_n, we see that if σ could be expressed as both an even and odd number of transpositions (giving row interchanges), then the matrix resulting from applying σ to I_n would have both determinant 1 and determinant -1.

26. If σ is a permutation and $\tau = (i,j)$ is a transposition in S_n, then by considering whether i and j are in the same or different orbits of σ, we can show that the number of orbits of σ and of $\tau\sigma$ differ by 1. Starting with the identity permutation ι which has n orbits and multiplying by transpositions to produce σ, we see that the number of transpositions can't be both even and odd, for σ has either an even or odd number of orbits, but not both.

27. a. Note that $(1,2)(1,2)$ is the identity permutation in S_n, and $2 \leq n-1$ if $n > 2$. Because $(1,2,3,4,\cdots,n) = (1,n)(1,n-1)\cdots(1,3)(1,2)$, we see that a cycle of length n can be written as a product of $n-1$ transpositions. Now a permutation in S_n can be written as a product of disjoint cycles, the sum of whose lengths is $\leq n$. If there are r disjoint cycles involved, we see the permutation can be written as a product of at most $n-r$ transpositions. Because $r \geq 1$, we can always write the permutation as a product of at most $n-1$ transpositions.

b. This follows from our proof of **a.**, because we must have $r \geq 2$.

c. Write the odd permutation σ as a product of s transpositions, where $s \leq n-1$ by Part(**a**). Then s is an odd number and $2n+3$ is an odd number, so $2n+3-s$ is an even number. Adjoin $2n+3-s$ transpositions (1,2) as factors at the right of the product of the s transpositions that comprise σ. The same permutation σ results because the product of an even number of factors (1,2) is the identity permutation. Thus σ can be written as a product of $2n+3$ permutations.

If σ is even, we proceed in exactly the same way, but this time s is even so $2n + 8 - s$ is also even. We tack the identity permutation, written as a product of the $2n + 8 - s$ factors $(1, 2)$, onto the end of σ and obtain σ as a product of $2n + 8$ transpositions.

28. LaTeX is unable to draw these figures the way I would like. Make the modifications listed in your own sketches. The final solid lines in your sketch will indicate the orbit after performing the additional transposition (i, j).

a. Consider the right circle to be drawn with a dashed rather than solid curve, and also the short arc from b to j on the left circle to be dashed.

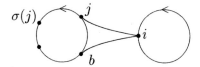

b. Consider the left and right circles both to be drawn with dashed curves, indicating the orbits before performing the additional transposition (i, j).

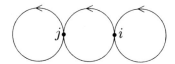

29. Suppose $\sigma \in H$ is an odd permutation. Let $\phi : H \to H$ be defined by $\phi(\mu) = \sigma\mu$ for $\mu \in H$. If $\phi(\mu_1) = \phi(\mu_2)$, then $\sigma\mu_1 = \sigma\mu_2$, so $\mu_1 = \mu_2$ by group cancellation. Also, for any $\mu \in H$, we have $\phi(\sigma^{-1}\mu) = \sigma\sigma^{-1}\mu = \mu$. This shows that ϕ is a one-to-one map of H onto itself. Because σ is an odd permutation, we see that ϕ maps an even permutation onto an odd one, and an odd permutation onto an even one. Because ϕ maps the set of even permutations in H one to one onto the set of odd permutations in H, it is immediate that H has the same number of even permutations as odd permutations. Thus we have shown that if H has one odd permutation, it has the same number of even permutations as odd permutations.

30. If the cycle has length 1, then no element is moved. If it has length $n > 1$, then n elements are moved, because elements not in the cycle are not moved.

31. *Closure:* Let $\sigma, \mu \in H$. If σ moves elements $s_1, s_2, \cdots, s_k$ of A and μ moves elements $r_1, r_2, \cdots, r_m$ of A, then $\sigma\mu$ can't move any elements not in the list $s_1, s_2, \cdots, s_k, r_1, r_2, \cdots, r_m$, so $\sigma\mu$ moves at most a finite number of elements of A, and hence is in H. Thus H is closed under the operation of S_A.

Identity: The identity permutation is in H because it moves no elements of A.

Inverses: Because the elements moved by $\sigma \in H$ are the same as the elements moved by σ^{-1}, we see that for each $\sigma \in H$, we have $\sigma^{-1} \in H$ also. Thus H is a subgroup of S_A.

32. No, K is not a subgroup. If $\sigma, \mu \in K$ and σ is a cycle of length 40 while μ is a cycle of length 30 and these two cycles are disjoint, then $\sigma\mu$ moves 70 elements of A, and is thus not in K. Thus K is not closed under permutation multiplication.

33. Let μ be any odd permutation in S_n. Because σ is an odd permutation, so is σ^{-1}, and consequently $\sigma^{-1}\mu$ is an even permutation, and thus is in A_n Because $\mu = \sigma(\sigma^{-1}\mu)$, we see that μ is indeed a product of σ and a permutation in A_n.

34. It is no loss of generality to assume that $\sigma = (1, 2, 3, \cdots, m)$ where m is odd. Because m is odd, we easily compute that

$$\sigma^2 = (1, 2, 3, \cdots, m)(1, 2, 3, \cdots, m)(1, 3, 5, \cdots, m, 2, 4, 6, \cdots, m-1),$$

which is again a cycle.

35. If σ is a cycle of length n, then σ^r is also a cycle if and only if n and r are relatively prime, that is, if and only if $\gcd(n, r) = 1$. To see why, let the cycle be $\sigma = (1, 2, 3, \cdots, n)$ Computing, we find that σ^r carries 1 into $1 + r$, or more precisely, into $1 + r$ modulo n in case $r \geq n$. Then $1 + r$ modulo n is carried in turn into $1 + 2r$ modulo n, etc. Thus the cycle in σ^r containing 1 is

$$(1, 1 + r, 1 + 2r, 1 + 3r, \cdots, 1 + mr)$$

where all entries are to be read modulo n, and m is the smallest positive integer such that $1 + mr \equiv 1 (\mathrm{mod}\ n)$, or equivalently, the smallest positive integer such that $mr \equiv 0 (\mathrm{mod}\ n)$. Thus the length of the cycle containing 1 in σ^r is the smallest positive integer m such that mr is divisible by n. In order for σ^r to be a cycle, this value of m must be n, which is the case if and only if $\gcd(n, r) = 1$.

36. We must show that λ_a is one to one and onto G. Suppose that $\lambda_a(g_1) = \lambda_a(g_2)$. Then $ag_1 = ag_2$. The group cancellation property then yields $g_1 = g_2$, so λ_a is one to one. Let $g \in G$. Then $\lambda_a(a^{-1}g) = a(a^{-1}g) = g$, so λ_a is onto G.

37. *Closure:* Let $\lambda_a, \lambda_b \in H$. For $g \in G$, we have $(\lambda_a \lambda_b)(g) = \lambda_a(\lambda_b(g)) = \lambda_a(bg) = (ab)g = \lambda_{ab}(g)$. Thus $\lambda_a \lambda_b = \lambda_{ab}$, so H is closed under permutation multiplication (function composition).

Identity: Cearly λ_e is the identity permutation of G.

Inverses: We have $\lambda_a \lambda_{a^{-1}} = \lambda_{aa^{-1}} = \lambda_e$, so $\lambda_a^{-1} = \lambda_{a^{-1}}$. Thus H is a subgroup of S_G.

38. We must show that for each $a, b \in G$, there exists some $\lambda_c \in H$ such that $\lambda_c(a) = b$. We need only choose c such that $ca = b$. That is, we take $c = ba^{-1}$.

39. We show that $(1, 2, 3, \cdots, n)^r (1, 2)(1, 2, 3, \cdots, n)^{n-r} = (1, 2)$ for $r = 0$, $(2, 3)$ for $r = 1$, $(3, 4)$ for $r = 2, \cdots, (n, 1)$ for $r = n - 1$. To see this, note that any number not mapped into 1 or 2 by $(1, 2, 3, \cdots, n)^{n-r}$ is left fixed by the given product. For $r = i$, we see that $(1, 2, 3, \cdots, n)^{n-i}$ maps $i + 1$ into 1, which is then mapped into 2 by $(1, 2)$, which is mapped into $i + 2$ by $(1, 2, 3, \cdots, n)^i$. Also $(1, 2, 3, \cdots, n)^{n-i}$ maps $i + 2 \bmod n$ into 2, which is then mapped into 1 by $(1, 2)$, which is mapped into $i + 1$ by $(1, 2, 3, \cdots, n)^i$.

Let (i, j) be any transposition, written with $i < j$. We easily compute that

$$(i, j) = (i, i + 1)(i + 1, i + 2) \cdots (j - 2, j - 1)(j - 1, j)(j - 2, j - 1) \cdots (i + 1, i + 2)(i, i + 1).$$

By Corollary 9.12, every permutation in S_n can be written as a product of transpositions, which we now see can each be written as a product of the special transpositions $(1, 2), (2, 3), \cdots, (n, 1)$ and we have shown that these in turn can be expressed as products of $(1, 2)$ and $(1, 2, 3, \cdots, n)$. This completes the proof.

10. Cosets and the Theorem of Lagrange

1. (See the answer in the text.)

2. $4\mathbb{Z} = \{\cdots, -8, -4, 0, 4, 8, \cdots\}$, $2 + 4\mathbb{Z} = \{\cdots, -6, -2, 2, 6, 10, \cdots\}$

3. $\langle 2 \rangle = \{0, 2, 4, 6, 8, 10\}$, $1 + \langle 2 \rangle = \{1, 3, 5, 7, 9, 11\}$

4. $\langle 4 \rangle = \{0, 4, 8\}$, $\quad 1 + \langle 4 \rangle = \{1, 5, 9\}$, $\quad 2 + \langle 4 \rangle = \{2, 6, 10\}$, $\quad 3 + \langle 4 \rangle = \{3, 7, 11\}$

5. $\langle 18 \rangle = \{0, 18\}$, $\quad 1 + \langle 18 \rangle = \{1, 19\}$, $\quad 2 + \langle 18 \rangle = \{2, 20\}$, $\quad \cdots$, $\quad 17 + \langle 18 \rangle = \{17, 35\}$

6. $\{\rho_0, \mu_2\}$, $\{\rho_1, \delta_2\}$, $\{\rho_2, \mu_1\}$, $\{\rho_3, \delta_1\}$

7. $\{\rho_0, \mu_2\}$, $\{\rho_1, \delta_1\}$, $\{\rho_2, \mu_1\}$, $\{\rho_3, \delta_2\}$ They are not the same.

8. We do not get a coset group. The 2×2 blocks in the table do not all have elements of just one coset.

	ρ_0	μ_2	ρ_1	δ_2	ρ_2	μ_1	ρ_3	δ_1
ρ_0	ρ_0	μ_2	ρ_1	δ_2	ρ_2	μ_1	ρ_3	δ_1
μ_2	μ_2	ρ_0	δ_1	ρ_3	μ_1	ρ_2	δ_2	ρ_1
ρ_1	ρ_1	δ_2	ρ_2	μ_1	ρ_3	δ_1	ρ_0	μ_2
δ_2	δ_2	ρ_1	μ_2	ρ_0	δ_1	ρ_3	μ_1	ρ_2
ρ_2	ρ_2	μ_1	ρ_3	δ_1	ρ_0	μ_2	ρ_1	δ_2
μ_1	μ_1	ρ_2	δ_2	ρ_1	μ_2	ρ_0	δ_1	ρ_3
ρ_3	ρ_3	δ_1	ρ_0	μ_2	ρ_1	δ_2	ρ_2	μ_1
δ_1	δ_1	ρ_3	μ_1	ρ_2	δ_2	ρ_1	μ_2	ρ_0

9. $\{\rho_0, \rho_2, \}$, $\{\rho_1, \rho_2\}$, $\{\mu_1, \mu_2\}$, $\{\delta_1, \delta_2, \}$

10. The same cosets are obtained as in Exercise 9, so the right cosets of $\{\rho_0, \rho_2\}$ are the same as the left cosets.

11. (See the answer in the text.)

12. $\langle 3 \rangle = \{1, 3, 6, 9, 12, 15, 18, 21\}$ has 8 elements, so its index (the number of cosets) is $24/8 = 3$.

13. $\langle \mu_1 \rangle = \{\rho_0, \mu_1\}$ has 2 elements, so its index (the number of left cosets) is $6/2 = 3$.

14. $\langle \mu_2 \rangle = \{\rho_0, \mu_2\}$ has 2 elements, so its index (the number of left cosets) is $8/2 = 4$.

15. $\sigma = (1, 2, 5, 4)(2, 3) = (1, 2, 3, 5, 4)$ generates a cyclic subgroup of S_5 of order 5, so its index (the number of left cosets) is $5!/5 = 4! = 24$.

16. $\mu = (1, 2, 4, 5)(3, 6)$ generates a cyclic subgroup of S_6 of order 4, (the cycles are disjoint) so its index (the number of left cosets) is $6!/4 = 720/4 = 180$.

17. The definition is incorrect; we have no concept of a left coset of an arbitrary subset of a group G.

Let G be a group and let $H \leq G$. The **left coset of H containing** a is $aH = \{ah \mid h \in H\}$.

18. The definition is correct.

19. T T T F T F T T F T   (i) See the last sentence in this section.

20. This is impossible. For a subgroup H of an abelian group G, we have $a + H = H + a$ for all $a \in G$.

21. For any group G, just take the subgroup $H = G$.   **22.** The subgroup $\{0\}$ of $\mathbb{Z}_6$.

23. This is impossible. Because the cells are disjoint and nonempty, their number cannot exceed the order of the group.

24. This is impossible. The number of cells must divide the order of the group, and 4 does not divide 6.

25. The left cosets of the subgroup H form a partition of G and each coset has the same number of elements as H has.

26. *Reflexive:* Let $a \in G$. then $aa^{-1} = e$ and $e \in H$ because H is a subgroup. Thus $a \sim_R a$.

    Symmetric: Suppose $a \sim_R b$. Then $ab^{-1} \in H$. Because H is a subgroup, $(ab^{-1})^{-1} = ba^{-1}$ is in H, so $b \sim_R a$.

    Transitive: Suppose $a \sim_R b$ and $b \sim_R c$. Then $ab^{-1} \in H$ and $bc^{-1} \in H$. Because H is a subgroup $(ab^{-1})(bc^{-1}) = ac^{-1}$ is in H, so $a \sim_R c$.

27. Let $\phi_g : H \to Hg$ by $\phi_g(h) = hg$ for all $h \in H$. If $\phi_g(h_1) = \phi_g(h_2)$ for $h_1, h_2 \in H$, then $h_1 g = h_2 g$ and $h_1 = h_2$ by group cancellation, so ϕ_g is one to one. Clearly ϕ_g is onto Hg, for if $hg \in Hg$, then $\phi_g(h) = hg$.

28. We show that $gH = Hg$ by showing that each coset is a subset of the other. Let $gh \in gH$ where $g \in G$ and $h \in H$. Then $gh = ghg^{-1}g = [(g^{-1})^{-1}hg^{-1}]g$ is in Hg because $(g^{-1})^{-1}hg^{-1}$ is in H by hypothesis. Thus gH is a subset of Hg.

    Now let $hg \in Hg$ where $g \in G$ and $h \in H$. Then $hg = gg^{-1}hg = g(g^{-1}hg)$ is in gH because $g^{-1}hg$ is in H by hypothesis. Thus Hg is a subset of gH also, so $gH = Hg$.

29. Let $h \in H$ and $g \in G$. By hypothesis, $Hg = gH$. Thus $hg = gh_1$ for some $h_1 \in H$. Then $g^{-1}hg = h_1$, showing that $g^{-1}hg \in H$.

30. It is false. Let $G = S_3, H = \{\rho_0, \mu_1\}, a = \rho_1$ and $b = \mu_3$. (See Table 8.8.) Then $aH = \{\rho_1, \mu_3\} = bH$, but $Ha = \{\rho_1, \mu_2\}$ while $Hb = \{\rho_2, \mu_3\}$.

31. It is true; $b = eb$ and $e \in H$ so $b \in Hb$. Because $Hb = Ha$, we have $b \in Ha$.

32. It is true. Because H is a subgroup, we have $\{h^{-1} \mid h \in H\} = H$. Therefore $Ha^{-1} = \{ha^{-1} \mid h \in H\} = \{h^{-1}a^{-1} \mid h \in H\} = \{(ah)^{-1} \mid h \in H\}$. That is, Ha^{-1} consists of all inverses of elements in aH. Similarly, Hb^{-1} consists of all inverses of elements in bH. Because $aH = bH$, we must have $Ha^{-1} = Hb^{-1}$.

33. It is False. Let H be the subgroup $\{\rho_0, \mu_2\}$ of D_4 in Table 8.12. Then $\rho_1 H = \delta_2 H = \{\rho_1, \delta_2\}$, and $\rho_1^2 H = \rho_2 H = \{\rho_2, \mu_1\}$ but $\delta_2^2 H = \rho_0 H = H = \{\rho_0, \mu_2\}$.

34. The possible orders for a proper subgroup are p, q, and 1. Now p and q are primes and every group of prime order is cyclic, and of course every group of order 1 is cyclic. Thus every proper subgroup of a group of order pq must be cyclic.

35. From the proof in Exercise 32, $Ha^{-1} = \{(ah)^{-1} \mid h \in H\}$ This shows that the map ϕ of the collection of left cosets into the collection of right cosets defined by $\phi(aH) = Ha^{-1}$ is well defined, for if $aH = bH$, then $\{(ah)^{-1} \mid h \in H\} = \{(bh)^{-1} \mid h \in H\}$. Because Ha^{-1} may be any right coset of H, the map is onto the collection of right cosets. Because elements in disjoint sets have disjoint inverses, we see that ϕ is one to one.

36. Let G be abelian of order $2n$ where n is odd. Suppose that G contains two elements, a and b, of order 2. Then $(ab)^2 = abab = aabb = ee = e$ and $ab \neq e$ because the inverse of a is a itself. Thus ab also has order 2. It is easily checked that then $\{e, a, b, ab\}$ is a subgroup of G of order 4. But this is impossible because n is odd and 4 does not divide $2n$. Thus there can't be two elements of order 2.

37. Let G be of order ≥ 2 but with no proper nontrivial subgroups. Let $a \in G, a \neq e$. Then $\langle a \rangle$ is a nontrivial subgroup of G, and thus must be G itself. Because every cyclic group not of prime order has proper subgroups, we see that G must be finite of prime order.

38. Following the hint and using the notation there, it suffices to prove $\{(a_i b_j)K \mid i = 1, \cdots, r, j = 1, \cdots, s\}$ is the collection of distinct left cosets of K in G. Let $g \in G$ and let g be in the left coset $a_i H$ of H. Then $g = a_i h$ for some $h \in H$. Let h be in the left coset $b_j K$ of K in H. Then $h = b_j k$ for some $k \in K$, so $g = a_i b_j k$ and $g \in a_i b_j K$. This shows that the collection given in the hint includes all left cosets of K in G. It remains to show the cosets in the collection are distinct. Suppose that $a_i b_j K = a_p b_q K$, so that $a_i b_j k_1 = a_p b_q k_2$ for some $k_1, k_2 \in K$. Now $b_j k_1 \in H$ and $b_q k_2 \in H$. Thus a_i and a_p are in the same left coset of H, and therefore $i = p$ and $a_i = a_p$. Using group cancellation, we deduce that $b_j k_1 = b_q k_2$. But this means that b_j and b_q are in the same left coset of K, so $j = q$.

39. The partition of G into left cosets of H must be H and $G - H = \{g \in G \mid g \notin H\}$, because G has finite order and H must have half as many elements as G. For the same reason, this must be the partition into right cosets of H. Thus every left coset is also a right coset.

40. Let $a \in G$. Then $\langle a \rangle$ has order d that must divide the order of G, so that $n = dq$. We know that $a^d = e$. Thus $a^n = (a^d)^q = e^q = e$ also.

41. Let $r + \mathbb{Z}$ be a left coset of $\mathbb{Z}$ in $\mathbb{R}$, where $r \in \mathbb{R}$. Let $[r]$ be the greatest integer less than or equal to r. Then $0 \leq r - [r] < 1$ and $r + (-[r])$ is in $r + \mathbb{Z}$. Because the difference of any two distinct elements in $r + \mathbb{Z}$ is at least 1, we see that $x = r - [r]$ must be the unique element $x \in r + \mathbb{Z}$ satisfying $0 \leq x < 1$.

42. Consider a left coset $r + \langle 2\pi \rangle$ of $\langle 2\pi \rangle$ in $\mathbb{R}$. Then every element of this coset is of the form $r + n(2\pi)$ for $n \in \mathbb{Z}$. We know that $\sin(r + n(2\pi)) = \sin r$ for all $n \in \mathbb{Z}$ because the function *sine* is periodic with period 2π. Thus *sine* has the same value at each elements of the coset $r + \langle 2\pi \rangle$.

43. a. *Reflexive:* We have $a = eae$ where $e \in H$ and $e \in K$, so $a \sim a$.

Symmetric: Let $a \sim b$ so $a = hbk$ for some $h \in H, k \in K$. Then $b = h^{-1}ak^{-1}$ and $h^{-1} \in H$ and $k^{-1} \in K$ because H and K are subgroups Thus $b \sim a$.

Transitive: Let $a \sim b$ and $b \sim c$ so $a = hbk$ and $b = h_1 c k_1$ for some $h, h_1 \in H$ and $k, k_1 \in K$. Then $a = hh_1 ck_1 k$ and $hh_1 \in H$ and $k_1 k \in K$ because H and K are subgroups. Thus $a \sim c$.

b. The equivalence class containing the element a is $HaK = \{hak \mid h \in H, k \in K\}$. It can be formed by taking the union of all right cosets of H that contain elements in the left coset aK.

44. a. *Closure:* If $\sigma(c) = c$ and $\mu(c) = c$, then $(\sigma\mu)(c) = \sigma(\mu(c)) = \sigma(c) = c$, so $S_{c,c}$ is closed under permutation multiplication.

Identity: The identity permutation leaves c fixed so it is in $S_{c,c}$.

Inverses: If σ leaves c fixed, then σ^{-1} does also. Thus $S_{c,c}$ is a subgroup of S_A.

b. No, $S_{c,d}$ is not closed under permutation multiplication. If $\sigma, \mu \in S_{c,d}$, then $(\sigma\mu)(c) = \sigma(\mu(c)) = \sigma(d)$. Because $\sigma(c) = d$ and σ is one to one, we know that $\sigma(d) \neq d$ unless $c = d$.

c. Let $\mu \in S_{c,d}$. Then we claim that $S_{c,d}$ is the coset $\mu S_{c,c}$ of $S_{c,c}$ in S_A. It is obvious that $\mu S_{c,c} \subseteq S_{c,d}$. Let $\sigma \in S_{c,d}$. Then $(\mu^{-1}\sigma)(c) = \mu^{-1}(\sigma(c)) = \mu^{-1}(d) = c$. Thus $\mu^{-1}\sigma \in S_{c,c}$ so $\sigma \in \mu S_{c,c}$ which means that $S_{c,d} \subseteq \mu S_{c,c}$. Hence $S_{c,d} = \mu S_{c,c}$.

45. We can work with $\mathbb{Z}_n$. Let d divide n. Then $\langle n/d \rangle = \{0, n/d, 2n/d, \cdots, (d-1)n/d\}$ is a subgroup of $\mathbb{Z}_n$ of order d. It consists precisely of all elements $x \in \mathbb{Z}_n$ such that $dx = x + x + \cdots + x$ for d summands is equal to 0. Because an element x of any subgroup of order d of $\mathbb{Z}_n$ must satisfy $dx = 0$, we see that $\langle n/d \rangle$ is the only such subgroup. Because the order of a subgroup must divide the order of the whole group, we see that these are the only subgroups that $\mathbb{Z}_n$ has.

46. Every element in $\mathbb{Z}_n$ generates a subgroup of some order d dividing n, and the number of generators of that subgroup is $\varphi(d)$ by Corollary 6.16. By the preceding exercise, there is a unique such subgroup of order d dividing n. Thus $\Sigma_{d|n}\varphi(d)$ counts each element of $\mathbb{Z}_n$ once and only once as a generator of a subgroup of order d dividing n. Hence $\Sigma_{d|n}\varphi(d) = n$.

47. Let d be a divisor of $n = |G|$. Now if G contains a subgroup of order d, then each element of that subgroup satisfies the equation $x^d = e$. By the hypothesis that $x^m = e$ has at most m solutions in G, we see that there can be at most one subgroup of each order d dividing n. Now each $a \in G$ has some order d dividing n, and $\langle a \rangle$ has exactly $\varphi(d)$ generators. Because $\langle a \rangle$ must be the only subgroup of order d, we see that the number of elements of order d for each divisor d of n cannot exceed $\varphi(d)$. Thus we have

$$n = \sum_{d|n} (\text{number of elements of } G \text{ of order } d) \leq \sum_{d|n} \varphi(d) = n.$$

This shows that G must have exactly $\varphi(d)$ elements of each order d dividing n, and thus must have $\varphi(n) \geq 1$ elements of order n. Hence G is cyclic.

11. Direct Products and Finitely Generated Abelian Groups

1. (See the answer in the text.)

2. The group is cyclic because there are elements of order 12.

Element	Order	Element	Order	Element	Order
(0,0)	1	(1,0)	3	(2,0)	3
(0,1)	4	(1,1)	12	(2,1)	12
(0,2)	2	(1,2)	6	(2,2)	6
(0,3)	4	(1,3)	12	(2,3)	12

3. $\text{lcm}(2, 2) = 2$. (The abbreviation lcm stands for *least common multiple*.)

4. $\text{lcm}(3, 5) = 15$. (The abbreviation lcm stands for *least common multiple*.)

5. $\text{lcm}(3, 9) = 9$. (The abbreviation lcm stands for *least common multiple*.)

6. $\text{lcm}(4, 6, 5) = 60$. (The abbreviation lcm stands for *least common multiple*.)

7. $\text{lcm}(4, 2, 5, 3) = 60$. (The abbreviation lcm stands for *least common multiple*.)

8. For $\mathbb{Z}_6 \times \mathbb{Z}_8$: the lcm (*least common multiple*) of 6 and 8 which is 24.
For $\mathbb{Z}_{12} \times \mathbb{Z}_{15}$: the lcm of 12 and 15 which is 60.

9. $\{(0,0),(1,0)\}$ $\{(0,0),(0,1)\}$ $\{(0,0),(1,1)\}$

10. There are 7 order 2 subgroups: $\langle(1,0,0)\rangle, \langle(0,1,0)\rangle, \langle(0,0,1)\rangle, \langle(1,1,0)\rangle, \langle(1,0,1)\rangle, \langle(0,1,1)\rangle, \langle(1,1,1)\rangle$.

There are 7 order 4 subgroups:
$\{(0,0,0),(1,0,0),(0,1,0),(1,1,0)\}$ $\{(0,0,0),(1,0,0),(0,0,1),(1,0,1)\}$
$\{(0,0,0),(1,0,0),(0,1,1),(1,1,1)\}$ $\{(0,0,0),(1,1,0),(0,0,1),(1,1,1)\}$
$\{(0,0,0),(1,1,0),(0,1,1),(1,0,1)\}$ $\{(0,0,0),(1,1,1),(0,1,0),(1,0,1)\}$
$\{(0,0,0),(0,1,1),(0,0,1),(0,1,0)\}$

11. (See the answer in the text.)

12. $\{(0,0,0),(1,0,0),(0,1,0),(1,1,0)\}$ $\{(0,0,0),(1,0,0),(0,0,2),(1,0,2)\}$
$\{(0,0,0),(1,0,0),(0,1,2),(1,1,2)\}$ $\{(0,0,0),(1,1,0),(0,0,2),(1,1,2)\}$
$\{(0,0,0),(1,1,0),(0,1,2),(1,0,2)\}$ $\{(0,0,0),(1,1,2),(0,1,0),(1,0,2)\}$
$\{(0,0,0),(0,1,2),(0,0,2),(0,1,0)\}$

13. $\mathbb{Z}_3 \times \mathbb{Z}_{20}$, $\mathbb{Z}_4 \times \mathbb{Z}_{15}$, $\mathbb{Z}_5 \times \mathbb{Z}_{12}$, $\mathbb{Z}_3 \times \mathbb{Z}_4 \times \mathbb{Z}_5$

14. a. 4   **b.** 12   **c.** 12   **d.** 2, 2   **e.** 8

15. The maximum possible order is $12 = \mathrm{lcm}(4, 6)$.

16. Yes. Both groups are isomorphic to $\mathbb{Z}_2 \times \mathbb{Z}_3 \times \mathbb{Z}_4$.

17. The maximum possible order is $120 = \mathrm{lcm}(8, 20, 24)$.

18. No. $\mathbb{Z}_8 \times \mathbb{Z}_{10} \times \mathbb{Z}_{24} \simeq \mathbb{Z}_2 \times \mathbb{Z}_8 \times \mathbb{Z}_8 \times \mathbb{Z}_3 \times \mathbb{Z}_5$ but $\mathbb{Z}_4 \times \mathbb{Z}_{12} \times \mathbb{Z}_{40} \simeq \mathbb{Z}_4 \times \mathbb{Z}_4 \times \mathbb{Z}_8 \times \mathbb{Z}_3 \times \mathbb{Z}_5$

19. The maximum possible order is $180 = \mathrm{lcm}(4, 18, 15)$.

20. Yes. Both groups are isomorphic to $\mathbb{Z}_2 \times \mathbb{Z}_4 \times \mathbb{Z}_3 \times \mathbb{Z}_9 \times \mathbb{Z}_5$.

21. $\mathbb{Z}_8$, $\mathbb{Z}_2 \times \mathbb{Z}_4$, $\mathbb{Z}_2 \times \mathbb{Z}_2 \times \mathbb{Z}_2$

22. $\mathbb{Z}_{16}$, $\mathbb{Z}_2 \times \mathbb{Z}_8$, $\mathbb{Z}_4 \times \mathbb{Z}_4$, $\mathbb{Z}_2 \times \mathbb{Z}_2 \times \mathbb{Z}_4$, $\mathbb{Z}_2 \times \mathbb{Z}_2 \times \mathbb{Z}_2 \times \mathbb{Z}_2$

23. (See the answer in the text.)

24. $\mathbb{Z}_{16} \times \mathbb{Z}_9 \times \mathbb{Z}_5$, $\mathbb{Z}_2 \times \mathbb{Z}_8 \times \mathbb{Z}_9 \times \mathbb{Z}_5$, $\mathbb{Z}_4 \times \mathbb{Z}_4 \times \mathbb{Z}_9 \times \mathbb{Z}_5$,
$\mathbb{Z}_2 \times \mathbb{Z}_2 \times \mathbb{Z}_4 \times \mathbb{Z}_9 \times \mathbb{Z}_5$, $\mathbb{Z}_2 \times \mathbb{Z}_2 \times \mathbb{Z}_2 \times \mathbb{Z}_2 \times \mathbb{Z}_9 \times \mathbb{Z}_5$,
$\mathbb{Z}_{16} \times \mathbb{Z}_3 \times \mathbb{Z}_3 \times \mathbb{Z}_5$, $\mathbb{Z}_2 \times \mathbb{Z}_8 \times \mathbb{Z}_3 \times \mathbb{Z}_3 \times \mathbb{Z}_5$,
$\mathbb{Z}_4 \times \mathbb{Z}_4 \times \mathbb{Z}_3 \times \mathbb{Z}_3 \times \mathbb{Z}_5$, $\mathbb{Z}_2 \times \mathbb{Z}_2 \times \mathbb{Z}_4 \times \mathbb{Z}_3 \times \mathbb{Z}_3 \times \mathbb{Z}_5$,
$\mathbb{Z}_2 \times \mathbb{Z}_2 \times \mathbb{Z}_2 \times \mathbb{Z}_2 \times \mathbb{Z}_3 \times \mathbb{Z}_3 \times \mathbb{Z}_5$

25. (See the answer in the text.)

26. There are 3 of order 24, arising from the subscript sequences 8, 3 and 2, 4, 3 and 2, 2, 2, 3 on the factors $\mathbb{Z}$. Similarly, there are 2 of order 25 arising from the subscript sequences 25 and 5, 5. There are $3 \cdot 2 = 6$ of order $24 \cdot 25$, because each of three for order 24 can be paired with each of the two of order 25.

27. Because there are no primes that divide both m and n, any abelian group of order mn is isomorphic to a direct product of cyclic groups of prime-power order where all cyclic groups given by primes dividing m appear before any of the primes dividing n. Thus any abelian group of order mn is isomorphic to a direct product of a group of order m with a group of order n, when $\gcd(m, n) = 1$. Because there are r choices for the group of order m and s choices for the group of order n, there are rs choices in all.

28. We have $10^5 = 2^5 5^5$. There are 7 groups of order 2^5, up to isomorphism, by Exercise 23. Replacing factors 2 by factors 5 in the answer to Exercise 23, we see that there are also 7 abelian groups of order 5^5, up to isomorphism. By Exercise 27, there are $7 \cdot 7 = 49$ abelian groups of order 10^5, up to isomorphism.

29. a. We just illustrate with the computation for groups of order p^8, to get the last entry 22 in the table. We try to be systematic, according as there is just one factor $\mathbb{Z}$, then two factors $\mathbb{Z}$, then three, etc. For each of these cases, we list the possible sequences of exponents i that appear on the subscripts p^i on the factors $\mathbb{Z}_{p^i}$.

Factors	Exponent Sequences	Total
1	8	1
2	1, 7 2, 6 3, 5 4, 4	4
3	1, 1, 6 1, 2, 5 1, 3, 4 2, 2, 4 2, 3, 3	5
4	1, 1, 1, 5 1, 1, 2, 4 1, 1, 3, 3 1, 2, 2, 3 2, 2, 2, 2	5
5	1, 1, 1, 1, 4 1, 1, 1, 2, 3 1, 1, 2, 2, 2	3
6	1, 1, 1, 1, 1, 3 1, 1, 1, 1, 2, 2	2
7	1,1,1,1,1,1,2	1
8	1,1,1,1,1,1,1,1	1

Thus there are a total of $1 + 4 + 5 + 5 + 3 + 2 + 1 + 1 = 22$ abelian groups of order p^8, up to isomorphism.

b. We use the entries from the table in the answer in the text.

i) $3 \cdot 5 \cdot 15 = 225$ ii) $5 \cdot 15 = 225$ iii) $q^5 r^4 q^3 = q^8 r^4$ so our computation becomes $22 \cdot 5 = 110$

30. Finish this diagram by a double arrow at the right end of each row looping around to the left end of the row, and a single arrow at the bottom of each column looping around to the top of the column. LaTeX can't do dashed arrows or the looping ones.

$$
\begin{array}{ccccccccc}
(0,0) & \Rightarrow & (0,1) & \Rightarrow & (0,2) & \Rightarrow & \cdots & \Rightarrow & (0,m) \\
\downarrow & & \downarrow & & \downarrow & & \cdots & & \downarrow \\
(1,0) & \Rightarrow & (1,1) & \Rightarrow & (1,2) & \Rightarrow & \cdots & \Rightarrow & (1,m) \\
\downarrow & & \downarrow & & \downarrow & & \cdots & & \downarrow \\
(2,0) & \Rightarrow & (2,1) & \Rightarrow & (2,2) & \Rightarrow & \cdots & \Rightarrow & (2,m) \\
\vdots & & \vdots & & \vdots & & \vdots & & \vdots \\
\downarrow & & \downarrow & & \downarrow & & \cdots & & \downarrow \\
(n,0) & \Rightarrow & (n,1) & \Rightarrow & (n,2) & \Rightarrow & \cdots & \Rightarrow & (n,m)
\end{array}
$$

$$
\begin{array}{cc}
(0,1) & (1,0) \\
\Rightarrow & \longrightarrow
\end{array}
$$

31. a. It is abelian if the two generators a and b representing the two arc types commute. From a diagram, we check that this is the case when the arrows on both n-gons have the same (clockwise or counterclockwise) direction.
b. $\mathbb{Z}_2 \times \mathbb{Z}_n$ **c.** $\mathbb{Z}_2 \times \mathbb{Z}_n$ is cyclic when n is odd.
d. It is isomorphic to the dihedral group D_n, for it is generated by an element ρ (a rotation) of order n and an element μ (a reflection) of order 2 satisfying $\rho\mu = \mu\rho^{-1}$.

32. T T F T F F F F F T   **33.** $\mathbb{Z}_p$ is an example for any prime p.

34. a. Cardinality considerations show that the only subgroup of $\mathbb{Z}_5 \times \mathbb{Z}_6$ that it isomorphic to $\mathbb{Z}_5 \times \mathbb{Z}_6$ is $\mathbb{Z}_5 \times \mathbb{Z}_6$ itself.

b. There are an infinite number of them. Subgroup $m\mathbb{Z} \times n\mathbb{Z}$ is isomorphic to $\mathbb{Z} \times \mathbb{Z}$ for all positive integers m and n.

35. S_3 is an example, for its nontrivial proper subgroups are all abelian, so any direct product of them would be abelian, and could not be isomorphic to nonabelian S_3.

36. T F F T T F T F T T   **37.** The numbers are the same.

38. a. Yes, it has just one subgroup of order 8 because $72 = 8 \cdot 9$ so the subgroup of order 8 consists of all elements having order that divides 8.

b. No. If the group is $\mathbb{Z}_8 \times \mathbb{Z}_9$, then it has just one subgroup $\{(0,0), (2,0), (4,0), (6,0)\}$ of order 4, but if it is isomorphic to $\mathbb{Z}_2 \times \mathbb{Z}_4 \times \mathbb{Z}_9$, it has more than one subgroup of order 4, namely

$$\{(0,0,0), (0,1,0), (0,2,0), (0,3,0)\} \text{ and } \{(0,0,0), (2,0,0), (0,2,0), (2,2,0)\}.$$

39. Let G be abelian and let $a, b \in G$ have finite order. Then $a^r = b^s = e$ for some positive integers r and s. Because G is abelian, we see that $(ab)^{rs} = (a^r)^s (b^s)^r = e^s e^r = ee = e$, so ab has finite order. This shows that the subset H of G consisting of all elements of finite order is closed under the group operation. Of course $e \in H$ because e has order 1. If $a^r = e$, then $a^{-r} = (a^{-1})^r = e$ also, showing that $a \in H$ implies $a^{-1} \in H$, and completing the demonstration that H is a subgroup of G.

40. The torsion subgroup of $\mathbb{Z}_4 \times \mathbb{Z} \times \mathbb{Z}_3$ has order $4 \cdot 3 = 12$. The torsion subgroup of $\mathbb{Z}_{12} \times \mathbb{Z} \times \mathbb{Z}_{12}$ has order $12 \cdot 12 = 144$.

41. $\{-1, 1\}$ **42.** $\{e^{q\pi i} \mid q \in \mathbb{Q}\}$

43. Let G be a finitely generated abelian group and write it (up to isomorphism) in the form described in Theorem 11.12. Put parentheses around the first portion, involving factors of the form $\mathbb{Z}_{p^r}$, and then put parenthese around the second part, containing the factors $\mathbb{Z}$. We have then exhibited G, up to isomorphism, in the form $H \times K$ where H is a torsion group and K is torsion free.

44. a. 36 ;   **b.** 2, 12, and 60 as explained in Part(**c**).

c. Find an isomorphic group that is a direct product of cyclic groups of prime-power order. For each prime divisor of the order of the group, write the subscripts in the direct product involving that prime in a row in order of increasing magnitude. Keep the right-hand ends of the rows aligned. Then take the product of the numbers down each column of the array. These are the torsion coefficients. Illustrating with the group in **b.**, we first form $\mathbb{Z}_2 \times \mathbb{Z}_3 \times \mathbb{Z}_3 \times \mathbb{Z}_4 \times \mathbb{Z}_4 \times \mathbb{Z}_5$. We now form the array and multiply columns, as in

$$
\begin{array}{rrr}
2 & 4 & 4 \\
  & 3 & 3 \\
  &   & 5 \\
\hline
2 & 12 & 60
\end{array}
$$

obtaining the torsion coefficients 2, 12, 60.

45. If m and n are relatively prime, then $(1, 1)$ has order mn so the group is cyclic of order mn. If m and n are not relatively prime, then no element has order exceeding the least common multiple of m and n, which has to be less than mn, so the group is not cyclic.

46. Computation in a direct product of n groups consists of computing using the individual group operations in each of the n components. In a direct product of abelian groups, the individual group operations are all commutative, and it follows at once that the direct product is an abelian group.

47. *Closure:* Let $a, b \in H$. Then $a^2 = b^2 = e$. Because G is abelian, we see that $(ab)^2 = abab = aabb = ee = e$, so $ab \in H$ also. Thus H is closed under the group operation.

   Identity: We are given that $e \in H$.

   Inverses: For all $a \in H$, the equation $a^2 = e$ means that $a^{-1} = a \in H$. Thus H is a subgroup.

48. Yes, H is a subgroup for order 3, by essentially the same proof as in the preceding exercise. No, H is not a subgroup for order 4, because the square of an element of order 4 has order 2, so H is not closed under the operation. For prime positive integers, H will be a subgroup.

49. S_3 is a counterexample.

50. a. $(h, k) = (h, e)(e, k)$ **b.** $(h, e)(e, k) = (h, k) = (e, k)(h, e)$. **c.** The only element of $H \times K$ of the form (h, e) and also of the form (e, k) is $(e, e) = e$.

51. *Uniqueness:* Suppose that $g = hk = h_1 k_1$ for $h, h_1 \in H$ and $k, k_1 \in K$. Then $h_1^{-1} h = k_1 k^{-1}$ is in both H and K, and we know that $H \cap K = \{e\}$. Thus $h_1^{-1} h = k_1 k^{-1} = e$, from which we see that $h = h_1$ and $k = k_1$.

   Isomorphic: Suppose $g_1 = h_1 k_1$ and $g_2 = h_2 k_2$. Then $g_1 g_2 = h_1 k_1 h_2 k_2 = h_1 h_2 k_1 k_2$ because elements of H and K commute by hypothesis **b.** Thus by uniqueness, $g_1 g_2$ is renamed $(h_1 h_2, k_1 k_2) = (h_1, k_1)(h_2, k_2)$ in $H \times K$.

52. Recall that every subgroup of a cyclic group is cyclic. Thus if a finite abelian group G contains a subgroup isomorphic to $\mathbb{Z}_p \times \mathbb{Z}_p$, which is not cyclic, then G cannot be cyclic.

   Conversely, suppose that G is a finite abelian group that is not cyclic. By Theorem 11.12, G contains a subgroup isomorphic to $\mathbb{Z}_{p^r} \times \mathbb{Z}_{p^s}$ for the same prime p, because if all components in the direct product correspond to distinct primes, then G would be cyclic by Theorem 11.5. The subgroup $\langle p^{r-1} \rangle \times \langle p^{s-1} \rangle$ of $\mathbb{Z}_{p^r} \times \mathbb{Z}_{p^s}$ is clearly isomorphic to $\mathbb{Z}_p \times \mathbb{Z}_p$.

53. By the Theorem of Lagrange, the order of an element of a finite group (that is, the order of the cyclic subgroup it generates) divides the order of the group. Thus if G has prime-power order, then the order of every element is also a power of the prime. The hypothesis of commutativity was not used.

54. By Theorem 11.12, the groups that appear in the decompositions of $G \times K$ and of $H \times K$ are unique except for the order of the factors. Because $G \times K$ and $H \times K$ are isomorphic, these factors in their decompositions must be the same. Because the decompositions of $G \times K$ and $H \times K$ can both be written in the order with the factors from K last, we see that G and H must have the same factors in their expression in the decomposition described in Theorem 11.12. Thus G and H are isomorphic.

12. Plane Isometries

1. (See the answer in the text.)   **2.**

	P	R
P	P	R
R	R	P

3. (See the answer in the text.)

4. A figure with a one-element group of plane symmetries.

5. A figure with a two-element group of plane symmetries.

6. A figure with a three-element group of plane symmetries.

7. (See the answer in the text.)

8. A figure with a four-element group of plane symmetries, isomorphic to $\mathbb{Z}_2 \times \mathbb{Z}_2$.

9. (See the answer in the text.)

10. Rotations and reflections can have fixed points. A translation slides all points by the same amount, and a glide reflection moves all points the same distance.

11. A rotations is the only type with just one fixed point.

12. No plane isometry has exactly two fixed points. If P and Q are left fixed, so are all points on the line through these two points.

13. Only the identity and reflections have an infinite number of fixed points.

14. If P, Q, and R are three non collinear points, then three circles with centers at P, Q, and R have at most one point in common. Namely, two circles intersect in two points, and if the center of the third circle does not lie on the line through the centers of the first two, then it can't pass through both points of intersection of the first two. An isometry ϕ that leaves P, Q, and R fixed must leave every other point S fixed because it must preserve its distance to P, Q, and R, so that both S and $\phi(S)$ must be the unique points of intersection of three circles with P, Q, and R as centers and the appropriate radii.

15. If $\phi(P_i) = \psi(P_i)$ for $i = 1, 2$, and 3, then $\phi^{-1}(\psi(P_i)) = P_i$ for $i = 1, 2$, and 3. Thus by Exercise 14, $\phi^{-1}\psi = \iota$, the identity map, so $\psi = \phi$.

16. No, the product of two rotations (about different points) may be a translation, so the set of rotations is not closed under multiplication.

17. (See the answer in the text.)

18. Yes, they do form a subgroup. Think of the fixed point as the origin in the plane of complex numbers. Rotations about that point correspond to multiplying by complex numbers z such that $|z| = 1$. The set $U = \{z \in \mathbb{C} \mid |z| = 1\}$ is a group under multiplication, and the multiplication corresponds to function composition of rotations. The number 1 corresponds to the identity map.

19. (See the answer in the text.)

20. No, the product of two glide reflections is orientation preserving, and hence is not a glide reflection.

21. (See the answer in the text.)

22. Because G is finite, it can contain no translations, so the orientation preserving isometries in G consist of the rotations in G and the identity map. Because the product of two orientation preserving isometries is orientation preserving, we see that the set H of all orientation preserving isometries in G is closed under multiplication (function composition). Because the inverse of a rotation is also a rotation, we see that H contains the inverse of each element, and is thus a subgroup of G. If $H \neq G$, let μ be an element of G that is not in H. If σ is another element of G not in H, then $\mu^{-1}\sigma \in H$, because the product of two orientation reversing isometries is order preserving. Thus $\sigma \in \mu H$. This shows that the coset μH contains all elements of G that are not in H. Because $|\mu H| = |H|$, we see that in this case $|G| = 2|H|$.

23. We can consider all the rotations in G to be clockwise. Let ρ be the rotation in G which rotates the plane clockwise through the smallest positive angle. Such a rotation exists because G is a finite group. We claim that G is cyclic, generated by ρ. Let α be the angle of rotation for ρ. Let σ be another rotation in G with angle of rotation β. Write $\beta = q\alpha + \theta$, according to the division algorithm. Then $\theta = \beta - q\alpha$, and the isometry $\rho^{-q}\sigma$ rotates the plane through the angle θ. By the division algorithm, either $\theta = 0$ or $0 < \theta < \alpha$. Because $0 < \theta < \alpha$ is impossible by our choice of α as the smallest nonzero angle of rotation, we see that $\theta = 0$. Hence $\beta = q\alpha$, so $\sigma = \rho^q$, showing that G is cyclic and generated by ρ.

24. a. No **b.** No **c.** No **d.** No **e.** $\mathbb{Z}$

25. a. No **b.** No **c.** Yes **d.** No **e.** D_∞

26. a. No **b.** Yes **c.** No **d.** No **e.** $\mathbb{Z} \times \mathbb{Z}_2$

27. a. Yes **b.** No **c.** No **d.** No **e.** D_∞

28. a. Yes **b.** Yes **c.** Yes **d.** No **e.** $D_\infty \times \mathbb{Z}_2$

29. a. No **b.** No **c.** No **d.** Yes **e.** $\mathbb{Z}$

30. a. Yes **b.** No **c.** Yes **d.** Yes **e.** D_∞

31. a. Yes, 90° and 180° **b.** Yes **c.** No **32. a.** Yes, 180° **b.** Yes **c.** No

33. a. No   **b.** No   **c.** No   **34. a.** No   **b.** Yes   **c.** No

35. a. Yes, 180°   **b.** Yes   **c.** No   **36. a.** Yes, 60°, 120°, and 180°   **b.** Yes   **c.** No

37. a. Yes, 120°   **b.** Yes   **c.** No   **38. a.** No   **b.** No   **c.** Yes   **d.** (1, 0) and (0, 1)

39. a. Yes, 90° and 180°   **b.** Yes   **c.** No   **d.** (1, 1) and (-1,1)

40. a. Yes, 120°   **b.** No   **c.** No   **d.** (1,0) and $(1, \sqrt{3})$

41. a. Yes, 120°   **b.** Yes   **c.** No   **d.** (0, 1) and $(\sqrt{3}, 1)$

42. Let us call the four diagonals of the cube through its center d_1, d_2, d_3, and d_4. By rotating the cube, any diagonal can be moved to fall on the line segment formerly occupied by any of the diagonals (including itself) in two ways. For example, if d_1 goes from point P to point Q and d_2 from point R to point S, then d_1 can be moved into the segment from R to S with the vertex formerly at P falling on either point R or point S. Thus diagonal d_1 can be moved onto a diagonal (including itself) in $4 \cdot 2 = 8$ ways. Once diagonal d_1 is in position, we can keep the ends of d_1 fixed and rotate the cube through a total of three positions, giving a total of $3 \cdot 8 = 24$ possible rotations of the cube. But the set $\{d_1, d_2, d_3, d_4\}$ admits only $4! = 24$ permutations. Thus, identifying each rotation with one of these permutations of the four diagonals, we see that the group of rotations must be isomorphic to the full symmetric group S_4 on four letters.

13. Homomorphisms

1. It is a homomorphism, because $\phi(m + n) = m + n = \phi(m) + \phi(n)$.

2. It is not a homomorphism, because $\phi(2.6 + 1.6) = \phi(4.2) = 4$ but $\phi(2.6) + \phi(1.6) = 2 + 1 = 3$.

3. It is a homomorphism, because $\phi(xy) = |xy| = |x||y| = \phi(x)\phi(y)$ for $x, y \in \mathbb{R}^*$.

4. It is a homomorphism. Let $m, n \in \mathbb{Z}_6$. In $\mathbb{Z}$, let $m + n = 6q + r$ by the division algorithm in $\mathbb{Z}$. Then $\phi(m +_6 n)$ is the remainder of r modulo 2. Because 2 divides 6, the remainder of $m + n$ in $\mathbb{Z}$ modulo 2 is also the remainder of r modulo 2. Now the map $\gamma : \mathbb{Z} \to \mathbb{Z}_2$ of Example 13.10 is a homomorphism, and we have just shown that $\phi(m +_6 n) = \gamma(m + n)$ for $m, n \in \mathbb{Z}_6$. Thus we have $\phi(m +_6 n) = \gamma(m + n) = \gamma(m) +_2 \gamma(n) = \phi(m) + \phi(n)$.

5. It is not a homomorphism, because $\phi(5 +_9 7) = \phi(3) = 1$ but $\phi(5) +_2 \phi(7) = 1 +_2 1 = 0$.

6. It is a homomorphism, because $\phi(x + y) = 2^{x+y} = 2^x 2^y = \phi(x)\phi(y)$ for $x, y \in \mathbb{R}^*$.

7. It is a homomorphism. Let $a, b \in G_i$. Then

$$
\begin{aligned}
\phi(ab) &= (e_1, e_2, \cdots, ab, \cdots, e_r) \\
&= (e_1, e_2, \cdots, a, \cdots, e_r)(e_1, e_2, \cdots, b, \cdots, e_r) \\
&= \phi(a)\phi(b).
\end{aligned}
$$

8. It is not a homomorphism if G is not abelian. We have $\phi(ab) = (ab)^{-1} = b^{-1}a^{-1} = \phi(b)\phi(a)$ which may not equal $\phi(a)\phi(b)$ if G is not abelian. For a specific example, let $G = S_3$ with our the notation in Section 8. Then $\phi(\rho_1\mu_1) = \phi(\mu_3) = \mu_3^{-1} = \mu_3$, but $\phi(\rho_1)\phi(\mu_1) = \rho_1^{-1}\mu_1^{-1} = \rho_2\mu_1 = \mu_2$.

9. Yes, it is a homomorphism. By calculus, $(f+g)'' = f'' + g''$. Then $\phi(f+g) = (f+g)'' = f'' + g'' = \phi(f) + \phi(g)$.

10. Yes, it is a homomorphism since we have $\int_a^b [f(x) + g(x)]\,dx = \int_a^b f(x)\,dx + \int_a^b g(x)\,dx$, so $\phi(f+g) = \int_0^4 [f(x) + g(x)]\,dx = \int_0^4 f(x)\,dx + \int_0^4 g(x)\,dx = \phi(f) + \phi(g)$.

11. Yes, it is a homomorphism. By definition, $3(f+g)(x) = 3[f(x) + g(x)] = 3 \cdot f(x) + 3 \cdot g(x) = (3f)(x) + (3g)(x) = (3f + 3g)(x)$, showing that $3(f+g)$ and $3f + 3g$ are the same function. Thus $\phi(f+g) = 3(f+g) = 3f + 3g = \phi(f) + \phi(g)$.

12. No, it is not a homomorphism. Let $n = 2$ and $A = \begin{bmatrix} 1 & 0 \\ 0 & 1 \end{bmatrix}$ and $B = \begin{bmatrix} 1 & 1 \\ 1 & 1 \end{bmatrix}$, so that $A + B = \begin{bmatrix} 2 & 1 \\ 1 & 2 \end{bmatrix}$. We see that $\phi(A + B) = \det(A + B) = 4 - 1 = 3$ but $\phi(A) + \phi(B) = \det(A) + \det(B) = 1 + 0 = 1$.

13. Yes, it is a homomorphism. Let $A = (a_{ij})$ and $B = (b_{ij})$ where the element with subscript ij is in the ith row and jth column. Then

$$
\begin{aligned}
\phi(A + B) &= \operatorname{tr}(A + B) = \sum_{i=1}^n (a_{ii} + b_{ii}) = \sum_{i=1}^n a_{ii} + \sum_{i=1}^n b_{ii} \\
&= \operatorname{tr}(A) + \operatorname{tr}(B) = \phi(A) + \phi(B).
\end{aligned}
$$

14. No, it is not a homomorphism. We see that $\phi(I_n I_n) = \phi(I_n) = \operatorname{tr}(I_n) = n$, but $\phi(I_n) + \phi(I_n) = \operatorname{tr}(I_n) + \operatorname{tr}(I_n) = n + n = 2n$.

15. No, it is not a homomorphism. Let $f(x) = x^2 + 1$. We have $\phi(f \cdot f) = \int_0^1 (x^2 + 1)^2\,dx = \int_0^1 (x^4 + 2x^2 + 1)\,dx = \frac{1}{5} + \frac{2}{3} + 1 = \frac{28}{15}$ but $\phi(f)\phi(f) = \left[\int_0^1 (x^2 + 1)\,dx\right]^2 = (\frac{1}{3} + 1)^2 = \frac{16}{9}$.

16. $\operatorname{Ker}(\phi)$ consists of the even permutations, so $\operatorname{Ker}(\phi) = A_3 = \{\rho_0, \rho_1, \rho_2\}$.

17. $\operatorname{Ker}(\phi) = 7\mathbb{Z}$ because 4 has order 7 in $\mathbb{Z}_7$. We have

$$
\begin{aligned}
\phi(25) &= \phi(21 + 4) = \phi(21) +_7 \phi(4) = 0 +_7 \phi(4) \\
&= \phi(1) +_7 \phi(1) +_7 \phi(1) +_7 \phi(1) \\
&= 4 +_7 4 +_7 4 +_7 4 = 1 +_7 1 = 2.
\end{aligned}
$$

18. $\operatorname{Ker}(\phi) = 5\mathbb{Z}$ because 6 has order 5 in $\mathbb{Z}_{10}$. We have

$$
\begin{aligned}
\phi(18) &= \phi(15 + 3) = \phi(15) +_{10} \phi(3) = 0 +_{10} \phi(3) \\
&= \phi(1) +_{10} \phi(1) +_{10} \phi(1) = 6 +_{10} 6 +_{10} 6 \\
&= 2 +_{10} 6 = 8.
\end{aligned}
$$

19. In S_8, we have $\sigma = (1,4,2,6)(2,5,7) = (1,4,2,5,7,6)$ which is of order 6, so $\operatorname{Ker}(\phi) = 6\mathbb{Z}$. Then $\phi(20) = \phi(18 + 2) = \phi(18)\phi(2) = \iota\,\sigma^2 = (1,2,7)(4,5,6)$.

20. $\operatorname{Ker}(\phi) = \langle 5 \rangle = \{0, 5\}$ because 8 has order 5 in $\mathbb{Z}_{20}$. We have $\phi(3) = 8 +_{20} 8 +_{20} 8 = 16 +_{20} 8 = 4$.

21. The element $\sigma = (2,5)(1,4,6,7)$ has order 4, so $\operatorname{Ker}(\phi) = \langle 4 \rangle = \{0, 4, 8, 12, 16, 20\}$. Then $\phi(14) = \phi(12 +_{24} 2) = \iota\,\sigma^2 = (1,6)(4,7)$.

22. Now $\phi(m,n) = 3m - 5n$ so $\text{Ker}(\phi) = \{(m,n) \mid 3m = 5n \text{ for } m, n \in \mathbb{Z}\}$. Then $\phi(-3, 2) = 3(-3) - 5(2) = -19$.

23. We have $\phi(m,n) = (2m - n, -3m + 5n)$ and the only simultaneous solution of the equations $2m - n = 0$ and $-3m + 5n = 0$ is $m = n = 0$, so $\text{Ker}(\phi) = \{(0,0)\}$. Also, $\phi(4,6) = (8 - 6, -12 + 30) = (2, 18)$.

24. Let $\sigma = (3,5)(2,4)$ and $\mu = (1,7)(6,10,8,9)$. Because σ has order 2 and μ has order 4, we see that $\text{Ker}(\phi) = 2\mathbb{Z} \times 4\mathbb{Z}$. Because our all the cycles are disjoint, we find that

$$\begin{aligned} \phi(3, 10) &= \sigma^3 \mu^{10} = (3,5)^3(2,4)^3(1,7)^{10}(6,10,8,9)^{10} \\ &= (3,5)(2,4)(6,10,8,9)^2 = (3,5)(2,4)(6,8)(9,10). \end{aligned}$$

25. Because the homomorphism ϕ must be *onto* $\mathbb{Z}$, $\phi(1)$ must be a generator of $\mathbb{Z}$. Thus there are only two such homomorphisms ϕ, one where $\phi(1) = 1$ so $\phi(n) = n$ for all $n \in \mathbb{Z}$, and one where $\phi(1) = -1$ so $\phi(n) = -n$ for all $n \in \mathbb{Z}$.

26. There are an infinite number of them. For any nonzero $n \in \mathbb{Z}$, we know that $\langle n \rangle$ is isomorphic to $\mathbb{Z}$, and that $\phi : \mathbb{Z} \to \mathbb{Z}$ given by $\phi(m) = mn$ is an isomorphism, and hence a homomorphism. Of course ϕ defined by $\phi(m) = 0$ for all $m \in \mathbb{Z}$ is also a homomorphism.

27. There are two of them; one where $\phi(1) = 1$ (see Example 13.10 with n = 2) and one where $\phi(1) = 0$.

28. Because we must have $\phi_g(e)$ by Theorem 13.12, we must have $ge = e$, so $g = e$ is the only possibility. Because $\phi_e(x) = ex = x$ is the identity map, it is indeed a homomorphism.

29. We have $\phi_g(xy) = g(xy)g^{-1} = (gxg^{-1})(gyg^{-1}) = \phi_g(x)\phi_g(y)$ for all $x, y \in G$, so ϕ_g is a homomorphism for all $g \in G$.

30. Incorrect. It should say what ϕ maps to what, what x and y are, and include the necessary quantifier, "for all".

A map ϕ of a group G into a group G' is a **homomorphism** if and only if $\phi(xy) = \phi(x)\phi(y)$ for all $x, y \in G$.

31. The definition is correct.   **32.** T T F T F F T T F F

33. There are no nontrivial homomorphisms. By Theorem 13.12, the image $\phi[\mathbb{Z}_{12}]$ would be a subgroup of $\mathbb{Z}_5$, and hence all of $\mathbb{Z}_5$ for a nontrivial ϕ. By Theorem 13.15, the number of cosets of $\text{Ker}(\phi)$ must then be 5. But the number of cosets of a subgroup of a finite group is a divisor of the order of the group, and 5 does not divide 12.

34. Let $\phi(n)$ be the remainder of n when divided by 4 for $n \in \mathbb{Z}_{12}$. Replacing 6 by 12 and 2 by 4 in the solution of Exercise 4 shows that ϕ is a homomorphism.

35. Let $\phi(m,n) = (m, 0)$ for all $(m,n) \in \mathbb{Z}_2 \times \mathbb{Z}_4$.

36. There are no nontrivial homomorphisms because $\mathbb{Z}$ has no finite subgroups other than $\{0\}$.

37. Let $\phi(n) = \rho_n$ for $n \in \mathbb{Z}_3$, using our notation in Section 8 for elements of S_3. Both $\mathbb{Z}_3$ and $\langle \rho_1 \rangle$ are cyclic of order 3.

38. Let $\phi(n)$ be the identity in S_3 for n even, and the transposition $(1,2)$ for n odd in $\mathbb{Z}$. Note that $\langle (1,2) \rangle$ is of order 2, isomorphic to $\mathbb{Z}_2$, and this homomorphism mirrors the homomorphism γ of Example 13.10 for $n = 2$.

39. Let $\phi(m, n) = 2m$. Then $\phi((m, n) + (r, s)) = \phi(m+r, n+s) = 2(m+r) = 2m+2r = \phi(m, n) + \phi(r, s)$.

40. Let $\phi(2n) = (2n, 0)$ for $n \in \mathbb{Z}$. Then $\phi(2m + 2n) = \phi(2(m + n)) = (2(m + n), 0) = (2m + 2n, 0) = (2m, 0) + (2n, 0) = \phi(2n) + \phi(2m)$.

41. Viewing D_4 as a group of permutations, let $\phi(\sigma) = (1, 2)$ for each odd permutation $\sigma \in D_4$, and let $\phi(\sigma)$ be the identity permutation for each even $\sigma \in D_4$. Note that $\langle (1, 2) \rangle$ is a subgroup of S_3 of order 2, isomorphic to $\mathbb{Z}_2$. This homomorphism mirrors the homomorphism for $n = 4$ in Example 13.3, restricted to the subgroup D_4 of S_4.

42. For each $\sigma \in S_3$, let $\phi(\sigma) = \mu$ where $\mu(i) = \sigma(i)$ for $i = 1, 2, 3$ and $\mu(4) = 4$. This is obviously a homomorphism.

43. Let $\phi(\sigma) = (1, 2)$ for each odd permutation $\sigma \in S_4$, and let $\phi(\sigma)$ be the identity permutation for each even $\sigma \in S_4$. Note that $\langle (1, 2) \rangle$ is a subgroup of S_3 of order 2, isomorphic to $\mathbb{Z}_2$. This homomorphism mirrors the homomorphism for $n = 4$ in Example 13.3.

44. Because $\phi[G] = \{\phi(g) \mid g \in G\}$, we see that $|\phi[G]| \le |G|$, so $|\phi[G]|$ must be finite also. By Theorem 13.15, there is a one-to-one correspondence between the elements of $\phi[G]$ and the cosets of $\mathrm{Ker}(\phi)$ in G. Thus $|\phi[G]| = |G|/|\mathrm{Ker}(\phi)|$, so $|\phi[G]|$ divides $|G|$.

45. By Theorem 13.14, $\phi[G]$ is a subgroup of G', so if $|G'|$ is finite, then $|\phi[G]|$ is finite. By the Theorem of Lagrange, we see that $|\phi[G]|$ is then a divisor of $|G'|$.

46. Let $x \in G$. By Theorem 7.6, there are (not necessarily distinct) indices $i_1, i_2, i_3, \cdots, i_m$ in I such that

$$x = a_{i_1}^{n_1} a_{i_2}^{n_2} a_{i_3}^{n_3} \cdots a_{i_m}^{n_m} \text{ where the } n_j \text{ are in } \mathbb{Z}.$$

Because $\phi(a_{i_j}) = \mu(a_{i_j})$ for $j = 1, 2, 3, \cdots, m$, it follows from Definition 13.1 (extended by induction) and Property 2 in Theorem 13.12 that $\phi(x) = \mu(x)$. Thus ϕ and μ are the same map of G into G'.

47. By Theorem 13.12, $\mathrm{Ker}(\phi)$ is a subgroup of G. By the Theorem of Lagrange, either $\mathrm{Ker}(\phi) = \{e\}$ or $\mathrm{Ker}(\phi) = G$ because $|G|$ is a prime number. If $\mathrm{Ker}(\phi) = \{e\}$, then ϕ is one to one by Corollary 13.18. If $\mathrm{Ker}(\phi) = G$, then ϕ is the trivial homomorphism, mapping everything into the identity element.

48. We see that $\mathrm{Ker}(\mathrm{sgn}_n) = A_n$. The multiplicative group $\{-1, 1\}$ is isomorphic to the group $\mathbb{Z}_2$, and if $1 \in \{-1, 1\}$ is renamed 0 and -1 is renamed 1, then this becomes the homomorphism of Example 13.3.

49. Let $a, b \in G$. For the composite function $\gamma\phi$, we have

$$\gamma\phi(ab) = \gamma(\phi(ab)) = \gamma(\phi(a)\phi(b)) = \gamma(\phi(a))\gamma(\phi(b)) = \gamma\phi(a)\gamma\phi(b)$$

where the first equality uses the definition of the composite map $\gamma\phi$, the second equality uses the fact that ϕ is a homomorphism, the third uses the fact that γ is a homomorphism, and the last uses the definition of $\gamma\phi$ again. This shows that $\gamma\phi$ is indeed a homomorphism.

50. Let $x', y' \in \phi[G]$ and let $\phi(x) = x'$ and $\phi(y) = y'$ where $x, y \in G$. Then $\phi[G]$ is abelian

$$\text{if and only if } x'y' = y'x',$$
$$\text{if and only if } (y'x')^{-1}x'y' = e',$$
$$\text{if and only if } x'^{-1}y'^{-1}x'y' = e',$$
$$\text{if and only if } \phi(x)^{-1}\phi(y)^{-1}\phi(x)\phi(y) = e',$$
$$\text{if and only if } \phi(x^{-1}y^{-1}xy) = e',$$
$$\text{if and only if } x^{-1}y^{-1}xy \in \mathrm{Ker}(\phi)$$

for all $x', y' \in \phi[G]$. Note that because x' and y' could be any elements of $\phi[G]$, x and y could be any elements of G.

51. Let $m, n \in \mathbb{Z}$. We have $\phi(m + n) = a^{m+n} = a^m a^n = \phi(m)\phi(n)$, showing that ϕ is a homomorphism. The image of ϕ is the cyclic subgroup $\langle a \rangle$ of G, and $\text{Ker}(\phi)$ is one of the subgroups of $\mathbb{Z}$, which must be cyclic and consist of all (positive, negative and zero) multiples of some integer j in $\mathbb{Z}$. If a has finite order in G, then j is the order of a; otherwise, $j = 0$.

52. We show that each of $S = \{x \in G \mid \phi(x) = \phi(a)\}$ and Ha is a subset of the other. Let $s \in S$. Using Theorem 13.12 and the homomorphism property, we have $\phi(sa^{-1}) = \phi(s)\phi(a^{-1}) = \phi(a)\phi(a^{-1}) = \phi(a)\phi(a)^{-1} = e'$ so $sa^{-1} = h \in H$. Then $s = ha$ so $s \in Ha$. Thus S is a subset of Ha.

Let $h \in H$ so that $ha \in Ha$. Then $\phi(ha) = \phi(h)\phi(a) = e'\phi(a) = \phi(a)$, showing that $ha \in S$. Thus Ha is a subset of S, so $Ha = S$.

53. We have $\phi(1, 0) = h^1 k^0 = h$ and $\phi(0, 1) = h^0 k^1 = k$. Let ϕ be a homomorphism. Using addition notation in $\mathbb{Z} \times \mathbb{Z}$ as usual, we have

$$\begin{aligned}
\phi(1,1) &= \phi((1,0) + (0,1)) = \phi(1,0) + \phi(0,1) = hk, \\
\phi(1,1) &= \phi((0,1) + (1,0)) = \phi(0,1) + \phi(1,0) = kh.
\end{aligned}$$

Thus if ϕ is a homomorphism, we must have $hk = kh$.

Conversely, suppose that $hk = kh$. Then for any (i, j) and (m, n) in $\mathbb{Z} \times \mathbb{Z}$, we have

$$\begin{aligned}
\phi((i,j) + (m,n)) &= \phi(i+m, j+n) = h^{i+m}k^{j+n} = h^i h^m k^j k^n \\
&= h^i k^j h^m k^n = \phi(i,j)\phi(m,n)
\end{aligned}$$

where the first equality in the second line follows from the commutativity of h and k. Thus ϕ is a homomorphism if and only if $hk = kh$.

54. The preceding exercise shows that ϕ is a homomorphism for all choices of h and k in G if and only if $hk - kh$ for all h and k in G, that is, if and only if G is an abelian group.

55. The map ϕ is a homomorphism if and only if $h^n = e$, the identity in G.

Proof: If ϕ is a homomorphism, then $\phi(0) = e$. Consequently

$$h^n = \phi(1)^n = \phi(\underbrace{1 + 1 + \cdots + 1}_{n \text{ summands}}) = \phi(0) = e.$$

Conversely, suppose that $h^n = e$, so that $\langle h \rangle \simeq \mathbb{Z}_m$ where m is a divisor of n. Let $i, j \in \mathbb{Z}_n$. Viewing i and j in $\mathbb{Z}$, write $i + j = qm + r$, $i = q_1 m + r_1$, and $j = q_2 m + r_2$, all by the division algorithm. Then $\phi(i + j) = h^{i+j} = h^{qm+r} = (h^m)^q h^r = e^q h^r = h^r$. Similarly, $\phi(i) = h^{r_1}$ and $\phi(j) = h^{r_2}$, so $\phi(i)\phi(j) = h^{r_1 + r_2}$. Because $i + j = (q_1 + q_2)m + r_1 + r_2$, the remainder $i + j$ when divided by m is the same as the remainder of $r_1 + r_2$ when divided by m. Thus $h^{r_1 + r_2} = h^r$ so $\phi(i)\phi(j) = \phi(i + j)$. Hence ϕ is a homomorphism.

14. Factor Groups

1. $\langle 3 \rangle$ has 2 elements, so $\mathbb{Z}_6/\langle 3 \rangle$ has $6/2 = 3$ elements.

2. $\langle 2 \rangle \times \langle 2 \rangle$ has $2 \cdot 6 = 12$ elements, so the factor group has $48/12 = 4$ elements.

3. $\langle (2,1) \rangle$ has 2 elements, so the factor group has $8/2 = 4$ elements.

4. $\{0\} \times \mathbb{Z}_5$ has 5 elements, so the factor group has $15/5 = 3$ elements.

5. $\langle (1,1) \rangle$ has 4 elements, so the factor group has $8/4 = 2$ elements.

6. $\langle (4,3) \rangle$ has 6 elements, so the factor group has $(12 \cdot 18)/6 = 36$ elements.

7. $\langle (1, \rho_1) \rangle$ has 6 elements, so the factor group has $12/6 = 2$ elements.

8. $(1, 1)$ generates the entire group so the factor group has just one element.

9. $\langle 4 \rangle = \{0, 4, 8\}$. Now $5 + 5 = 10$, $5 + 5 + 5 = 3$, and $5 + 5 + 5 + 5 = 8$. Because $5 + 5 + 5 + 5 = 8$ is the first repeated sum of 5 in $\langle 4 \rangle$, we see that the coset $5 + \langle 4 \rangle$ is of order 4 in this factor group.

10. $\langle 12 \rangle = \{0, 12, 24, 36, 48\}$. We prefer to compute sums of the element 2 in the coset $26 + \langle 12 \rangle$, rather than the element 26. Computing, $2 + 2 = 4$, $4 + 2 = 6$, $6 + 2 = 8$, $8 + 2 = 10$, and $10 + 2 = 12 \in \langle 12 \rangle$. Thus $26 + \langle 12 \rangle$ has order 6 in the factor group.

11. $\langle (1,1) \rangle = \{(0,0), (1,1), (2,2), (0,3), (1,4), (2,5)\}$. Computing, $(2, 1) + (2, 1) = (1, 2)$, $(1, 2) + (2, 1)$ $= (0, 3) \in \langle (1,1) \rangle$. Thus $(2,1) + \langle (1,1) \rangle$ has order 3 in the factor group.

12. $\langle (1,1) \rangle = \{(0,0), (1,1), (2,2), (3,3)\}$. Computing, we find that $(3, 1) + (3, 1) = (2, 2) \in \langle (1,1) \rangle$. Thus $(3,1) + \langle (1,1) \rangle$ has order 2 in the factor group.

13. $\langle (0,2) \rangle = \{(0,0), (0,4), (0,6)\}$. Computing, $(3, 1) + (3, 1) = (2, 2)$, $(2, 2) + (3, 1) = (1, 3)$, $(1, 3) +$ $(3, 1) = (0, 4) \in \langle (0,2) \rangle$. Thus $(3,1) + \langle (0,2) \rangle$ has order 4 in the factor group.

14. $\langle (1,2) \rangle = \{(0,0), (1,2), (2,4), (3,6)\}$. We have $(3, 3) + (3, 3) = (2, 6)$, $(2, 6) + (3, 3) = (1, 1)$, $(1, 1) +$ $(3, 3) = (0, 4)$, $(0, 4) + (3, 3) = (3, 7)$, $(3, 7) + (3, 3) = (2, 2)$, $(2, 2) + (3, 3) = (1, 5)$, $(1, 5) + (3, 3) =$ $(0, 0) \in \langle (1,2) \rangle$. Thus $(3,3) + \langle (1,2) \rangle$ has order 8 in the factor group. It generates the entire factor group.

15. $\langle (4,4) \rangle = \{(0,0), (4,4), (2,0), (0,4), (4,0), (2,4)\}$. We see that $(2,0) \in \langle (4,4) \rangle$ so $(2,0) + \langle (4,4) \rangle$ has order 1 in the factor group.

16. Because $\rho_1 \mu_1 \rho_1^{-1} = \rho_1 \mu_1 \rho_2 = \mu_2$ and $\rho_1 \rho_0 \rho_1^{-1} = \rho_1 \rho_0 \rho_2 = \rho_0$, we see that $i_{\rho_1}(H) = \{\rho_0, \mu_2\}$.

17. The definition is incorrect.

    A **normal subgroup** H of a group G is a subgroup satisfying $gH = Hg$ for all $g \in G$.

18. The definition is correct.

19. The definition is incorrect. Change "homomorphism" to "isomorphism" and "into" to "onto".

    An **automorphism** of a group G is an isomorphism mapping G onto G.

20. Normal subgroups are those whose cosets can be used to form a factor group, because multiplication of left cosets by multiplying representatives is a well-defined binary operation.

21. (See the answer in the text for Part(**a**) and Part(**b**)).

c. Taking a and b as representatives of the cosets aH and bH respectively, we see that $(aH)(bH) = (ab)H$. Because G is abelian, $ab = ba$, so $(ab)H = (ba)H = (bH)(aH)$. Thus $(aH)(bH) = (bH)(aH)$ so G/H is abelian.

22. a. When working with a factor group G/H, one would let x be an element of G, not an element of G/H. The student probably does not understand what elements of G/H look like and can write nothing sensible concerning them.

b. We must show that each element of G/H is of finite order. Let $xH \in G/H$.

c. Because G is a torsion group, we know that $x^m = e$ in G for some positive integer m. Computing $(xH)^m$ in G/H using the representative x, we have $(xH)^m = x^m H = eH = H$, so xH is of finite order. Because xH can be any element of G/H, we see that G/H is a torsion group.

23. T T T T T F T F T F

24. If $n \geq 2$, then $|A_n| = |S_n|/2$, so the only cosets of A_n are A_n and the set of all odd permutations in S_n. Thus the left and right cosets must be the same, and A_n is a normal subgroup of S_n. Because S_n/A_n has order 2, it is isomorphic to $\mathbb{Z}_2$. If $n = 1$, then $A_n = S_n$ so S_n/A_n is the trivial group of one element.

25. Let $h \in H$ and $a \in G$. Suppose left coset multiplication $(aH)(bH)$ by choosing representatives is well defined. Then $(a^{-1}H)(aH) = eH = H$. Choosing the representatives $a^{-1}h$ from $a^{-1}H$ and a from aH, we see that $a^{-1}ha = h_1$ for some $h_1 \in H$. Thus $ha = ah_1$, so $ha \in aH$. This shows that $Ha \subseteq aH$.

26. Exercise 39 of Section 11 proves that the elements of G of finite order do form a subgroup T of the abelian group G. Because G is abelian, every subgroup of G is a normal subgroup, so T is normal in G. Suppose that xT is of finite order in G/T; in particular, suppose that $(xT)^m = T$. Then $x^m \in T$. Because T is a torsion group, we must have $(x^m)^r = x^{mr} = e$ in G for some positive integer r. Thus x is of finite order in G, so that $x \in T$. This means that $xT = T$. Thus the only element of finite order in G/T is the identity T, so G/T is a torsion free group.

27. *Reflexive:* Because $i_e[H] = H$ for every subgroup H of G, we see that every subgroup is conjugate to itself.

Symmetric: Suppose that $i_g[H] = K$, so that for each $k \in K$, we have $k = ghg^{-1}$ for exactly one $h \in H$. Then $h = (g^{-1})kg = (g^{-1})k(g^{-1})^{-1}$, and we see that $i_{g^{-1}}[K] = H$, so K is also conjugate to H.

Transitive: Suppose that $i_a[H] = K$ and $i_b[K] = S$ for elements $a, b \in G$ and subgroups H, K, and S of G. Then each $s \in S$ can be written as $s = bkb^{-1}$ for a unique $k \in K$. But $k = aha^{-1}$ for a unique $h \in H$. Substituting, we have $s = b(aha^{-1})b^{-1} = (ba)h(a^{-1}b^{-1}) = (ba)h(ba)^{-1}$, so $i_{ba}[H] = S$ and H is conjugate to S.

28. We have $\overline{H} = \{H\}$ if and only if $gHg^{-1} = H$ for all $g \in G$, which is true if and only if H is a normal subgroup of G. We see that the normal subgroups of G are precisely the subgroups in the one-element cells of the conjugacy partition of the subgroups of G.

29. We see that $\rho_1\mu_2\rho_1^{-1} = \rho_1\mu_2\rho_2 = \mu_3$, and $\rho_2\mu_2\rho_2^{-1} = \rho_2\mu_2\rho_1 = \mu_1$, and conjugation by other elements of S_3 again yield either μ_1, μ_2, or μ_3. Thus the subgroups of S_3 conjugate to $\{\rho_0, \mu_2\}$ are $\{\rho_0, \mu_2\}, \{\rho_0, \mu_1\}$, and $\{\rho_0, \mu_3\}$.

30. We have $|G/H| = m$. Because the order of each element of a finite group divides the order of the group, we see that $(aH)^m = H$ for all elements aH of G/H. Computing using the representative a of aH, we see that $a^m \in H$ for all $a \in G$.

31. Let $\{H_i \mid i \in I\}$ be a set of normal subgroups of a group G. Let $K = \bigcap_{i \in I} H_i$. If $a, b \in K$, then $a, b \in H_i$ for each $i \in I$, and $ab \in H_i$ for each $i \in I$ because H_i is a subgroup of G. Thus $ab \in K$ and K is closed under the group operation of G. We see that $e \in K$ because $e \in H_i$ for each $i \in I$. Because $a^{-1} \in H_i$ for each $i \in I$, we see that $a^{-1} \in K$ also. Thus K is a subgroup of G. Let $g \in G$ and $k \in K$. Then $k \in H_i$ for $i \in I$ and $gkg^{-1} \in H_i$ for each $i \in I$ because each H_i is a normal subgroup of G. Hence $gkg^{-1} \in K$, and K is a normal subgroup of G.

32. Let $\{H_i \mid i \in I\}$ be the set of all normal subgroups of G containing S. Note that G is such a subgroup of G, so I is nonempty. Let $K = \bigcap_{i \in I} H_i$. By Exercise 31, we know that K is a normal subgroup of G, and of course K contains S because H_i contains S for each $i \in I$. By our constructions, we see that K is contained in *every* normal subgroup H_i of G containing S, so K must be the smallest normal subgroup of G containing S.

33. Consider two elements aC and bC in G/C. Now $(aC)^{-1} = a^{-1}C$ and $(bC)^{-1} = b^{-1}C$. Conseqently, choosing representatives, we see that $(aC)(bC)(aC)^{-1}(bC)^{-1} = aba^{-1}b^{-1}C$. However, $aba^{-1}b^{-1} \in C$ because C contains all commutators in G, so $(aC)(bC)(aC)^{-1}(bC)^{-1} = C$. Thus $(aC)(bC) = C(bC)(aC) = (bC)(aC)$ which shows that G/C is abelian.

34. Let $g \in G$. Because the inner automorphism $i_g : G \to G$ is a one-to-one map, we see that $i_g[H]$ has the same order as H. Because H is the only subgroup of G of that order, we find that $i_g[H] = H$ for all $g \in G$. Therefore H is invariant under all inner automorphisms of G, and hence is a normal subgroup of G.

35. By Exercise 54 of Section 5, we know that $H \cap N$ is a subgroup of G, and is contained in H, so it is a subgroup of H. Let $h \in H$ and $x \in H \cap N$. Then $x \in N$ and because N is a normal subgroup of G, we find that $hxh^{-1} \in N$, and of course $hxh^{-1} \in H$ because $h, x \in H$. Thus $hxh^{-1} \in H \cap N$, so $H \cap N$ is a normal subgroup of H.

Let $G = D_4$, let $N = \{\rho_0, \rho_2, \mu_1, \mu_2\}$, and let $H = \{\rho_0, \mu_1\}$, using the notation in Section 8. Then N is normal in G, but $H \cap N = H$ is not normal in G.

36. Let H be the intersection of all subgroups of G that are of order s. We are told that this intersection is nonempty. By Exercise 31, H is a subgroup of G. Let $x \in H$ and $g \in G$. Let K be any subgroup of G of order s. To show that $gxg^{-1} \in H$, we must show that $gxg^{-1} \in K$. Now $g^{-1}Kg$ is a subgroup of G of order s, so $x \in g^{-1}Kg$. Let $x = g^{-1}kg$ where $k \in K$. Then $k = gxg^{-1}$, so gxg^{-1} is indeed in K. Because K can be any subgroup of G of order s, we see that $gxg^{-1} \in H$, so H is a normal subgroup of G.

37. a. By Exercise 49 of Section 13, the composition of two automorphisms of G is a homomorphism of G into G. Because each automorphism is a one-to-one map of G onto G, their composition also has this property, and is thus an automorphism of G. Thus compostion gives a binary operation on the set of all automorphisms of G. The identity map acts as identity automorphism, and the inverse map of an automorphism of G is again an automorphism of G. Thus the automorphisms form a group under function composition.

b. For $a, b, x \in G$, we have $i_a(i_b(x)) = i_a(bxb^{-1}) = a(bxb^{-1})a^{-1} = (ab)x(b^{-1}a^{-1}) = (ab)x(ab)^{-1} = i_{ab}(x)$, so the composition of two inner automorphisms is again an inner automorphism. Clearly, i_e

acts as identity and the equation $i_a i_b = i_{ab}$ shows that $i_a i_{a^{-1}} = i_e$, so $i_{a^{-1}}$ is the inverse of i_a. Thus the set of inner automorphisms is a group under function composition.

Let $a \in G$ and let ϕ be any automorphism of G. We must show that $\phi i_a \phi^{-1}$ is an inner automorphism of G in order to show that the inner automorphisms are a normal subgroup of the entire automorphism group of G. For any $x \in G$, we have $(\phi i_a \phi^{-1})(x) = \phi(i_a(\phi^{-1}(x))) = \phi(a\phi^{-1}(x)a^{-1}) = \phi(a)\phi(\phi^{-1}(x))\phi(a^{-1}) = \phi(a)x(\phi(a))^{-1} = i_{\phi(a)}(x)$, so $\phi i_a \phi^{-1} = i_{\phi(a)}$ which is indeed an inner automorphism of G.

38. Let $H = \{g \in G \mid i_g = i_e\}$. Let $a, b \in H$. Then for $x \in G$, we have $(ab)x(ab)^{-1} = a(bxb^{-1})a^{-1} = axa^{-1} = x$, so $i_{ab} = i_e$ and $ab \in H$. Of course $e \in H$, and $axa^{-1} = x$ yields $x = a^{-1}xa = a^{-1}x(a^{-1})^{-1}$, so $a^{-1} \in H$. Thus H is a subgroup of G.

To show that H is a normal subgroup of G, let $a \in H$ and $x \in G$. We must show that $xax^{-1} \in H$, that is, that $i_{xax^{-1}} = i_e$. For any $y \in G$, we have $i_{xax^{-1}}(y) = (xax^{-1})y(xax^{-1})^{-1} = x[a(x^{-1}yx)a^{-1}]x^{-1} = x(x^{-1}yx)x^{-1} = y = i_e(y)$, so $i_{xax^{-1}} = i_e$.

39. For $gH \in G/H$, let $\phi_*(gH) = \phi(g)H'$. Because we defined ϕ_* using the representative g of gH, we must show that ϕ_* is well defined. Let $h \in H$, so that gh is another representative of gH. Then $\phi(gh) = \phi(g)\phi(h)$. Because we are told that $\phi[H]$ is contained in H', we know that $\phi(h) = h' \in H'$, so $\phi(g)\phi(h) = \phi(g)h' \in \phi(g)H'$. This shows that ϕ_* is well defined, for the same coset $\phi(g)H'$ was obtained using the representatives g and gh.

For the homomorphism property, let $aH, bH \in G/H$. Because ϕ is a homomorphism, we obtain $\phi_*((aH)(bH)) = \phi_*((ab)H) = \phi(ab)H' = (\phi(a)\phi(b))H' = (\phi(a)H')(\phi(b)H') = \phi_*(aH)\phi_*(bH)$. Thus ϕ_* is a homomorphism.

40. a. Let H be the subset of $GL(n, \mathbb{R})$ consisting of the $n \times n$ matrices with determinant 1. The property $\det(AB) = \det(A) \cdot \det(B)$ shows that the set H is closed under matrix multiplication. Now $\det(I_n) = 1$ and every matrix in $GL(n, \mathbb{R})$ has a nonzero determinant and is invertible. From $1 = \det(I_n) = \det(AA^{-1}) = \det(A) \cdot \det(A^{-1})$, it follows that $\det(A^{-1}) = 1/\det(A)$, so if $A \in H$, then $A^{-1} \in H$. Thus H is a subgroup of $GL(n, \mathbb{R})$. Let $A \in H$ and let $X \in GL(n, \mathbb{R})$. Because X is invertible, $\det(X) \neq 0$. Then $\det(XAX^{-1}) = \det(X) \cdot \det(A) \cdot \det(X^{-1}) = \det(X) \cdot \det(A) \cdot (1/\det(X)) = \det(A) = 1$, so $XAX^{-1} \in H$. Thus H is a normal subgroup of $GL(n, \mathbb{R})$.

b. Let K be the subset of $GL(n, \mathbb{R})$ consisting of the $n \times n$ matrices with determinant ± 1. Note from Part(**a**) that if $\det(A) = -1$, then $\det(A^{-1}) = 1/(-1) = -1$. The same arguments as in Part(**a**) then show that if K is the subset of $n \times n$ matrices with determinant ± 1, then K is a subgroup of $GL(n, \mathbb{R})$. Part(**a**) shows that if $A \in K$ and $X \in GL(n, \mathbb{R})$, then $\det(XAX^{-1}) = \det(A) = \pm 1$, so that again $XAX^{-1} \in K$ and K is a normal subgroup of $GL(n, \mathbb{R})$.

41. a. Let A, B, and C be subsets of G. Then

$$\begin{aligned}(AB)C &= \{(ab)c \mid a \in A, b \in B, c \in C\} \\ &= \{a(bc) \mid a \in A, b \in B, c \in C\} = A(BC)\end{aligned}$$

by the associativity of multiplication in G. The subset $\{e\}$ acts as identity for this multiplication. Let $a, b \in G$ with $a \neq b$. Then the set $\{a, b\}$ has no multiplicative inverse, because the product of $\{a, b\}$ with any other nonempty subset of G yields a set with at least two elements, and hence not $\{e\}$. The product of any subset with the empty subset is the empty subset, so even if $G = \{e\}$, we still do not have a group, for $\varnothing$ has no inverse.

b. The proof that if N is a normal subgrop and $a, b \in G$, then the subset product $(aN)(bN)$ is contained in the coset $(ab)N$ would just repeat the last paragraph of the proof of Theorem 14.4. To show that $(ab)N$ is contained in $(aN)(bN)$, we let $n \in N$. Then $(ab)n = (ae)(bn)$, and this equation exhibits an element of $(ab)N$ as a product of elements in (aN) and (bN).

c. Associativity was proved for all subsets of G in Part(**a**), so it is surely true for the cosets of the normal subgroup N. Because $N = eN$, we see that $(aN)N = (aN)(eN) = (ae)N = aN$, and similarly $(eN)(aN) = aN$. Thus the coset N acts as identity element. The equation $(a^{-1}N)(aN) = eN = (aN)(a^{-1}N)$ shows that each coset has an inverse, so these cosets of N do form a group under this set multiplication. The identity of the coset group is N, while the identity for the multiplication of all subsets of G is $\{e\}$. If $N \neq \{e\}$, these identities are different.

15. Factor Group Computations and Simple Groups

1. Because $\langle (0, 1) \rangle$ has order 4, the factor group has order 2 and must be isomorphic to $\mathbb{Z}_2$. This is also obvious because this factor group essentially collapses everything in the second factor of $\mathbb{Z}_2 \times \mathbb{Z}_4$ to the identity, leaving just the first factor.

2. In this factor group,the first factor is not touched, but in the second factor, the element 2 is collapsed to 0. Because $\mathbb{Z}_4/\langle 2 \rangle$ is isomorphic to $\mathbb{Z}_2$, we see the factor group isomorphic to $\mathbb{Z}_2 \times \mathbb{Z}_2$. Alternatively, we can argue that the factor group has order four but no element of order greater than two.

3. We have $\langle (1, 2) \rangle = \{(0, 0), (1, 2)\}$, so the factor group is of order $8/2 = 4$. We easily see that $(1, 1) + \langle (1, 2) \rangle$ has order 4 in this factor group, which must then be isomorphic to $\mathbb{Z}_4$.

4. We have $\langle (1, 2) \rangle = \{(0, 0), (1, 2), (2, 4), (3, 6)\}$, so the factor group has order $32/4 = 8$. Because $(0, 1)$ must be added to itself eight times for the sum to lie in $\langle (1, 2) \rangle$, we see that $(0, 1) + \langle (1, 2) \rangle$ is of order 8 in this factor group, which is thus cyclic and isomorphic to $\mathbb{Z}_8$.

5. We have $\langle (1, 2, 4) \rangle = \{(0, 0, 0), (1, 2, 4), (2, 0, 0), (3, 2, 4)\}$ so the factor group has order $(4 \cdot 4 \cdot 8)/4 = 32$. The factor group can have no element of order greater than 8 because $\mathbb{Z}_4 \times \mathbb{Z}_4 \times \mathbb{Z}_8$ has no elements of order greater than 8. Because $(0, 0, 1)$ must be added to iself eight times for the sum to lie in $\langle (1, 2, 4) \rangle$, we see that the factor group has an element $(0, 0, 1) + \langle (1, 2, 4) \rangle$ of order 8, and is thus either isomorphic to $\mathbb{Z}_4 \times \mathbb{Z}_8$ or to $\mathbb{Z}_2 \times \mathbb{Z}_2 \times \mathbb{Z}_8$. The first group has only three elements of order 2, while the second one has seven elements of order 2. We count the elements of $\mathbb{Z}_4 \times \mathbb{Z}_4 \times \mathbb{Z}_8$ not in $\langle (1, 2, 4) \rangle$ but which, when added to themselves, yield an element of $\langle (1, 2, 4) \rangle$. No element added to itself yields $(1, 2, 4)$ or $(3, 2, 4)$. There are six such elements that yield $(0, 0, 0)$ when added to themselves. There are another six such elements that yield $(2, 0, 0)$ when added to themselves. These twelve elements are only enough to form three 4-element cosets in the factor group, which must be ismorphic to $\mathbb{Z}_4 \times \mathbb{Z}_8$.

6. Factoring out by $\langle (0, 1) \rangle$ collapses the second factor of $\mathbb{Z} \times \mathbb{Z}$ to zero without touching the first factor, so the factor group is isomorphic to $\mathbb{Z}$. (For those who objcct to this "collapsing" argument, the projection map $\pi_1 : \mathbb{Z} \times \mathbb{Z} \to \mathbb{Z}$ where $\pi_1(m, n) = m$ has $\langle (0, 1) \rangle$ as its kernel.)

7. The 1 in the generator $(1, 2)$ of $\langle (1, 2) \rangle$ shows that each coset of $\langle (1, 2) \rangle$ contains a unique element of the form $(0, m)$, and of course, every such element of $\mathbb{Z} \times \mathbb{Z}$ is in some coset of $\langle (1, 2) \rangle$. We can choose these representatives $(0, m)$ to compute in the factor group, which must therefore be isomorphic to $\mathbb{Z}$.

8. The 1 in the generator $(1, 1, 1)$ of $\langle (1, 1, 1) \rangle$ shows that each coset of $\langle (1, 1, 1) \rangle$ contains a unique element of the form $(0, m, n)$, and of course, every such element of $\mathbb{Z} \times \mathbb{Z} \times \mathbb{Z}$ is in some coset of

$\langle (1,1,1) \rangle$. We can choose these representatives $(0,m,n)$ to compute in the factor group, which must therefore be isomorphic to $\mathbb{Z} \times \mathbb{Z}$.

9. We conjecture that $(\mathbb{Z} \times \mathbb{Z} \times \mathbb{Z}_4)/\langle (3,0,0) \rangle$ is isomorphic to $\mathbb{Z}_3 \times \mathbb{Z} \times \mathbb{Z}_4$, because only the multiples of 3 in the first factor are collapsed to zero. It is easy to check that $\phi : \mathbb{Z} \times \mathbb{Z} \times \mathbb{Z}_4 \to \mathbb{Z}_3 \times \mathbb{Z} \times \mathbb{Z}_4$ defined by $\phi(n,m,s) = (r,m,s)$, where r is the remainder of n when divided by 3 in the division algorithm, is an onto homomorphism with kernel $\langle (3,0,0) \rangle$. By Theorem 14.6, such a check proves our conjecture.

10. We conjecture that $(\mathbb{Z} \times \mathbb{Z} \times \mathbb{Z}_8)/\langle (0,4,0) \rangle$ is isomorphic to $\mathbb{Z} \times \mathbb{Z}_4 \times \mathbb{Z}_8$, because only the multiples of 4 in the second factor are collapsed to zero. It is easy to check that $\phi : \mathbb{Z} \times \mathbb{Z} \times \mathbb{Z}_8 \to \mathbb{Z} \times \mathbb{Z}_4 \times \mathbb{Z}_8$ defined by $\phi(n,m,s) = (n,r,s)$, where r is the remainder of m when divided by 4 in the division algorithm, is an onto homomorphism with kernel $\langle (0,4,0) \rangle$. By Theorem 14.6, such a check proves our conjecture.

11. Note that $(1,1) + \langle (2,2) \rangle$ is of order 2 in the factor group and $(0,1) + \langle (2,2) \rangle$ generates an infinite cyclic subgroup of the factor group. This suggests that the factor group is isomorphic to $\mathbb{Z}_2 \times \mathbb{Z}$. We construct a homomophism ϕ mapping $\mathbb{Z} \times \mathbb{Z}$ onto $\mathbb{Z}_2 \times \mathbb{Z}$ having kernel $\langle (2,2) \rangle$. By Theorem 14.6, we will then know that $(\mathbb{Z} \times \mathbb{Z})/\langle (2,2) \rangle$ is isomorphic to $\mathbb{Z}_2 \times \mathbb{Z}$.

   We want to have $\phi(1,1) = (1,0)$ and $\phi(0,1) = (0,1)$. Because $(m,n) = m(1,1) + (n-m)(0,1)$, we try to define ϕ by

$$\phi(m,n) = (m \cdot 1, n - m).$$

Here $m \cdot 1$ means $1 + 1 + \cdots 1$ for m summands in $\mathbb{Z}_2$, in other words, the remainder of m modulo 2. Because

$$
\begin{aligned}
\phi[(m,n) + (r,s)] &= \phi(m+r, n+s) \\
&= ((m+r) \cdot 1, n+s-m-r) \\
&= (m \cdot 1, n-m) + (r \cdot 1, s-r) \\
&= \phi(m,n) + \phi(r,s),
\end{aligned}
$$

we see that ϕ is indeed a homomorphism. For $(r,s) \in \mathbb{Z}_2 \times \mathbb{Z}$, we see that $\phi(r, s+r) = (r,s)$, so ϕ is onto $\mathbb{Z}_2 \times \mathbb{Z}$. If $\phi(m,n) = (0,0)$, then $m \cdot 1 = 0$ in $\mathbb{Z}_2$ and $n - m = 0$ in $\mathbb{Z}$. Thus m is even and $m = n$, so $(m,n) = (m,m)$ lies in $\langle (2,2) \rangle$. Thus $\text{Ker}(\phi)$ is contained in $\langle (2,2) \rangle$. It is easy to see that $\langle (2,2) \rangle$ is contained in $\text{Ker}(\phi)$, so $\text{Ker}(\phi) = \langle (2,2) \rangle$. As we observed above, Theorem 14.6 shows that our factor group is isomorphic to $\mathbb{Z}_2 \times \mathbb{Z}$.

12. Clearly $(1,1,1) + \langle (3,3,3) \rangle$ is of order 3 in the factor group, while $(0,1,0) + \langle (3,3,3) \rangle$ and $(0,0,1) + \langle (3,3,3) \rangle$ both generate infinite subgroups of the factor group. We conjecture that the factor group is isomorphic to $\mathbb{Z}_3 \times \mathbb{Z} \times \mathbb{Z}$. As in the solution to Exercise 11, we show that by defining a homomorphism ϕ of $\mathbb{Z} \times \mathbb{Z} \times \mathbb{Z}$ onto $\mathbb{Z}_3 \times \mathbb{Z} \times \mathbb{Z}$ having kernel $\langle (3,3,3) \rangle$. Just as in Exercise 11, we are motivated to let

$$\phi(m,n,s) = (m \cdot 1, n - m, s - m).$$

We easily check that ϕ is a homomorphism with the onto property and kernel that we desire, completing the proof. Just follow the arguments in the solution of Exercise 11.

13. Checking the Table 8.12 for D_4, we find that only ρ_0 and ρ_2 commute with every element of D_4. Thus $Z(D_4) = \{\rho_0, \rho_2\}$. It follows that $\{\rho_0, \rho_2\}$ is a normal subgroup of D_4. Now $D_4/Z(D_4)$ has order 4 and is hence abelian. Therefore the commutator subgroup C is contained in $Z(D_4)$. Because D_4 is not abelian, we see that $C \neq \{\rho_0\}$, so $C = Z(D_4) = \{\rho_0, \rho_2\}$.

14. Because ρ_0 is the only element of S_3 that commutes with every element of S_3 (see Table 8.8), we see that $Z(\mathbb{Z}_3 \times S_3) = \mathbb{Z}_3 \times \{\rho_0\}$. Because A_3 is the commutator subgroup of S_3, we see that the commutator subgroup of $\mathbb{Z}_3 \times S_3$ is $\{0\} \times A_3$.

15. From Tables 8.8 and 8.12, $Z(S_3 \times D_4) = \{(\rho_0, \rho_0), (\rho_0, \rho_2)\}$. From Exercise 13, we see that the commutator subgroup is $A_3 \times \{\rho_0, \rho_2\}$.

16. We present the answers in tabular form. The order of the factor group is easy to determine, as we did in Exercises 1 through 12. It is clear the factor groups listed are the only ones possible. Which of the possibilities is the correct one for the given subgroup can easily be determined by taking into account the order, and checking whether there is an element of order 4 in the factor group. We leave the "up to isomorphism" label off the "Factor Group" heading to conserve space.

Subgroup	Factor Group	Subgroup	Factor Group
$\langle (1,0) \rangle$	$\mathbb{Z}_4$	$\langle 2 \rangle \times \langle 2 \rangle$	$\mathbb{Z}_2 \times \mathbb{Z}_2$
$\langle (0,1) \rangle$	$\mathbb{Z}_4$	$\langle (2,0) \rangle$	$\mathbb{Z}_2 \times \mathbb{Z}_4$
$\langle (1,1) \rangle$	$\mathbb{Z}_4$	$\langle (0,2) \rangle$	$\mathbb{Z}_4 \times \mathbb{Z}_2$
$\langle (1,2) \rangle$	$\mathbb{Z}_4$	$\langle (2,2) \rangle$	$\mathbb{Z}_2 \times \mathbb{Z}_4$
$\langle (2,1) \rangle$	$\mathbb{Z}_4$	$\langle (0,0) \rangle$	$\mathbb{Z}_4 \times \mathbb{Z}_4$
$\langle (1,3) \rangle$	$\mathbb{Z}_4$		

17. The definition is incorrect. Replace "contains" by "consists of".

    The **center** of a group G consists of all elements of G that commute with every element of G.

18. The definition is correct.

19. T F F T F T F F T

20. F/K is isomorphic to $H = \{f \in F \mid f(0) = 0\}$, because every coset of K in F contains a unique function in H, and H is a subgroup of F. There is nothing special about 0 as the choice of a point in the domain of the functions. F/K is also isomorphic to $H_a = \{f \in F \mid f(a) = 0\}$ for the same reason.

21. F^*/K^* is isomorphic to $H^* = \{f \in F^* \mid f(1) = 1\}$, because every coset of K^* in F^* contains a unique function in H^*, and H^* is a subgroup of F^*. There is nothing special about 1 as the choice of a point in the domain of the functions. F^*/K^* is also isomorphic to $H_a^* = \{f \in F^* \mid f(a) = 1\}$ for the same reason.

22. No, if $f + K$ has order 2 in F/K, then we would have to have $f \notin K$ but $g = f + f \in K$. Thus we would have to have g be continuous, but have $f = \frac{1}{2}g$ be not continuous. This is impossible.

23. (See the answer in the text.)

24. $U/z_0 U$ is isomorphic to $\{e\}$, for $z_0 U = U$.

25. $U/\langle -1 \rangle$ is isomorphic to U, for the map $\phi : U \to U$ given by $\phi(z) = z^2$ is a homomorphism of U onto U with kernel $\{-1, 1\}$. By Theorem 14.6, $U/\langle -1 \rangle$ is isomorphic to U.

26. $U/\langle \zeta_n \rangle$ is isomorphic to U, for the map $\phi : U \to U$ given by $\phi(z) = z^n$ is a homomorphism of U onto U with kernel $\langle \zeta_n \rangle$. By Theorem 14.6, $U/\langle \zeta_n \rangle$ is isomorphic to U.

27. The factor group $\mathbb{R}/\mathbb{Z}$ is isomorphic to U, the multiplicative group of complex numbers having absolute value 1. The map $\phi : \mathbb{R} \to U$ given by $\phi(r) = e^{(2\pi r)i} = \cos(2\pi r) + i\sin(2\pi r)$ is a homomorphism of $\mathbb{R}$ onto U with kernel $\mathbb{Z}$. By Theorem 14.6, $\mathbb{R}/\mathbb{Z}$ is isomorphic to U.

28. The group $\mathbb{Z}$ is an example, for $\mathbb{Z}/\langle 2 \rangle$ has only elements of finite order.

29. Let $G = \mathbb{Z}_2 \times \mathbb{Z}_4$. Then $H = \langle (1,0) \rangle$ is isomorphic to $K = \langle (0,2) \rangle$, but G/H is isomorphic to $\mathbb{Z}_4$ while G/K is isomorphic to $\mathbb{Z}_2 \times \mathbb{Z}_2$.

30. a. The center is a whole group.

b. The center is $\{e\}$, because the center is a normal subgroup and the group is simple.

31. a. The commutator subgroup of an abelian group is $\{e\}$.

b. The whole group G is simple. Because C is a normal subgroup, $C = \{e\}$ or $C = G$. Because G is nonabelian and G/C is abelian, we must have $C = G$.

32. Every coset of a factor group G/H that contains a generator of the cyclic group G will generate the factor group.

33. If M and L are normal subgroups of G and $M < L < G$, then L/M is a proper nontrivial normal subgroup of G/M. If $\gamma : G \to G/M$ is the canonical homomorphism and K is a proper nontrivial normal subgroup of G/M, then $\gamma^{-1}[K]$ is a normal subgroup of G and $M < \gamma^{-1}[K] < G$.

34. Every subgroup H of index 2 is normal, because both left and right cosets of H are H itself and $\{g \in G \mid g \notin H\}$. Thus G cannot be simple if it has a subgroup H of index 2.

35. We know that $\phi[N]$ is a subgroup of $\phi[G]$ by Theorem 13.12. We need only show that $\phi[N]$ is normal in $\phi[G]$. Let $g \in G$ and $x \in N$. Because ϕ is a homomophism, Theorem 13.12 tells us that $\phi(g)\phi(x)\phi(g)^{-1} = \phi(g)\phi(x)\phi(g^{-1}) = \phi(gxg^{-1})$. Because N is normal, we know that $gxg^{-1} \in N$, so $\phi(gxg^{-1})$ is in $\phi[N]$, and we are done.

36. We know that $\phi^{-1}[N']$ is a subgroup of G by Theorem 13.12. We need only show that $\phi^{-1}[N']$ is normal in G. Let $x \in \phi^{-1}[N']$, so that $\phi(x) \in N'$. For each $g \in G$, we have $\phi(gxg^{-1}) = \phi(g)\phi(x)\phi(g^{-1}) = \phi(g)\phi(x)\phi(g)^{-1} \in N'$ because N' is a normal subgroup of G'. Thus $gxg^{-1} \in \phi^{-1}[N']$, showing that $\phi^{-1}[N']$ is a normal subgroup of G. (Note that N' need only be normal in $\phi[G]$ for the conclusion to hold.)

37. Suppose that $G/Z(G)$ is cyclic and is generated by the coset $aZ(G)$. Let $x, y \in G$. Then x is a member of a coset $a^m Z(G)$ and y is a member of a coset $a^n Z(G)$ for some $m, n \in \mathbb{Z}$. We can thus write $x = a^m z_1$ and $y = a^n z_2$ where $z_1, z_2 \in Z(G)$. Because z_1 and z_2 commute with every element of G, we have $xy = a^m z_1 a^n z_2 = a^{m+n} z_1 z_2 = a^n z_2 a^m z_1 = yx$, showing that G is abelian. Therefore, if G is not abelian, then $G/Z(G)$ is not cyclic.

38. Let G be nonabelian of order pq. Suppose that $Z(G) \neq \{e\}$. Then $|Z(G)|$ is a divisor of pq greater than 1, but less than pq because G is nonabelian, and hence $|Z(G)|$ is either p or q. But then $|G/Z(G)|$ is either q or p, and hence is cyclic, which contradicts Exercise 37. Therefore $Z(G) = \{e\}$.

39. a. Because $(a, b, c) = (a, c)(a, b)$, we see that every 3-cycle is an even permutation, and hence is in A_n.

b. Let $\sigma \in A_n$ and write σ as a product of transpositions. The number of transpositions in the product will be even by definition of A_n. The product of the first two transpositions will be either of the form $(a, b)(c, d)$ or of the form $(a, b)(a, c)$ or of the form $(a, b)(a, b)$, depending on repetition of letters in the transpositions. If the form is $(a, b)(a, b)$, it can be deleted from the product altogether. As the hint shows, either of the other two forms can be expressed as a 3-cycle. We then proceed with the next pair of transpositions in the product, and continue until we have expressed σ as a product of 3-cycles. Thus the 3-cycles generate A_n.

c. Following the hint, we find that

$$(r, s, i)^2 = (r, i, s),$$
$$(r, s, j)(r, s, i)^2 = (r, s, j)(r, i, s) = (r, i, j),$$
$$(r, s, j)^2(r, s, i) = (r, j, s)(r, s, i) = (s, i, j),$$
$$(r, s, i)^2(r, s, k)(r, s, j)^2(r, s, i) = (r, i, s)(r, s, k)(s, i, j) = (i, j, k).$$

Now every 3-cycle either contains neither r nor s and is of the form (i, j, k), or just one of r or s and is of the form (r, i, j) or (s, i, j), or both r and s and is of the form (r, s, i) or $(r, i, s) = (s, r, i)$. Beause all of these forms can be obtained from our special 3-cycles, we see that the special 3-cycles generate A_n.

d. Following the hint and using Part(**c**) we find that

$$((r, s)(i, j))(r, s, i)^2((r, s)(i, j))^{-1} = (r, s)(i, j)(r, i, s)(i, j)(r, s) = (r, s, j).$$

Thus if N is a normal subgroup of A_n and contains a 3-cycle, which we can consider to be (r, s, i) because r and s could be any two numbers from 1 to n in Part(**c**), we see that N must contain all the special 3-cycles and hence be all of A_n by Part(**c**).

e. Before making the computations in the hints of the five cases, we observe that one of the cases must hold. If Case 1 is not true and Case 2 is not true, then when elements of N are written as a product of disjoint cycles, no cycle of length greater than 3 occurs, and no element of N is a single 3-cycle. The remaining cases cover the possibilities that at least one of the products of disjoint cycles involves two cycles of length 3, involves one cycle of length 3, or involves no cycle of length 3. Thus all possibilities are covered, and we now turn to the computations in the hints.

Case 1. By Part(**d**), if N contains a 3-cycle, then $N = A_n$ and we are done.

Case 2. Note that $a_1, a_2, \cdots, a_r$ do not appear in μ because the product contained disjoint cycles. We have

$$\sigma^{-1}[(a_1, a_2, a_3)\sigma(a_1, a_2, a_3)^{-1}]$$
$$= (a_r, \cdots, a_2, a_1)\mu^{-1}(a_1, a_2, a_3)\mu(a_1, a_2, \cdots, a_r)(a_1, a_3, a_2)$$
$$= (a_1, a_3, a_r),$$

and this element is in N because it is the product of σ^{-1} and a conjugate of σ by an element of A_n. Thus in this case, N contains a 3-cycle and is equal to A_n by Part(**d**).

Case 3. Note that $a_1, a_2, \cdots, a_6$ do not appear in μ. As in Case 2, we see that

$$\sigma^{-1}[(a_1, a_2, a_4)\sigma(a_1, a_2, a_4)^{-1}] =$$
$$(a_1, a_3, a_2)(a_4, a_6, a_5)\mu^{-1}(a_1, a_2, a_4)\mu(a_4, a_5, a_6)(a_1, a_2, a_3)(a_1, a_4, a_2)$$
$$= (a_1, a_4, a_2, a_6, a_3)$$

is in N. Thus N contains a cycle of length greater than 3, and $N = A_n$ by Case 2.

Case 4. Note that $a_1, a_2,$ and a_3 do not appear in μ. Of course $\sigma^2 \in N$ because $\sigma \in N$, so $\sigma^2 = \mu(a_1, a_2, a_3)\mu(a_1, a_2, a_3) = (a_1, a_3, a_2) \in N$, so N contains a 3-cycle and hence $N = A_n$ as shown by Part(**d**).

Case 5. Note that $a_1, a_2, a_3,$ and a_4 do not appear in μ. As in Case 2, we see that

$$\sigma^{-1}[(a_1, a_2, a_3)\sigma(a_1, a_2, a_3)^{-1}] =$$
$$(a_1, a_2)(a_3, a_4)\mu^{-1}(a_1, a_2, a_3)\mu(a_3, a_4)(a_1, a_2)(a_1, a_3, a_2) =$$
$$(a_1, a_3)(a_2, a_4)$$

is in N. Continuing with the hint given, we let $\alpha = (a_1, a_3)(a_2, a_4)$ and $\beta = (a_1, a_3, i)$ where i is different from a_1, a_2, a_3, and a_4. Then $\beta \in A_n$ and $\alpha \in N$ and N a normal subgroup of A_n imply that $(\beta^{-1}\alpha\beta)\alpha \in N$. Computing, we find that

$$(\beta^{-1}\alpha\beta)\alpha = (a_1, i, a_3)(a_1, a_3)(a_2, a_4)(a_1, a_3, i)(a_1, a_3)(a_2, a_4) = (a_1, a_3, i).$$

Thus $N = A_n$ in this case also, by Part(**d**).

40. *Closure:* Let $h_1n_1, h_2n_2 \in HN$ where $h_1, h_2 \in H$ and $n_1, n_2 \in N$. Because N is a normal subgroup, left cosets are right cosets so $Nh_2 = h_2N$; in particular, $n_1h_2 = h_2n_3$ for some $n_3 \in N$. Then $(h_1n_1)(h_2n_2) = (h_1h_2)(n_3n_2) \in HN$, so HN is closed under the group operation.

Identity: Because $e \in H$ and $e \in N$, we see that $e = ee \in HN$.

Inverses: Now $(h_1n_1)^{-1} = n_1^{-1}h_1^{-1} \in Nh_1^{-1}$ and $Nh_1^{-1} = h_1^{-1}N$ because N is normal. Thus $n_1^{-1}h_1^{-1} = h_1^{-1}n_4$ for some $n_4 \in N$, so $(h_1n_1)^{-1} \in HN$, and we see that HN is a subgroup of G.

Clearly HN is the smallest subgroup of G containing both H and N, because any such subgroup must contain all the products hn for $h \in H$ and $n \in N$.

41. Exercise 40 shows that NM is a subgroup of G. We must show that $g(nm)g^{-1} \in NM$ for all $g \in G, n \in N$, and $m \in M$. We have $g(nm)g^{-1} = (gng^{-1})(gmg^{-1})$. Because N and M are both normal, we know that $gng^{-1} \in N$ and $gmg^{-1} \in M$. Thus $g(nm)g^{-1} \in NM$ so NM is a normal subgroup of G.

42. The fact that K is normal shows that $hkh^{-1} \in K$, so $(hkh^{-1})k^{-1} \in K$. The fact that H is normal shows that $kh^{-1}k^{-1} \in H$, so $h(kh^{-1}k^{-1}) \in H$. Thus $hkh^{-1}k^{-1} \in H \cap K$, so $hkh^{-1}k^{-1} = e$. It follows that $hk = kh$.

16. Group Action on a Set

1. (See the answer in the text.)

2. $G_1 = G_3 = \{\rho_0, \delta_2\}$, $G_2 = G_4 = \{\rho_0, \delta_1\}$,
$G_{s_1} = G_{s_3} = \{\rho_0, \mu_1\}$, $G_{s_2} = G_{s_4} = \{\rho_0, \mu_2\}$,
$G_{m_1} = G_{m_2} = \{\rho_0, \rho_2, \mu_1, \mu_2\}$, $G_{d_1} = G_{d_2} = \{\rho_0, \rho_2, \delta_1, \delta_2\}$,
$G_C = G$, $G_{P_1} = G_{P_3} = \{\rho_0, \mu_1\}$, $G_{P_2} = G_{P_4} = \{\rho_0, \mu_2\}$

3. $\{1, 2, 3, 4\}$, $\{s_1, s_2, s_3, s_4\}$, $\{m_1, m_2\}$, $\{d_1, d_2\}$, $\{C\}$, $\{P_1, P_2, P_3, P_4\}$

4. The definition is incorrect. We need a universal quantifier.

A group G **acts faithfully** on X if and only if $gx = x$ for all $x \in G$ implies that $g = e$.

5. The definition is incorrect, and is an example of a nonsense definition.

A group G is **transitive** on a G-set X if and only if for each $a, b \in X$, there exists some $g \in G$ such that $ga = b$.

6. Every sub-G-set of a G-set X consists of a union of orbits in X under G.

7. A G-set if transitive if and only if it has only one orbit.

8. F T F T F T T F T T

9. a. $\{P_1, P_2, P_3, P_4\}$ and $\{s_1, s_2, s_3, s_4\}$ are isomorphic sub-D_4-sets. Note that if you change each P to an s in Table 16.10, you get a duplication of the four columns for s_1, s_2, s_3, and s_4.

b. δ_1 leaves two elements, 2 and 4, of $\{1, 2, 3, 4\}$ fixed, but δ_1 leaves no elements of $\{s_1, s_2, s_3, s_4\}$ fixed.

c. Yes, for after Part(**b**), the only other conceivable choice for an isomorphism is $\{m_1, m_2\}$ with $\{d_1, d_2\}$. However, μ_1 leaves the elements of $\{m_1, m_2\}$ fixed and moves both elements of $\{d_1, d_2\}$ so they are not isomorphic.

10. a. Yes, for ρ_0 is the only element of G that leaves every element of X fixed.

b. $\{1, 2, 3, 4\}$, $\{s_1, s_2, s_3, s_4\}$, and $\{P_1, P_2, P_3, P_4\}$

11. Let $g_1, g_2 \in G$. Now suppose that $g_1 a = g_2 a$ for all $a \in X$, which is true if and only if and only if $g_2^{-1} g_1 a = a$ for all $a \in X$. If $g_1 \neq g_2$, then $g_2^{-1} g_1 \neq e$, and the action of G on X is not faithful. If the action of G on X is faithful, then we must have $g_2^{-1} g_1 = e$ and $g_1 = g_2$, that is, two distinct elements of G cannot act the same on each $a \in G$.

12. *Closure:* Let $g_1, g_2 \in G_Y$. Then for each $y \in Y$, we have $(g_1 g_2)y = g_1(g_2 y) = g_1 y = y$, so $g_1 g_2 \in G_Y$, and G_Y is closed under the group operation.

Identity: Because $ey = y$ for all $y \in Y$, we see that $e \in G_Y$.

Inverses: From $y = g_1 y$ for all $y \in Y$, it follows that $g_1^{-1} y = g_1^{-1}(g_1 y) = (g_1^{-1} g_1)y = ey = y$ for all $y \in Y$, so $g_1^{-1} \in G_Y$ also, and consequently $G_Y \leq G$.

13. a. Because rotation through 0 radians leaves each point of the plane fixed, the first requirement of Definition 16.1 is satisfied. The second requirement $(\theta_1 + \theta_2)P = \theta_1(\theta_2)P$ is also valid, because a rotation counterclockwise through $\theta_1 + \theta_2$ radians can be achieved by sequentially rotating through θ_2 radians and then through θ_1 radians.

b. The orbit containing P is a circle with center at the origin $(0, 0)$ and radius the distance from P to the origin.

c. The group G_P is the cyclic subgroup $\langle 2\pi \rangle$ of G.

14. a. Let $X = \bigcup_{i \in I} X_i$ and let $x \in X$. Then $x \in X_i$ for precisely one index $i \in I$ because the sets are disjoint, and we define gx for each $g \in G$ to be the value given by the action of G on X_i. Conditions (1) and (2) in Definition 16.1 are satisfied because X_i is a G-set by assumption.

b. We have seen that each orbit in X is a sub-G-set. The G-set X can be regarded as the union of these sub-G-sets because the action gx of $g \in G$ on $x \in X$ coincides with the sub-G-set action gx of $g \in G$ on the same element x viewed as an element of its orbit.

15. Let $\phi : X \to L$ be defined by $\phi(x) = gG_{x_0}$ where $gx_0 = x$. Because G is transitive on X, we know that such a g exists. We must show that ϕ is well defined. Suppose that $g_1 x_0 = x$ and $g_2 x_0 = x$. Then $g_2 x_0 = g_1 x_0$ so $(g_1^{-1} g_2)x_0 = x_0$. But then $g_1^{-1} g_2 \in G_{x_0}$ so $g_2 \in g_1 G_{x_0}$ and $g_1 G_{x_0} = g_2 G_{x_0}$. This shows that the definition of $\phi(x)$ is independent of the choice of g such that $gx = x_0$, that is, ϕ is well defined.

It remains to show that ϕ is one to one and onto L, and that $g\phi(x) = \phi(gx)$ for all $x \in X$ and $g \in G$. Suppose that $\phi(x_1) = \phi(x_2)$ for $x_1, x_2 \in X$, and let $g_1 x_0 = x_1$ and $g_2 x_0 = x_2$. Then $\phi(x_1) = \phi(x_2)$ implies that $g_1 G_{x_0} = g_2 G_{x_0}$ so $g_2 = g_1 g_0$ for some $g_0 \in G_{x_0}$. The equation $g_2 x_0 = x_2$ then yields $g_1 g_0 x_0 = x_2$. Because $g_0 \in G_{x_0}$, we then obtain $g_1 x_0 = x_2$ so $x_1 = x_2$ and ϕ is one to one. If $g \in G$, then $\phi(gx_0) = gG_{x_0}$ shows that ϕ maps X onto L. Finally, to show that $g\phi(x) = \phi(gx)$, let $x = g_1 x_0$. Then $gx = g(g_1 x_0) = (gg_1)x_0$ so $\phi(gx) = (gg_1)G_{x_0} = g(g_1 G_{x_0}) = g\phi(x)$.

16. By Exercise 14, each G-set X is the union of its G-set orbits X_i for $i \in I$. By Exercise 15, each G-set orbit X_i is isomorphic to a G-set consisting of left cosets of $G_{x_{i,0}}$ where $x_{i,0}$ is any point of X_i. It is possible that the group $G_{x_{i,0}}$ may the the same as the group $G_{x_{j,0}}$ for some $j \neq i$ in I, but by attaching the index i to each coset of $G_{x_{i,0}}$ and j to each coset of $G_{x_{j,0}}$ as indicated in the statement of the exercise, we can consider these ith and jth coset G-sets to be disjoint. Identifying X_i with this isomorphic ith coset G-set, we see that X is isomorphic to a disjoint union of left coset G-sets.

17. a. If $g \in K$ so that $g(g_0 x_0) = g_0 x_0$, then $(g_0^{-1} g g_0)x_0 = x_0$, which means that $g_0^{-1} g g_0 \in H$, so $g \in g_0 H g_0^{-1}$. Because g may be any element of K, this shows that $K \subseteq g_0 H g_0^{-1}$. Making a symmetric argument, starting with $g \in H$, $g_0 x_0$ as initial base point, and obtaining x_0 as second base point by g_0^{-1} acting on $g_0 x_0$, we see that $H \subseteq g_0^{-1} K g_0$, or equivalently, $g_0 H g_0^{-1} \subseteq K$. Thus $K = g_0 H g_0^{-1}$.

b. *Conjecture:* The G-set of left cosets of H is isomorphic to the G-set of left cosets of K if and only if H and K are conjugate subgroups of G.

c. We first show that if H and K are conjugate subgroups of G, then the G-set L_H of left cosets of H is isomorphic to the G-set L_K of left cosets of K. Let $g_0 \in G$ be chosen such that $K = g_0 H g_0^{-1}$. Note that for $aH \in L_H$, we have $aHg_0^{-1} = ag_0^{-1}g_0 H g_0^{-1} = ag_0^{-1}K \in L_K$. We define $\phi : L_H \to L_K$ by $\phi(aH) = ag_0^{-1}K$. We just saw that $ag_0^{-1}K = (aH)g_0^{-1}$ so ϕ is independent of the choice of $a \in H$, that is, ϕ is well defined. Because ag_0^{-1} assumes all values in G as a varies through G, we see that ϕ is onto L_K. If $\phi(aH) = \phi(bH)$, then $(aH)g_0^{-1} = ag_0^{-1}K = bg_0^{-1}K = (bH)g_0^{-1}$, so $aH = bH$ and ϕ is one to one. To show ϕ is an isomorphism of G-sets, it only remains to show that $\phi(g(aH)) = g\phi(aH)$ for all $g \in G$ and $aH \in L_H$. But $\phi(g(aH)) = \phi((ga)H) = (ga)g_0^{-1}K = g(ag_0^{-1}K) = g\phi(aH)$, and we are done.

Conversely, suppose that $\phi : L_H \to L_K$ is an isomorphism of the G-set of left cosets of H onto the G-set of left cosets of K. Because ϕ is an onto map, there exists $g_0 \in G$ such that $\phi(g_0 H) = K$. Because ϕ commutes with the action of G, we have $(g_0 h g_0^{-1})K = (g_0 h g_0^{-1})\phi(g_0 H) = \phi(g_0 h g_0^{-1}g_0 H) = \phi(g_0 H) = K$, so $g_0 h g_0^{-1} \in K$ for all $h \in H$, that is, $g_0 H g_0^{-1} \subseteq K$. From $\phi(g_0 H) = K$, we easily see that $\phi^{-1}(g_0^{-1}K) = H$, and an argument similar to the one just made then shows that $g_0^{-1}K g_0 \subseteq H$. Thus $g_0 H g_0^{-1} = K$, that is, the subgroups are indeed conjugate.

18. There are three of them; call them X, Y, and $\mathbb{Z}_4$ corresponding to the three subgroups $\{0, 1, 2, 3\}, \{0, 2\}$ and $\{0\}$, respectively, of $\mathbb{Z}_4$, no two of which are conjugate. The tables for them are

	X	Y	
	a	a	b
0	a	a	b
1	a	b	a
2	a	a	b
3	a	b	a

and essentially the group table for $\mathbb{Z}_4$ itself corresponding to $\{0\}$. (Conceptually, entries in the body of that group table should be in braces, like $\{3\}$ to denote the coset, rather than the element.)

19. There are four of them; call them X, Y, Z, and $\mathbb{Z}_6$ corresponding respectively to subgroups $\mathbb{Z}_6, \langle 2 \rangle, \langle 3 \rangle$, and $\{0\}$. See the text answer for the action tables for X, Y, and Z. The group table for $\mathbb{Z}_6$ is essentially the action table for $\{0\}$. (Conceptually, the entries in the body of that group table should in braces, like $\{3\}$, to denote the coset rather than the element.)

20. There are four of them; using the notation for S_3 in Section 8, call them X, Y, Z, and S_3 corresponding respectively to the subgroup S_3, the subgroup $\{\rho_0, \rho_1, \rho_2\}$, the three conjugate subgroups $\{\rho_0, \mu_1\}, \{\rho_0, \mu_2\}$, and $\{\rho_0, \mu_3\}$, and the trivial subgroup $\{\rho_0\}$. We choose the subgroup $\{\rho_0, \mu_1\}$ to illustrate the action on the 3-element set. The action tables for these subgroups are

	X	Y		Z		
	a	a	b	a	b	c
ρ_0	a	a	b	a	b	c
ρ_1	a	a	b	b	c	a
ρ_2	a	a	b	c	a	b
μ_1	a	b	a	a	c	b
μ_2	a	b	a	c	b	a
μ_3	a	b	a	b	a	c

and essentially the group table for S_3 itself corresponding to the subgroup $\{\rho_0\}$. (Conceptually, the entries in the body of that group table should in braces, like $\{\mu_1\}$, to denote the coset rather than the element.)

17. Applications of G-Sets to Counting

1. $G = \langle (1,3,5,6) \rangle$ has order 4. Let $X = \{1,2,3,4,5,6,7,8\}$. We have $|X_{(1)}| = 8$, $|X_{(1,3,5,6)}| = |\{2,4,7,8\}| = 4$, $|X_{(1,5)(3,6)}| = |\{2,4,7,8\}| = 4$, $|X_{(1,6,5,3)}| = |\{2,4,7,8\}| = 4$. Therefore we have $\sum_{g \in G} |X_g| = 8 + 4 + 4 + 4 = 20$. The number of orbits under G is then $(1/4)(20) = 5$.

2. The group G generated by $(1, 3)$ and $(2, 4, 7)$ has order 6. Let $X = \{1,2,3,4,5,6,7,8\}$. We have

$$|X_{(1)}| = 8 \qquad |X_{(1,3)}| = 6 \qquad |X_{(2,4,7)}| = 5$$
$$|X_{(2,7,4)}| = 5 \quad |X_{(1,3)(2,4,7)}| = 3 \quad |X_{(1,3)(2,7,4)}| = 3.$$

Thus $\sum_{g \in G} |X_g| = 8 + 6 + 5 + 5 + 3 + 3 = 30$. The number of orbits under G is then $(1/6)(30) = 5$.

3. The group of rigid motions of the tetrahedron has 12 elements because any one of four triangles can be on the bottom and the tetrahedron can then be rotated though 3 positions, keeping the same face on the bottom. We see that $|X_g| = 0$ unless g is the identity ι of this group G, and $|X_\iota| = 4! = 24$. Thus there are $(1/12)(24) = 2$ distinguishable tetrahedral dice.

4. The total number of ways such a block can be painted with different colors on each face is $8 \cdot 7 \cdot 6 \cdot 5 \cdot 4 \cdot 3$. The group of rigid motions of the cube has 24 elements. The only rigid motion leaving unchanged a block with different colors on all faces is the identity, which leaves all such blocks fixed. Thus the number of distinguishable blocks is $(1/24)(8 \cdot 7 \cdot 6 \cdot 5 \cdot 4 \cdot 3) = 8 \cdot 7 \cdot 5 \cdot 3 = 40 \cdot 21 = 840$.

5. There are 8^6 ways of painting the faces of a block, allowing for repetition of the 8 colors. Following the breakdown of the group G of rotations given in the hint, and using sublabels to suggest the categories in this breakdown, we have

$|X_\iota| = 8^6$ where ι is the identity,

$|X_{\text{opp face, } 90° \text{ or } 270° \text{ rotation}}| = 8 \cdot 8 \cdot 8$; there are 6 such,

$|X_{\text{opp face, } 180° \text{ rotation}}| = 8 \cdot 8 \cdot 8 \cdot 8$; there are 3 such,

$|X_{\text{opp vertices}}| = 8 \cdot 8$; there are 8 such,

$|X_{\text{opp edges}}| = 8 \cdot 8 \cdot 8$; there are 6 such.

Thus $\sum_{g \in G} |X_g| = 8^6 + 6 \cdot 8^3 + 3 \cdot 8^4 + 8 \cdot 8^2 + 6 \cdot 8^3 = 8^3(8^3 + 37)$. The number of distinguishable blocks is thus $(1/24)[8^3(8^3 + 37)] = 11,712$.

6. Proceding as in Exercise 5 using the same group G acting on the set X of 4^8 ways of coloring the eight vertices, we obtain

$|X_\iota| = 4^8$ where ι is the identity,

$|X_{\text{opp face, } 90° \text{ or } 270° \text{ rotation}}| = 4 \cdot 4$; there are 6 such,

$|X_{\text{opp face, } 180° \text{ rotation}}| = 4 \cdot 4 \cdot 4 \cdot 4$; there are 3 such,

$|X_{\text{opp vertices}}| = 4 \cdot 4 \cdot 4 \cdot 4$; there are 8 such,

$|X_{\text{opp edges}}| = 4 \cdot 4 \cdot 4 \cdot 4$; there are 6 such.

Thus $\sum_{g \in G} |X_g| = 4^8 + 6 \cdot 4^2 + 3 \cdot 4^4 + 8 \cdot 4^4 + 6 \cdot 4^4 = 4^4(273) + 96$. The number of distinguishable blocks is thus $(1/24)[4^4(273) + 96] = 2,916$.

7. **a.** The group is $G = D_4$ and has eight elements. We label them as in Section 8. There are $6 \cdot 5 \cdot 4 \cdot 3$ ways of painting the edges of the square, and we let X be this set of 360 elements. We have $|X_{\rho_0}| = 6 \cdot 5 \cdot 4 \cdot 3$ and $|X_g| = 0$ for $g \in D_4, g \neq \rho_0$. Thus the number of distinguishable such painted squares is $(1/8)(360) = 45$.

b. We let G be as in Part(**a**), and let X be the set of 6^4 ways of painting the edges of the square, allowing repetition of colors. This time, we have

$$|X_{\rho_0}| = 6^4, \quad |X_{\rho_1}| = |X_{\rho_3}| = 6, \quad |X_{\rho_2}| = 6 \cdot 6,$$
$$|X_{\mu_1}| = |X_{\mu_2}| = 6 \cdot 6 \cdot 6, \quad \text{and} \quad |X_{\delta_1}| = |X_{\delta_2}| = 6 \cdot 6.$$

Thus $\sum_{g \in G} |X_g| = 6^4 + 2 \cdot 6 + 6^2 + 2 \cdot 6^3 + 2 \cdot 6^2 = 6^2(51) + 12$. The number of distinguishable blocks is thus $(1/8)[6^2(51) + 12] = 231$.

8. The group of rigid motions of the tetrahedron is a subgroup G of the group of permutations of its vertices. The order of G is 12 because, viewing the tetrahedron as sitting on a table, any of the four faces may be on the bottom, and then the base can be rotated repreatedly through 120° to give three possible positions. If we call the vertex at the top of the tetrahedron number 1 and number the vertices on the table as 2, 3, and 4 counterclockwise when viewed from above, we can write the 12 group elements in cyclic notation as

#1 on top	#2 on top	#3 on top	#4 on top
(1)	(1, 2)(3, 4)	(1, 3)(2, 4)	(1, 4)(2, 3)
(2, 3, 4)	(1, 3, 2)	(1, 2, 3)	(1, 2, 4)
(2, 4, 3)	(1, 4, 2)	(1, 4, 3)	(1, 3, 4).

Let X be the 2^6 ways of placing either a 50-ohm resistor or 100-ohm resistor in each edge of the tetrahedron. Now the elements of G that are 3-cycles correspond to rotating, holding a single vertex fixed. These carry the three edges of the triangle opposite that vertex cyclically into themselves, and carry the three edges emenating from that vertex cyclically into themselves. Thus $|X_{3\text{-cycle}}| = 2 \cdot 2$. The element $(1, 2)(3, 4)$ of G carries the edge joining vertex 1 to vertex 2 and the edge joining vertex 3 to vertex 4 into themselves, swaps the edge joining vertices 1 and 3 with the one joining vertices 2 and 4, and swaps the edge joining vertices 1 and 4 with the one joining vertices 2 and 3. Thus we see that $|X_{(1,2)(3,4)}| = 2 \cdot 2 \cdot 2 \cdot 2$, and of course the analogous count can be made for the group elements $(1, 3)(2, 4)$ and $(1, 4)(2, 3)$. Thus we obtain

$$|X_\iota| = 2^6, \quad |X_{3\text{-cycle}}| = 2^2, \quad \text{and} \quad |X_{\text{other type}}| = 2^4.$$

Consequently $\sum_{g \in G} |X_g| = 2^6 + 8 \cdot 2^2 + 3 \cdot 2^4 = 144$. The number of distinguishable blocks is thus $(1/12)(144) = 12$.

9. **a.** The group G of rigid motions of the prism has order 8, four positions leaving the end faces in the same position and four positions with the end faces swapped. There are $6 \cdot 5 \cdot 4 \cdot 3 \cdot 2 \cdot 1$ ways of painting the faces different colors. We let X be the set of these 6! possibilities. Then $|X_\iota| = 6!$ and $|X_{\text{other}}| = 0$. Thus there are $(1/8)(6!) = 6 \cdot 5 \cdot 3 = 90$ distinguishable painted prisms using six different colors.

b. This time the set X of possible ways of painting the prism has 6^6 elements. We have
$|X_\iota| = 6^6$ where ι is the identity element,
$|X_{\text{same ends, rotate } 90° \text{ or } 270°}| = 6 \cdot 6 \cdot 6$,
$|X_{\text{same ends, rotate } 180°}| = 6 \cdot 6 \cdot 6 \cdot 6$,
$|X_{\text{swap ends, keeping top face on top}}| = 6 \cdot 6 \cdot 6 \cdot 6$,
$|X_{\text{swap ends, as above, rotate } 90° \text{ or } 270°}| = 6 \cdot 6$,
$|X_{\text{swap ends, as above, rotate } 180°}| = 6 \cdot 6 \cdot 6$.
Thus $\sum_{g \in G} |X_g| = 6^6 + 2 \cdot 6^3 + 6^4 + 6^4 + 2 \cdot 6^2 + 6^3 = 6^2(6^4 + 92)$. The number of distinguishable blocks is thus $(1/8)[6^2(6^4 + 82)] = 6,246$.

18. Rings and Fields

1. 0   **2.** 16   **3.** 1   **4.** 22   **5.** $(1, 6)$ **6.** $(2, 2)$

7. Yes, $n\mathbb{Z}$ for $n \in \mathbb{Z}^+$ is a commutative ring, but without unity unless $n = 1$, and is not a field.

8. No. $\mathbb{Z}^+$ is not a ring; there is no identity for addition.

9. Yes, $\mathbb{Z} \times \mathbb{Z}$ is a commutative ring with unit $(1, 1)$, but is not a field because $(2, 0)$ has no multiplicative inverse.

10. Yes. $2\mathbb{Z} \times \mathbb{Z}$ is a commutative ring, but without unity, and is not a field.

11. Yes, $\{a + b\sqrt{2} \mid a, b \in \mathbb{Z}\}$ is a commutative ring with unity, but is not a field because 2 has no multiplicative inverse.

12. Yes, $\{a + b\sqrt{2} \mid a, b \in \mathbb{Q}\}$ is a commutative ring with unity and is a field because

$$\frac{1}{a + b\sqrt{2}} = \frac{1}{a + b\sqrt{2}} \cdot \frac{a - b\sqrt{2}}{a - b\sqrt{2}} = \frac{a}{a^2 - 2b^2} + \frac{-b}{a^2 - 2b^2}\sqrt{2}.$$

13. No, $\mathbb{R}i$ is not closed under multiplication.   **14.** The units in $\mathbb{Z}$ are 1 and -1.

15. The units in $\mathbb{Z} \times \mathbb{Z}$ are (1, 1), (1, -1), (-1, 1), and (-1, -1).

16. The units in $\mathbb{Z}_5$ are 1, 2, 3, and 4 because $1 \cdot 1 = 2 \cdot 3 = 4 \cdot 4 = 1$.

17. All nonzero elements of $\mathbb{Q}$ are units.

18. The units in $\mathbb{Z} \times (Q) \times \mathbb{Z}$ are $(1, q, 1), (-1, q, 1), (1, q, -1)$ and $(-1, q, -1)$ for any nonzero $q \in \mathbb{Q}$.

19. The units in $\mathbb{Z}_4$ are 1 and 3; $1 \cdot 1 = 3 \cdot 3 = 1$.

20. a. Each of the four entries in the matrix can be either of two elements, so there are $2^4 = 16$ matrices in $M_2(\mathbb{Z}_2)$.

    b. $\begin{bmatrix} 1 & 1 \\ 1 & 0 \end{bmatrix} \begin{bmatrix} 0 & 1 \\ 1 & 1 \end{bmatrix} = I_2$, and $\begin{bmatrix} 1 & 0 \\ 1 & 1 \end{bmatrix}$, $\begin{bmatrix} 1 & 1 \\ 0 & 1 \end{bmatrix}$, $\begin{bmatrix} 1 & 0 \\ 0 & 1 \end{bmatrix}$ and $\begin{bmatrix} 0 & 1 \\ 1 & 0 \end{bmatrix}$ are all their own inverse. These six matrices are the units.

21. (See the answer in the text.)

22. Because $\det(A + B)$ need not equal $\det(A) + \det(B)$, we see that det is not a ring homomorphism. For example, $\det(I_n + I_n) = 2^n$ but $\det(I_n) + \det(I_n) = 1 + 1 = 2$.

23. Let $\phi : \mathbb{Z} \to \mathbb{Z}$ be a ring homomorphism. Because $1^2 = 1$, we see that $\phi(1)$ must be an integer whose square is itself, namely either 0 or 1. If $\phi(1) = 1$ then $\phi(n) = \phi(n \cdot 1) = n$, so ϕ is the identity map of $\mathbb{Z}$ onto itself which is a homomorphism. If $\phi(1) = 0$, then $\phi(n) = \phi(n \cdot 1) = 0$, so ϕ maps everthing onto 0, which also yields a homomorphism.

24. As in the preceding solution, we see that for a ring homomorphism $\phi : \mathbb{Z} \to \mathbb{Z} \times \mathbb{Z}$, we must have $\phi(1)^2 = \phi(1^2) = \phi(1)$. The only elements of $\mathbb{Z} \times \mathbb{Z}$ that are their own squares are (0, 0), (1, 0), (0, 1), and (1, 1). Thus the possibilities for ϕ are $\phi_1(n) = (0, 0), \phi_2(n) = (n, 0), \phi_3(n) = (0, n)$, and $\phi_4(n) = (n, n)$. It is easily checked that these four maps are ring homomorphisms.

25. Because both (1, 0) and (0, 1) are their own squares, their images under a ring homomorphism $\phi : \mathbb{Z} \times \mathbb{Z} \to \mathbb{Z}$ must also have this property, and thus must be either 0 or 1. Because (1, 0) and (0, 1) generate $\mathbb{Z} \times \mathbb{Z}$ as an additive group, this determines the possible values of the homomorphism on $(n, m) \in \mathbb{Z} \times \mathbb{Z}$. Thus the possibilities are given by $\phi_1(n, m) = 0, \phi_2(n, m) = n, \phi_3(n, m) = m$, and $\phi_4(n, m) = n + m$. It is easily checked that ϕ_1, ϕ_2, and ϕ_3 are homomorpisms. However, ϕ_4 is not a homomorphism because $n + m = \phi_4(n, m) = \phi_4((1, 1)(n, m)) \neq \phi_4(1, 1)\phi_4(n, m) = (1 + 1)(n + m) = 2(n + m)$.

26. As in Exercise 25, we see that the images of additive generators (1, 0, 0), (0, 1, 0), and (0, 0, 1) under a ring homomorphism $\phi : \mathbb{Z} \times \mathbb{Z} \times \mathbb{Z} \to \mathbb{Z}$ can only be 0 and 1. As shown in the argument for ϕ_4 in that exercise, mapping more than one of these generators into 1 will not give a ring homomorphism. Thus either they are all mapped into 0 giving the trivial homomorphism, or we have a projection homomorphism where one of the generators is mapped into 1 and the other two are mapped into 0. Thus there are four such ring homomorphisms ϕ.

27. (See the answer in the text)

28. We have $x^2 + x - 6 = (x + 3)(x - 2)$. In $\mathbb{Z}_{14}$, it is possible to have a product of two nonzero elements be 0. Trying all elements x from -6 to 7 in $\mathbb{Z}_{14}$ to see if $(x + 3)(x - 2)$ is zero, we find that this happens for $x = -5, -3, 2$, and 4. Thus the elements 2, 4, 9, and 11 in $\mathbb{Z}_{14}$ are solutions of the quadratic equation.

29. The definition is incorrect. Insert the word "commutative" before either "ring" or "group".

   A **field** F is a commutative ring with nonzero unity such that the set of nonzero elements of F is a group under multiplication.

30. The definition is incorrect. We have not defined any concept of magnitude for elements of a ring.

   A **unit** in a ring with nonzero unity is an element that has a multiplicative inverse.

31. In the ring $\mathbb{Z}_6$, we have $2 \cdot 3 = 0$, so we take $a = 2$ and $b = 3$.

32. $\mathbb{Z} \times \mathbb{Z}$ has unity $(1, 1)$; however the subring $\mathbb{Z} \times \{0\}$ has unity $(1, 0)$. Also, $\mathbb{Z}_6$ has unity 1 while the subring $\{0, 2, 4\}$ has unity 4, and the subring $\{0, 3\}$ has unity 3.

33. T F F F T F T T T T

34. Let $f, g, h \in F$. Now $[(fg)h](x) = [(fg)(x)]h(x) = [f(x)g(x)]h(x)$. Because multiplication in $\mathbb{R}$ is associative, we continue with $[f(x)g(x)]h(x) = f(x)[g(x)h(x)] = f(x)[(gh)(x)] = [f(gh)](x)$. Thus $(fg)h$ and $f(gh)$ have the same value on each $x \in \mathbb{R}$, so they are the same function and axiom 2 holds. For axiom 3, we use the distributive laws in $\mathbb{R}$ and we have $[f(g + h)](x) = f(x)[(g + h)(x)] = f(x)[g(x) + h(x)] = f(x)g(x) + g(x)h(x) = (fg)(x) + (fh)(x) = (fg + fh)(x)$ so $f(g + h)$ and $fg + fh$ are the same function and the left distributive law holds. The right distributive law is proved similarly.

35. For $f, g \in F$, we have $\phi_a(f + g) = (f + g)(a) = f(a) + g(a) = \phi_a(f) + \phi_a(g)$. Turning to the multiplication, we have $\phi_a(fg) = (fg)(a) = f(a)g(a) = \phi_a(f)\phi_a(g)$. Thus ϕ_a is a homomorphism.

36. We need check only the multiplicative property.

   Reflexive: The identity map ι of a ring R into itself satisfies $\iota(ab) = ab = \iota(a)\iota(b)$, so the reflexive property is satisfied.

   Symmetric: Let $\phi : R \to R'$ be an isomorphism. We know from group theory that $\phi^{-1} : R' \to R$ is an isomorphism of the additive group of R' with the additive group of R. For $\phi(a), \phi(b) \in R'$, we have $\phi^{-1}(\phi(a)\phi(b)) = \phi^{-1}(\phi(ab)) = ab = \phi^{-1}(\phi(a))\phi^{-1}(\phi(b))$.

   Transitive: Let $\phi : R \to R'$ and $\psi : R' \to R''$ be ring isomorphisms. Exercise 27 of Section 3 shows that $\psi\phi$ is an isomorphism of both the additive binary structure and the multiplicative binary stucture. Thus $\psi\phi$ is again a ring isomorphism.

37. Let $u, v \in U$. Then there exists $s, t \in R$ such that $us = su = 1$ and $vt = tv = 1$. These equations show that s and t are also units in U. Then $(ts)(uv) = t(su)v = t1v = tv = 1$ and $(uv)(ts) = u(vt)s = u1s = 1$, so uv is again a unit and U is closed under multiplication. Of course multiplication in U is associative because multiplication in R is associative. The equation $(1)(1) = 1$ shows that 1 is a unit. We showed above that a unit u in U has a multiplicative inverse s in U. Thus U is a group under multiplication.

38. Now $(a + b)(a - b) = a^2 + ba - ab - b^2$ is equal to $a^2 - b^2$ if and only if $ba - ab = 0$, that is, if and only if $ba = ab$. But $ba = ab$ for all $a, b \in R$ if and only if R is commutative.

39. We need only check the second and third ring axioms. For axiom 2, we have $(ab)c = 0c = 0 = a0 = a(bc)$. For axiom 3, we have $a(b + c) = 0 = 0 + 0 = ab + ac$, and $(a + b)c = 0 = 0 + 0 = ac + bd$.

40. If $\phi : 2\mathbb{Z} \to 3\mathbb{Z}$ is an isomorphism, then by group theory for the additive groups we know that either $\phi(2) = 3$ or $\phi(2) = -3$, so that either $\phi(2n) = 3n$ or $\phi(2n) = -3n$. Suppose that $\phi(2n) = 3n$. Then

$\phi(4) = 6$ while $\phi(2)\phi(2) = (3)(3) = 9$. Thus $\phi(2n) = 3n$ does not give an isomorphism, and a similar computation shows that $\phi(2n) = -3n$ does not give an isomorphism either.

$\mathbb{R}$ and $\mathbb{C}$ are not isomorphic because every element in the field $\mathbb{C}$ is a square while -1 is not a square in $\mathbb{R}$.

41. In a commutative ring, we have $(a+b)^2 = a^2 + ab + ba + b^2 = a^2 + ab + ab + b^2 = a^2 + 2 \cdot ab + b^2$. Now the binomial theorem simply counts the number of each type of product $a^i b^{n-i}$ appearing in $(a+b)^n$. As long as our ring is commutative, every summand of $(a+b)^n$ can be written as a product of factors a and b with all the factors a written first, so the usual binomial expansion is valid in a commutative ring.

In $\mathbb{Z}_p$, the coefficient $\binom{p}{i}$ of $a^i b^{p-i}$ in the expansion of $(a+b)^p$ is a multiple of p if $1 \leq i \leq p-1$. Because $p \cdot a = 0$ for all $a \in \mathbb{Z}_p$, we see that the only nonzero terms in the expansion are those corresponding to $i = 0$ and $i = p$, namely b^p and a^p.

42. Let F be a field, and suppose that $u^2 = u$ for nonzero $u \in F$. Multiplying by u^{-1}, we find that $u = 1$. This shows that 0 and 1 are the only solutions of the equation $x^2 = x$ in a field. Now let K be a subfield of F. The unity of K satisfies the equation $x^2 = x$ in K, and hence also in F, and thus must be the unity 1 of F.

43. Let u be a unit in a ring R. Suppose that $su = us = 1$ and $tu = ut = 1$. Then $s = s1 = s(ut) = (su)t = 1t = t$. Thus the inverse of a unit is unique.

44. a. If $a^2 = a$ and $b^2 = b$ and if the ring is commutative, then $(ab)^2 = abab = aabb = a^2 b^2 = ab$, showing that the idempotents are closed under multiplication.

b. By trying all elments, we find that the idempotents in $\mathbb{Z}_6$ are 0, 1, 3, and 4 while the idempotents in $\mathbb{Z}_{12}$ are 0, 1, 4, and 9. Thus the idempotents in $\mathbb{Z}_6 \times \mathbb{Z}_{12}$ are

$$
\begin{array}{llll}
(0,0) & (0,1) & (0,4), & (0,9) \\
(1,0) & (1,1) & (1,4), & (1,9) \\
(3,0) & (3,1) & (3,4), & (3,9) \\
(4,0) & (4,1) & (4,4), & (4,9).
\end{array}
$$

45. We have

$$
\begin{aligned}
P^2 &= [A(A^T A)^{-1} A^T][A(A^T A)^{-1} A^T] = A[(A^T A)^{-1}(A^T A)](A^T A)^{-1} A \\
&= AI_n(A^T A)^{-1} A^T = A(A^T A)^{-1} A^T = P.
\end{aligned}
$$

46. As explained in the answer to Exercise 41, the binomial expansion is valid in a commutative ring. Suppose that $a^n = 0$ and $b^m = 0$ in R. Now $(a+b)^{m+n}$ is a sum of terms containing as a factor $a^i b^{m+n-i}$ for $0 \leq i \leq m+n$. If $i \geq n$, then $a^i = 0$ so each term with a factor $a^i b^{m+n-i}$ is zero. On the other hand, if $i < n$, then $m+n-i > m$ so $b^{m+n-i} = 0$ and each term with a factor $a^i b^{m+n-i}$ is zero. Thus $(a+b)^{m+n} = 0$, so $a+b$ is nilpotent.

47. If R has no nonzero nilpotent element, then the only solution of $x^2 = 0$ is 0, for any nonzero solution would be a nilpotent element. Conversely, suppose that the only solution of $x^2 = 0$ is 0, and suppose that $a \neq 0$ is nilpotent. Let n be the smallest positive integer such that $a^n = 0$. If n is even, then $a^{n/2} \neq 0$ but $(a^{n/2})^2 = a^n = 0$ so $a^{n/2}$ is a nonzero solution of $x^2 = 0$, contrary to assumption. If n is odd, then $(a^{(n+1)/2})^2 = a^{n+1} = a^n a = 0a = 0$ so $a^{(n+1)/2}$ is a nonzero solution of $x^2 = 0$, contrary to assumption. Thus R has no nonzero nilpotent elements.

48. It is clear that if S is a subring of R, then all three of the conditions must hold. Conversely, suppose the conditions hold. The first two conditions and Exercise 45 of Section 5 show that $\langle S, + \rangle$ is an additive group. The final condition shows that multiplication is closed on S. Of course the associative and distributive laws hold for elements of S, because they actually hold for all elements in R. Thus S is a subring of R.

49. a. Let R be a ring and let $H_i \leq R$ for $i \in I$. Theorem 7.4 shows that $H = \bigcap_{i \in I} H_i$ is an additive group. Let $a, b \in H$. Then $a, b \in H_i$ for $i \in I$, so $ab \in H_i$ for $i \in I$ because H_i is a subring of R. Therefore $ab \in H$ so H is closed under multiplication. Clearly the associative and distributive laws hold for elements from H, because they actually hold for all elements in R. Thus H is a subring of R.

b. Let F be a field, and let $K_i \leq F$ for $i \in I$. Part(**a**) shows that $K = \bigcap_{i \in I} K_i$ is a ring. Let $a \in K, a \neq 0$. Then $a \in K_i$ for $i \in I$ so $a^{-1} \in K_i$ for $i \in I$ because Exercises 42 and 43 show that the unity in each K_i is the same as in F and that inverses are unique. Therefore $a^{-1} \in K$. Of course multiplication in K is commutative because multiplication in F is commutative. Therefore K is a subfield of F.

50. We show that I_a satisfies the conditions of Exercise 48. Because $a0 = 0$ we see that $0 \in I_a$. Let $c, d \in I_a$. Then $ac = ad = 0$ so $a(c - d) = ac - ad = 0 - 0 = 0$; thus $(c - d) \in I_a$. Also $a(cd) = (ac)d = 0d = 0$ so $cd \in I_a$. This completes the check of the properties in Exercise 48.

51. Clearly a^n is in every subring containing a, so R_a contains a^n for every positive integer n. Thus $\langle R_a, + \rangle$ contains the additive group G generated by $S = \{a^n \mid n \in \mathbb{Z}^+\}$. We claim that $G = R_a$. We need only show that G is closed under multiplication. Now G consists of zero and all finite sums of terms of the form a^n or $-a^m$. By the distributive laws, the product of two elements that are finite sums of positive powers and inverses of postive powers of a can again be written as such a sum, and is thus again in G. Therefore G is actually a subring containing a and contained in R_a so we must have $G = R_a$.

52. Example 18.15 shows that the map $\phi : \mathbb{Z}_{rs} \to \mathbb{Z}_r \times \mathbb{Z}_s$ where $\phi(a) = a \cdot (1, 1)$ is an isomorphism. Let $b = \phi^{-1}(m, n)$. Computing $b \cdot (1, 1)$ by components, we see that $1 + 1 + \cdots + 1$ for b summands yields m in $\mathbb{Z}_r$ and yields n in $\mathbb{Z}_s$. Thus, viewing b as an integer in $\mathbb{Z}$, we see that $b \equiv m \pmod{r}$ and $b \equiv n \pmod{s}$.

53. a. Statement: Let $b_1, b_2, \cdots, b_n$ be integers such that $\gcd(b_i, b_j) = 1$ for $i \neq j$. Then $\mathbb{Z}_{b_1 b_2 \cdots b_n}$ is isomorphic to $\mathbb{Z}_{b_1} \times \mathbb{Z}_{b_2} \times \cdots \mathbb{Z}_{b_n}$ with an isomorphism ϕ where $\phi(1) = (1, 1, \cdots, 1)$.

Proof: By the hypothesis that $\gcd(b_i, b_j) = 1$ for $i \neq j$, we know that the image group is cyclic and that $(1, 1, \cdots, 1)$ generates the group. Because the domain group is cyclic generated by 1, we know that ϕ is an additive group isomorphism. It remains to show that $\phi(ms) = \phi(m)\phi(s)$ for m and s in the domain group. This follows from the fact that the ith component of $\phi(ms)$ in the image group is $(ms) \cdot 1$ which is equal to the product of m summands of 1 times s summands of 1 by the distributive laws in a ring.

b. Let $c = \phi^{-1}(a_1, a_2, \cdots, a_n)$ where ϕ be the isomorphism in Part(**a**). Computing $\phi(c) = \phi(c \cdot 1)$ in its ith component, we see that $1 + 1 + \cdots + 1$ for c summands in the ring $\mathbb{Z}_{b_i}$ yields a_i. Viewing c as an integer, this means that $c \equiv a_i \pmod{b_i}$.

54. Note that $a0 = 0$ for all $a \in S$ follows from the distributive laws, so associativity of multiplication for products containing a factor 0 holds, and associativity in the group $\langle S^*, \cdot \rangle$ takes care of associativity for other products. All of the other axioms needed to verify that S is a division ring follow at once from the two given group statements and the given distribtive laws, except for the commutativity of addition.

The left followed by the right distributive laws yield $(1+1)(a+b) = (1+1)a + (1+1)b = a+a+b+b$. The right followed by the left distributive laws yield $(1+1)(a+b) = 1(a+b) + 1(a+b) = a+b+a+b$. Thus $a+a+b+b = a+b+a+b$ and by cancellation in the additive group, we obtain $a+b = b+a$.

55. Let $a, b \in R$ where R is a Boolean ring. We have $a+b = (a+b)^2 = a^2 + ab + ba + b^2 = a + ab + ba + b$. Thus in a Boolean ring, $ab = -ba$. Taking $b = a$, we see that $aa = -aa$, so $a = -a$. Thus every element is its own additive inverse, so $-ba = ba$. Combining our equations $ab = -ba$ and $-ba = ba$, we obtain $ab = ba$, showing that R is commutative.

56. a.

+	$\varnothing$	$\{a\}$	$\{b\}$	S
$\varnothing$	$\varnothing$	$\{a\}$	$\{b\}$	S
$\{a\}$	$\{a\}$	$\varnothing$	S	$\{b\}$
$\{b\}$	$\{b\}$	S	$\varnothing$	$\{a\}$
S	S	$\{b\}$	$\{a\}$	$\varnothing$

$\cdot$	$\varnothing$	S	$\{a\}$	$\{b\}$
$\varnothing$	$\varnothing$	$\varnothing$	$\varnothing$	$\varnothing$
S	$\varnothing$	S	$\{a\}$	$\{b\}$
$\{a\}$	$\varnothing$	$\{a\}$	$\{a\}$	$\varnothing$
$\{b\}$	$\varnothing$	$\{b\}$	$\varnothing$	$\{b\}$

b. Let $A, B \in \mathcal{P}(S)$. Then $A + B = (A \cup B) - (A \cap B) = (B \cup A) - (B \cap A) = B + A$, so addition is commutative.

We check associativity of addition; it is easiest to think in terms of the elements in $(A+B)+C$ and the elements in $A+(B+C)$. By definition, the sum of two sets contains the elements in precisely one of the sets. Thus $A + B$ consists of the elements that are in either one of the sets A or B, but not in the other. Therefore $(A+B)+C$ consists of the elements that are in precisely one of the three sets A, B, C. Clearly $A + (B+C)$ yields this same set, so addition is associative.

The empty set $\varnothing$ acts a additive identity, for $A + \varnothing = (A \cup \varnothing) - (A \cap \varnothing) = A - \varnothing = A$ for all $A \in \mathcal{P}(S)$.

For $A \in \mathcal{P}(S)$, we have $A + A = (A \cup A) - (A \cap A) = A - A = \varnothing$, so each element of $\mathcal{P}(S)$ is its own additive inverse. This shows that $\langle \mathcal{P}(S), + \rangle$ is an abelian group.

For associativity of multiplication, we see that $(A \cdot B) \cdot C = (A \cap B) \cap C = A \cap (B \cap C) = A \cdot (B \cdot C)$.

For the left distributive law, we again think in terms of the elements in the sets. The set $A \cdot (B + C) = A \cap (B + C)$ consists of all elements of A that are in precisely one of the two sets B, C. This set thus contains all the elements in $A \cap B$ or in $A \cap C$, but not in both sets. This is precisely the set $(A \cap B) + (A \cap C) = (A \cdot B) + (A \cdot C)$. The right distributive law can be demonstrated by a similar argument.

We have shown that $\langle \mathcal{P}(S), +, \cdot \rangle$ is a ring. Because $A \cdot A = A \cap A = A$, we see from the definition in Exercise 55 that it is a Boolean ring.

19. Integral Domains

1. We rewrite the equation as $x(x-3)(x+1) = 0$, and simply try all the elements, -5 -4, -3 -2, -1, 0, 1, 2, 3, 4, 5, 6 of $\mathbb{Z}_{12}$, obtaining the solutions 0, 3, 5, 8, 9, and 11.

2. The solution in $\mathbb{Z}_7$ is 3 and the solution in $\mathbb{Z}_{23}$ is 16.

3. Trying all possibilities -2, -1, 0, 1, 2, and 3, we find no solutions.

4. Trying all possibilities -2, -1, 0, 1, 2, and 3, we find $x = 2$ as the only solution.

5. 0   **6.** 0   **7.** 0   **8.** 3   **9.** 12   **10.** 30

11. $(a+b)^4 = a^4 + 4 \cdot a^3 b + 6 \cdot a^2 b^2 + 4 \cdot ab^3 + b^4 = a^4 + 2 \cdot a^2 b^2 + b^4$

12. $(a+b)^9 = [(a+b)^3]^3 = [a^3 + 3 \cdot a^2 b + 3 \cdot ab^2 + b^3]^3 = (a^3 + b^3)^3 = a^9 + 3 \cdot a^6 b^3 + 3 \cdot a^3 b^6 + b^9 = a^9 + b^9.$

13. $(a+b)^6 = [(a+b)^3]^2 = [a^3 + 3 \cdot a^2 b + 3 \cdot ab^2 + b^3]^2 = (a^3 + b^3)^2 = a^6 + 2 \cdot a^3 b^3 + b^6.$

14. We have $\begin{bmatrix} 2 & -1 \\ 2 & -1 \end{bmatrix} \begin{bmatrix} 1 & 2 \\ 2 & 4 \end{bmatrix} = \begin{bmatrix} 0 & 0 \\ 0 & 0 \end{bmatrix}.$

15. The definition is incorrect. We must state that $a \neq 0$ and $b \neq 0$.

     If, in a ring R, nonzero elements a and b are such that $ab = 0$, then a and b are **divisors of zero**.

16. The definition is incorrect; n must be minimal in $\mathbb{Z}^+$.

     If for some $n \in \mathbb{Z}^+, n \cdot a = 0$ for all a in a ring R, the smallest such n is the **characteristic** of R. If no such n exists, then 0 is the **characteristic** of R.

17. F T F F T T F T F F

18. Specifying the regions by their number, we have as examples:

         1. $\mathbb{Q}$ 2. $\mathbb{Z}$ 3. $\mathbb{Z}_4$ 4. $2\mathbb{Z}$ 5. $M_2(\mathbb{R})$ and

         6. Upper-triangular matrices with integer entries and all zeros on the main diagonal.

19. (See the answer in the text.)

20. We need add only one region, No. 7, to the existing figure. Unfortunately, we can't draw big circles, so we had to use quadrilaterals which are a bit confusing. For example, the commutative ring quadrilateral extends into the high rectangle, and includes regions numbered 1, 2, 3, and 4. This would be apparent if we could have made it a large circle. We suggest that you draw circles, as in the text. Just add another region 7 inside the existing region 5 of the text, and label it "Strictly skew fields".

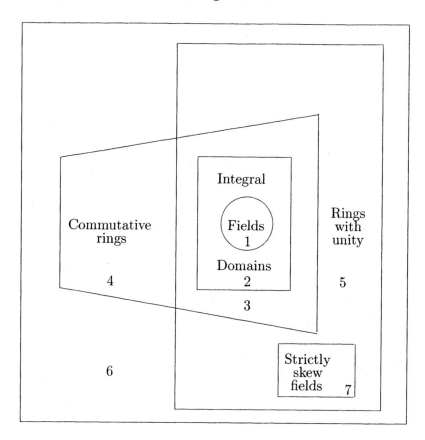

21. Rewriting $ab = ac$ as $a(b - c) = 0$, deduce $b = c$ from $a \neq 0$ and the absence of zero divisors.

22. If $a \neq 0$ is in the finite integral domain D, use the cancellation law and cardinality to deduce that the map of D into D where x is mapped into ax is one-to-one and onto D.

23. If $a^2 = a$, then $a^2 - a = a(a - 1) = 0$. If $a \neq 0$, then a^{-1} exists in R and we have $a - 1 = (a^{-1}a)(a - 1) = a^{-1}[a(a - 1)] = a^{-1}0 = 0$, so $a - 1 = 0$ and $a = 1$. Thus 0 and 1 are the only two idempotent elements in a division ring.

24. Exercise 49(a) in Section 18 showed that an intersection of subrings of a ring R is again a subring of R. Thus an intersection of subdomains D_i for $i \in I$ of an integral domain D is at least a ring. The preceding exercise shows that unity in an integral domain can be characterized as the nonzero idempotent. This shows that the unity in each D_i must be the unity 1 in D, so 1 is in the intersection of the D_i. Of course multiplication is commutative in the intersection because it is commutative in D and the operation is induced. Finally, if $ab = 0$ in the intersection, then $ab = 0$ in D so either $a = 0$ or $b = 0$, that is, the intersection has no divisors of zero, and is a subdomain of D.

25. Because R has no divisors of zero, multiplicative cancellation of nonzero elements is valid. The construction in the proof of Theorem 18.11 is valid and shows that each nonzero $a \in R$ has a right inverse, say a_i. A similar construction where the elements of R are all multiplied on the *right* by a shows that a has a left inverse, say a_j. By associativity of multiplication, we have $a_j = a_j(aa_i) = (a_ja)a_i = a_i$. Thus every nonzero $a \in R$ is a unit, so R is a division ring.

26. a. Let $a \neq 0$. We wish to show that a is not a divisor of zero. Let b be the unique element such that $aba = a$. Suppose $ac = 0$ or $ca = 0$. Then $a(b + c)a = aba + aca = a + 0 = a$. By uniqueness $b + c = b$ so $c = 0$.

b. From $aba = a$, we know that $b \neq 0$ also. Multiplying on the left by b, we obtain $baba = ba$. Because R has no divisors of zero by part **a**, multiplicative cancellation is valid and we see that $bab = b$.

c. We claim that ab is unity for nonzero a and b given in the statement of the exercise. Let $c \in R$. From $aba = a$, we see that $ca = caba$. Cancelling a, we obtain $c = c(ab)$. From part **b**, we have $bc = babc$, and cancelling b yields $c = (ab)c$. Thus ab satisfies $(ab)c = c(ab)$ for all $c \in R$, so ab is unity.

d. Let a be a nonzero element of the ring. By part **a**, $aba = a$. By part **c** , $ab = 1$ so b is a right inverse of a. Because the elements a and b behave in a symmetric fashion by part **b**, an argument symmetric to that in part **c**, starting with the equation $ac = abac$, shows that $ba = 1$ also. Thus b is also a left inverse of a, so a is a unit. This shows that R is a division ring.

27. By Exercise 23, we see that the unity in an integral domain can be characterized as the unique nonzero idempotent. The unity element in D must then also be the unity in every subdomain. Recall that the characteristic of a ring with unity is the minimum $n \in \mathbb{Z}^+$ such that $n \cdot 1 = 0$, if such an n exists, and is 0 otherwise. Because unity is the same in the subdomain, this computation will lead to the same result there as in the original domain.

28. Let $R = \{n \cdot 1 \mid n \in \mathbb{Z}\}$. We have $n \cdot 1 + m \cdot 1 = (n+m) \cdot 1$ so R is closed under addition. Taking $n = 0$, we see that $0 \in R$. Because the inverse of $n \cdot 1$ is $(-n) \cdot 1$, we see that R contains all additive inverses of elements, so $\langle R, + \rangle$ is an abelian group. The distributive laws show that $(n \cdot 1)(m \cdot 1) = (nm) \cdot 1$, so R is closed under multiplication. Because $1 \cdot 1 = 1$, we see that $1 \in R$. Thus R is a commutative ring with unity. Because a product $ab = 0$ in R can also be viewed as a product in D, we see that R also has no divisors of zero. Thus R is a subdomain of D.

29. Suppose the characteristic is mn for $m > 1$ and $n > 1$. Following the hint, the distributive laws show that $(m \cdot 1)(n \cdot 1) = (nm) \cdot 1 = 0$. Because we are in an integral domain, we must have either $m \cdot 1 = 0$ or $n \cdot 1 = 0$. But if $m \cdot 1 = 0$ then Theorem 19.15 shows that the characteristic of D is at most m. If $n \cdot 1 = 0$, the chracteristic of D is at most n. Thus the characteristic can't be a composite positive integer, so it must either be 0 or a prime p.

30. **a.** From group theory, we know that S is an abelian group under addition. We check the associativity of multiplication, using the facts that, for all $m, n \in \mathbb{Z}$ and $r, s \in R$, we have $n \cdot (m \cdot r) = (nm) \cdot r, n \cdot (r+s) = n \cdot r + n \cdot s, r(n \cdot s) = n \cdot (rs)$, and $(n \cdot r)s = n \cdot (rs)$, which all follow from commutativity of addition and the distributive laws in R. We have, for $r, s, t \in R$ and $k, m, n \in \mathbb{Z}$,

$(r,k)[(s,m)(t,n)] = (r,k)(st + m \cdot t + n \cdot s, mn)$
$= (r(st + m \cdot t + n \cdot s) + k \cdot (st + m \cdot t + n \cdot s) + mn \cdot r, kmn)$
$= (rst + k \cdot st + m \cdot rt + n \cdot rs + km \cdot t + kn \cdot s + mn \cdot r, kmn)$

and

$[(r,k)(s,m)](t,n) = (rs + k \cdot s + m \cdot r, km)(t,n)$
$= ((rs + k \cdot s + m \cdot r)t + km \cdot t + n \cdot (rs + k \cdot s + m \cdot r), kmn)$
$= (rst + k \cdot st + m \cdot rt + n \cdot rs + km \cdot t + kn \cdot s + mn \cdot r, kmn).$

Thus multiplication is associative. For the left distributive law, we obtain

$$
\begin{aligned}
(r,k)[(s,m) + (t,n)] &= (r,k)(s+t, m+n) \\
&= (r(s+t) + k \cdot (s+t) + (m+n) \cdot r, k(m+n)) \\
&= (rs + k \cdot s + m \cdot r, km) + (rt + k \cdot t + n \cdot r, kn) \\
&= (r,k)(s,m) + (r,k)(t,n).
\end{aligned}
$$

Proof of the right distributive law is a similar computation. Thus S is a ring.

b. We have $(0,1)(r,n) = (0r + 1 \cdot r + n \cdot 0, 1n) = (r,n) = (r0 + n \cdot 0 + 1 \cdot r, n1) = (r,n)(0,1)$, so $(0,1) \in S$ is unity.

c. Using Theorem 19.15 and part **b**, the ring S either has characteristic 0 or the smallest positive integer n such that $(0,0) = n \cdot (0,1) = (0,n)$. Clearly n has this property if and only if $S = R \times \mathbb{Z}_n$. Because we chose $\mathbb{Z}$ or $\mathbb{Z}_n$ to form S acccording as R has characteristic 0 or n, we see that R and S have the same characteristic.

d. We have $\phi(r_1 + r_2) = (r_1 + r_2, 0) = (r_1, 0) + (r_2, 0) = \phi(r_1) + \phi(r_2)$. Also, $\phi(r_1 r_2) = (r_1 r_2, 0) = (r_1 r_2 + 0 \cdot r_2 + 0 \cdot r_1, 00) = (r_1, 0)(r_2, 0) = \phi(r_1)\phi(r_2)$. Thus ϕ is a homomorphism. If $\phi(r_1) = \phi(r_2)$, then $(r_1, 0) = (r_2, 0)$ so $r_1 = r_2$; Thus ϕ is one to one. Therefore ϕ maps R isomorphically onto the subring $\phi[R]$ of S.

20. Fermat's and Euler's Theorems

1. Either 3 or 5   **2.** Either 2, 6, 7, or 8   **3.** Either 3, 5, 6, 7, 10, 11, 12, or 14

4. $3^{47} \equiv (3^{22})^2 \cdot 3^3 \equiv 1^2 \cdot 27 \equiv 27 \equiv 4 \pmod{23}$ **5.** $37^{49} \equiv 2^{49} \equiv (2^6)^8 \cdot 2 \equiv 1^8 \cdot 2 \equiv 2 \pmod 7$

6. $2^{17} \equiv (2^4)^4 \cdot 2 \equiv (-2)^4 \cdot 2 \equiv 16 \cdot 2 \equiv 14 \pmod{18}$. Thus $2^{17} = 18q + 14$. Then $2^{(2^{17})} \equiv 2^{18q+14} \equiv (2^{18})^q \cdot 2^{14} \equiv 1^q \cdot 2^{14} \equiv (2^4)^3 \cdot 2^2 \equiv (-3)^3 \cdot 2^2 \equiv -27 \cdot 4 \equiv -8 \cdot 4 \equiv 6 \pmod{19}$ so the answer is $6 + 1 = 7$.

7. (See the answer in the text.)

8. All positive integers less than p^2 that are not divisible by p are relatively prime to p. Thus we delete from the $p^2 - 1$ integers less than p^2 the integers $p, 2p, 3p, \cdots, (p-1)p$. There are $p-1$ integers deleted, so $\phi(p^2) = (p^2 - 1) - (p - 1) = p^2 - p$.

9. We delete from the $pq - 1$ integers less than pq those that are mltiples of p or of q to obtain those relatively prime to pq. The multiples of p are $p, 2p, 3p, \cdots, (q-1)p$ and the multiples of q are $q, 2q, 3q, \cdots, (p-1)q$. Thus we delete a total of $(q-1) + (p-1) = p + q - 2$ elements, so $\phi(pq) = (pq - 1) - (p + q + 2) = pq - p - q + 1 = (p-1)(q-1)$.

10. From Exercise 7, we find that $\phi(24) = 8$, so $7^8 \equiv 1 \pmod{24}$. Then $7^{1000} \equiv (7^8)^{125} \equiv 1^{125} \equiv 1 \pmod{24}$.

11. We can reduce the congruence to $2x \equiv 2 \pmod 4$. The gcd of 4 and 2 is $d = 2$ which divides $b = 2$. We divide by 2 and solve instead the congruence $x \equiv 1 \pmod 2$. Of course $x = 1$ is a solution. Another incongruent $\pmod 4$ solution is $x = 1 + 2 = 3$. Thus the solutions are the numbers in $1 + 4\mathbb{Z}$ and $3 + 4\mathbb{Z}$.

12. We can reduce the congruence to $7x \equiv 5 \pmod{15}$. The gcd of 15 and 7 is $d = 1$ which divides $b = 5$. By inspection, $x = 5$ is a solution, and all solutions must be congruent to 5. Thus the solutions are the numbers in $5 + 15\mathbb{Z}$.

13. The congruence can be reduced to $12x \equiv 15 \pmod{24}$. The gcd of 24 and 12 is $d = 12$ which does not divide $b = 15$, so there are no solutions.

14. The congruence can be reduced to $21x \equiv 15 \pmod{24}$. The gcd of 24 and 21 is $d = 3$ which divides $b = 15$. We divide by 3 and solve instead the congruence $7x \equiv 5 \pmod{8}$. By inspection, $x = 3$ is a solution. Other incongruent $\pmod{24}$ solutions are given by $x = 3 + 8 = 11$ and $x = 3 + 2 \cdot 8 = 19$. Thus the solutions are the numbers in $3 + 24\mathbb{Z}, 11 + 24\mathbb{Z}$, or $19 + 24\mathbb{Z}$.

15. The congruence can be reduced to $3x \equiv 8 \pmod{9}$. The gcd of 9 and 3 is $d = 3$ which does not divide $b = 8$, so there are no solutions.

16. The congruence can be reduced to $5x \equiv 8 \pmod{9}$. The gcd of 9 and 5 is 1 which divides 8. By inspection, $x = 7$ is a solution, and there are no other incongruent $\pmod{9}$ solutions, so solutions are the numbers in $7 + 9\mathbb{Z}$.

17. The congruence can be reduced to $25x \equiv 10 \pmod{65}$. The gcd of 65 and 25 is $d = 5$ which divides $b = 10$. We divide by 5 and solve instead the congruence $5x \equiv 2 \pmod{13}$. By inspection $x = 3$ is one solution. The other solutions that are incongruent $\pmod{65}$ are $3 + 13 = 16, 3 + 2 \cdot 13 = 29, 3 + 3 \cdot 13 = 42$, and $3 + 4 \cdot 13 = 55$. Thus the solutions are the numbers in $3 + 65\mathbb{Z}, 16 + 65\mathbb{Z}, 29 + 65\mathbb{Z}, 42 + 65\mathbb{Z}$, and $55 + 65\mathbb{Z}$.

18. The gcd of 130 and 39 is $d = 13$ which divides $b = 52$. We divide by 13 and solve instead $3x \equiv 4 \pmod{10}$. By inspection, $x = 8$ is a solution. Repeatedly adding 10 eleven times, we see that the solutions are the numbers in $8 + 130\mathbb{Z}, 18 + 130\mathbb{Z}, 28 + 130\mathbb{Z}, 38 + 130\mathbb{Z}, 48 + 130\mathbb{Z}, 58 + 130\mathbb{Z}, 68 + 130\mathbb{Z}, 78 + 130\mathbb{Z}, 88 + 130\mathbb{Z}, 98 + 130\mathbb{Z}, 108 + 130\mathbb{Z}, 118 + 130\mathbb{Z}$, or $128 + 130\mathbb{Z}$.

19. Because $(p - 1)! = (p - 1)(p - 2)!$, Exercise 28 shows that we have $-1 \equiv (p - 1) \cdot (p - 2)! \pmod{p}$. Reducing $\pmod{p}$, we have the congruence $-1 \equiv (-1) \cdot (p - 2)! \pmod{p}$, so we must have $(p - 2)! \equiv 1 \pmod{p}$.

20. Taking $p = 37$ and using Exercise 28, we have

$$36! \equiv (36)(35)(34!) \equiv -1 \pmod{37}$$

so $(-1)(-2)(34!) \equiv -1 \pmod{37}$ and $2(34!) \equiv 36 \pmod{37}$. Thus $34! \equiv 18 \pmod{37}$.

21. Taking $p = 53$ and using Exercise 28, we have

$$52! \equiv (52)(51)(50)(49!) \equiv -1 \pmod{53}$$

so $(-1)(-2)(-3)(49!) \equiv -1 \pmod{53}$ and $6(49!) \equiv 1 \pmod{53}$. By inspection, we see that $49! \equiv 9 \pmod{53}$.

22. Taking $p = 29$ and using Exercise 28, we have

$$28! \equiv (28)(27)(26)(25)(24!) \equiv -1 \pmod{29}$$

so $(-1)(-2)(-3)(-4)(24!) \equiv -1 \pmod{29}$ and $(-5)(24!) \equiv -1 \pmod{29}$. By inspection, we see that $24! \equiv 6 \pmod{29}$.

23. F T T F T T F T F T

24.

$\cdot_{12}$	1	5	7	11
1	1	5	7	11
5	5	1	11	7
7	7	11	1	5
11	11	7	5	1

This group is isomorphic to $\langle \mathbb{Z}_2 \times \mathbb{Z}_2, + \rangle$.

25. The nonzero elements of $\mathbb{Z}_p$ form a group of order $p-1$ under multiplication modulo p, and the order of an element of a finite group divides the order of the group.

26. The elements of $\mathbb{Z}_n$ that are integers relative prime to n form a group of order $\phi(n)$ under multiplication modulo n, and the order of an element of a finite group divides the order of the group.

27. If $a^2 = 1$, then $a^2 - 1 = (a-1)(a+1) = 0$. Because a field has no divisors of 0, either $a - 1 = 0$ or $a + 1 = 0$. Thus either $a = 1$ or $a = p - 1$.

28. Because $\mathbb{Z}_p$ is a field, for each factor in $(p-1)!$, its inverse in $\mathbb{Z}_p$ is also a factor. In two cases, namely for the factors 1 and $p - 1$, the inverse is the same factor (see Exercise 27), while in the other cases the inverse is a different factor. For $p \geq 3$ we see that

$$(p-1)! \equiv (p-1) \cdot \underbrace{(1)(1)\cdots(1)}_{\frac{p-3}{2}(1)'s} \cdot (1) \pmod{p}$$

so $(p-1)! \equiv p-1 \equiv -1 \pmod{p}$. When $p = 2$, we have $p - 1 \equiv 1 \equiv -1 \pmod 2$.

29. We show that $n^{37} - n$ is divisible by each of the primes 37, 19, 13, 7, 3, and 2 for every positive integer n. By Corollary 20.2, $a^p \equiv a \pmod p$ so of course $n^{37} \equiv n \pmod{37}$ for all n, so $n^{37} - n \equiv 0 \pmod{37}$ for all n, that is $n^{37} - n$ is divisible by 37 for all $n \in \mathbb{Z}+$.

Working modulo 19, we have $n^{37} - n \equiv n[(n^{18})^2 - 1]$. If n is divisible by 19, then so is $n^{37} - n$. If n is not divisible by 19, then by Fermat's theorem, $(n^{18})^2 - 1 \equiv 1^2 - 1 \equiv 0 \pmod{19}$, so again $n^{37} - n$ is divisible by 19. Notice that the reason this argumnet works is that $36 = 37 - 1$ is a multiple of $18 = 19 - 1$.

Divisiblity by 13, 7, 3, and 2 are handled in the same way, and the computatations are successful because

36 is a multiple of 13 - 1 = 12,
36 is a multiple of 7 - 1 = 6,
36 is a multiple of 3 - 1 = 2,
36 is a multiple of 2 - 1 = 1.

30. Looking at the argument in Exercise 29, we try to find still another prime p less than 37 such that 36 is divisible by $p - 1$. We see that $p = 5$ fills the bill, so $n^{37} - n$ is actually divisible by $5(383838) = 1919190$ for all integers n.

21. The Field of Quotients of an Integral Domain

1. The field of quotients of D is $\{q_1 + q_2 i \mid q_1, q_2 \in \mathbb{Q}\}$.

2. Because
$$\frac{1}{a+b\sqrt{2}} = \frac{1}{a+b\sqrt{2}} \cdot \frac{a - b\sqrt{2}}{a - b\sqrt{2}} = \frac{a}{a^2 - 2b^2} + \frac{-b}{a^2 - 2b^2}\sqrt{2},$$
we see that $\{q_1 + q_2\sqrt{2} \mid q_1, q_2 \in \mathbb{Q}\}$ is a field, and must be the field of quotients.

3. The definition is incorrect. We should think of the embedding as having taken place, so that $D \subseteq F$, and every element of F must be a quotient of elements of D.

A **field of quotients** of an integral domain D is a field F containing D as a subdomain and with the property that every $x \in F$ is equal to some quotient a/b for $a, b \in D$.

4. T F T F T T F T T T

5. Let $D = \{q \in \mathbb{Q} \mid q = m/2^n \text{ for } m, n \in \mathbb{Z}\}$, that is, the set of all rational numbers that can be written as a quotient of integers with denominator a power of 2. It is easy to see that D is an integral domain. Let $D' = \mathbb{Z}$. Then $\mathbb{Q}$ is a field of quotients of both D and D'.

6. We have

$$
\begin{aligned}
[(a,b)] + ([(c,d)] + [(e,f)]) &= [(a,b)] + [(cf + de, df)] \\
&= [(adf + bcf + bde, bdf)] \\
&= [(ad + bc, bd)] + [(e,f)] \\
&= ([(a,b)] + [(c,d)]) + [(e,f)].
\end{aligned}
$$

Thus addition is associative.

7. We have $[(0,1)] + [(a,b)] = [(0b + 1a, 1b] = [(a,b)]$. by Part 1 of Step 3, we also have $[(a,b)] + [(0,1)] = [(a,b)]$.

8. We have $[(-a,b)] + [(a,b)] = [(-ab + ba, b^2)] = [(0, b^2)]$. But $[(0, b^2)] \sim [(0,1)]$ because $(0)(1) = (b^2)(0) = 0$. Thus $[(-a,b)] + [(a,b)] = [(0,1)]$. By Part 1 of Step 3, $[(a,b)] + [(-a,b)] = [(0,1)]$ also.

9. Now $[(a,b)]([(c,d)][(e,f)]) = [(a,b)][(ce, df)] = [(ace, bdf)] = [(ac, bd)][(e,f)] = ([(a,b)][(c,d)])[(e,f)]$. Thus multiplication is associative.

10. We have $[(a,b)][(c,d)] = [(ac, bd)] = [(ca, db)] = [(c,d)][(a,b)]$ so multiplication is commutative.

11. For the left distributive law, we have $[(a,b)]([(c,d)] + [(e,f)]) = [(a,b)][(cf + de, df)] = [(acf + ade, bdf)]$. Also, $[(a,b)][(c,d)] + [(a,b)][(e,f)] = [(ac, bd)] + [(ae, bf)] = [(acbf + bdae, bdbf)] \sim [(acf + ade, bdf)]$ because $(acbf + bdae)bdf = acbfbdf + bdaebdf = bdbf(acf + ade)$, for multiplication in D is commutative. The right distributive law then follws from Part 6.

12. a. Because T is nonempty, there exists $a \in T$. Then $[(a,a)]$ is unity in $Q(R,T)$, because $[(a,a)][(b,c)] = [(ab, ac)] \sim [(b,c)]$ since $abc = acb$ in the commutative ring R.

b. A nonzero element $a \in T$ is identified with $[(aa, a)]$ in $Q(R,T)$. Because T has no divisors of zero, $[(a, aa)] \in Q(R,T)$, and we see that $[(aa, a)][(a, aa)] = [(aaa, aaa)] \sim [(a,a)]$ because $aaaa = aaaa$. We saw in part **a** that $[(a,a)]$ is unity in $Q(R,T)$. Commutativity of $Q(R,T)$ shows that $[(a, aa)][(aa, a)]$ is unity also, so $a \in T$ has an inverse in $Q(R,T)$ if $a \neq 0$.

13. We need only takt $T = \{a^n \mid n \in \mathbb{Z}^+\}$ in Exercise 12. This construction is entirely different from the one in Exercise 30 of Section 19.

14. There are four elements, for 1 and 3 are already units in $\mathbb{Z}_4$.

15. It is isomorphic to the ring D of all rational numbers that can be expressed as a quotient of integers with denominator a power of 2, as described in the answer to Exercise 5.

16. It is isomorphic to the ring of all rational numbers that can be expressed as a quotient of integers with denominator a power of 6. The 3 in the $3\mathbb{Z}$ does not restrict the numerator, because 1 can be recovered as $[(6,6)]$, 2 as $[(12,6)]$, etc.

17. It runs into trouble when we try to prove the transitive property in the proof of Lemma 21.2, for multiplicative cancellation may not hold. For $R = \mathbb{Z}_6$ and $T = \{1, 2, 4\}$ we have $(1,2) \sim (2,4)$ because $(1)(4) = (2)(2) = 4$ and $(2,4) \sim (2,1)$ because $(2)(1) = (4)(2)$ in $\mathbb{Z}_6$, but $(1,2) \nsim (2,1)$ because $(1)(1) \neq (2)(2)$ in $\mathbb{Z}_6$.

22. Rings of Polynomials

1. $f(x) + g(x) = 2x^2 + 5$,　$f(x)g(x) = 6x^2 + 4x + 6$

2. $f(x) + g(x) = 0$,　$f(x)g(x) = x^2 + 1$

3. $f(x) + g(x) = 5x^2 + 5x + 1$,　$f(x)g(x) = x^3 + 5x$

4. $f(x) + g(x) = 3x^4 + 2x^3 + 4x^2 + 1$,
$f(x)g(x) = x^7 + 2x^6 + 4x^5 + x^3 + 2x^2 + x + 3$

5. Such a polynomial is of the form $ax^3 + bx^2 + cx + d$ where each of a, b, c, d may be either 0 or 1. Thus there are $2 \cdot 2 \cdot 2 \cdot 2 = 16$ such polynomials in all.

6. Such a polynomial is of the form $ax^2 + bx + c$ where each of a, b, c maybe either 0, 1, 2, 3, or 4. Thus there are $5 \cdot 5 \cdot 5 = 125$ such polynomials in all.

7. $\phi_2(x^2 + 3) = 2^2 + 3 = 7$

8. $\phi_i(2x^3 - x^2 + 3x + 2) = 2i^3 - i^2 + 3i + 2 = -2i + 1 + 3i + 2 = i + 3$

9. $\phi_3[(x^4 + 2x)(x^3 - 3x^2 + 3)] = \phi_3(x^4 + 2x)\phi_3(x^3 - 3x^2 + 3) = (3^4 + 6)(3^3 - 3^3 + 3) = (4 + 6)(3) = 2$

10. $\phi_5[(x^3 + 2)(4x^2 + 3)(x^7 + 3x^2 + 1)] = \phi_5(x^3 + 2)\phi_5(4x^2 + 3)\phi_5(x^7 + 3x^2 + 1) =$
$(5^3 + 2)(4 \cdot 5^2 + 3)(5^7 + 3 \cdot 5^2 + 1) = (6 + 2)(2 + 3)(5 + 5 + 1) = (1)(5)(4) = 6$

11. $\phi_4(3x^{106} + 5x^{99} + 2x^{53}) = 3(4)^{106} + 5(4)^{99} + 2(4)^{53} = 3(4^6)^{17}4^4 + 5(4^6)^{16}4^3 + 2(4^6)^84^5$
$= 3(1)4 + 5(1)1 + 2(1)2 = 5 + 5 + 4 = 0$

12. $1^2 + 1 = 0$ but $0^2 + 1 \neq 0$, so 1 is the only zero.

13. Let $f(x) = x^3 + 2x + 2$. Then $f(0) = 2, f(1) = 5, f(2) = 0, f(3) = 0, f(-3) = 4, f(-2) = 4$, and $f(-1) = 6$ so 2 and 3 are the only zeros.

14. Let $f(x) = x^5 + 3x^3 + x^2 + 2x$. Then $f(0) = 0, f(1) = 2, f(2) = 4, f(-2) = 4$, and $f(-1) = 0$ so 0 and 4 are the only zeros.

15. Because $\mathbb{Z}_7$ is a field, $f(a)g(a) = 0$ if and only if either $f(a) = 0$ or $g(a) = 0$. Let $f(x) = x^3 + 2x^2 + 5$ and $g(x) = 3x^2 + 2x$. Then $f(0) = 5, f(1) = 1, f(2) = 0, f(3) = 1, f(-3) = 3, f(-2) = 5$, and $f(-1) = 6$ while $g(0) = 0, g(1) = 5, g(2) = 2, g(3) = 5, g(-3) = 0, g(-2) = 1$, and $g(-1) = 1$. Thus $f(x)g(x)$ has 0, 2, and 4 as its only zeros.

16. $\phi_3(x^{231} + 3x^{117} - 2x^{53} + 1) = 3^{231} + 3^{118} - 2(3^{53}) + 1 = (3^4)^{57}3^3 + (3^4)^{29}3^2 - 2(3^4)^{13}3 + 1 = 3^3 + 3^2 - 2(3) + 1 = 2 + 4 - 1 + 1 = 1$.

17. Let $f(x) = 2x^{219} + 3x^{74} + 2x^{57} + 3x^{44} = 2(x^4)^{54}x^3 + 3(x^4)^{18}x^2 + 2(x^4)^{14}x + 3(x^4)^{11}$. Then $f(0) = 0, f(1) = 2+3+2+3 = 0, f(2) = 1+2+4+3 = 0, f(-2) = 4+2+1+3 = 0$ and $f(-1) = 3+3+3+3 = 2$. Thus 0, 1, 2, and 3 are zeros of $f(x)$.

18. The definition is incorrect. All but a finite number of the a_i must be zero.

　　A polynomial with coefficients in a ring R is an infinite formal sum

$$\sum_{i=1}^{\infty} a_i x^i = a_0 + a_1 x + a_2 x^2 + \cdots + a_n x^n + \cdots$$

where $a_i \in R$ for $i = 0, 1, 2, \cdots$ and all but a finite number of the a_i are 0.

19. The definition is incorrect. The zero α may be in a field E containing F.

Let F be a subfield of a field E and let $f(x) \in F[x]$. A **zero of** $f(x)$ **in** E is an $\alpha \in E$ such that $\phi_\alpha(f(x)) = 0$, where $\phi_\alpha : F[x] \to E$ is the evaluation homomorphism mapping x into α.

20. $f(x,y) = (3x^3 + 2x)y^3 + (x^2 - 6x + 1)y^2 + (x^4 - 2x)y + (x^4 - 3x^2 + 2) =$
$(y + 1)x^4 + 3y^3x^3 + (y^2 - 3)x^2 + (2y^3 - 6y^2 - 2y)x + (y^2 + 2)$

21. (See the answer in the text.)

22. $2x + 1$ is a unit because $(2x + 1)^2 = 1$ in $\mathbb{Z}_4[x]$.

23. T T T T F F T T T F

24. Let $f(x) = a_nx^n + a_{n-1}x^{n-1} + \cdots + a_1x + a_0$ and $g(x) = b_mx^m + b_{m-1}x^{m-1} + \cdots + b_1x + b_0$ be polynomials in $D[x]$ with a_n and b_m both nonzero. Because D is an integral domain, we know that $a_nb_m \neq 0$, so $f(x)g(x)$ is nonzero because its term of highest degree has coefficient a_nb_m. As stated in the text, $D[x]$ is a commutative ring with unity, and we have shown it has no divisors of zero, so it is an integral domain.

25. a. The units in $D[x]$ are the units in D because a polnomial of degree n times a polynomial of degree m is a polynomial of degree nm, as proved in the preceding exercise. Thus a polynomial of degree 1 cannot be multiplied by anything in $D[x]$ to give 1, which is a polynomial of degree 0.

b. They are the units in $\mathbb{Z}$, namely 1 and -1.

c. They are the units in $\mathbb{Z}_7$, namely 1, 2, 3, 4, 5, and 6.

26. Let $f(x) = \sum_{i=0}^{\infty} a_ix^i$, $g(x) = \sum_{i=0}^{\infty} b_ix^i$, and $h(x) = \sum_{i=0}^{\infty} c_ix^i$. Then

$$
\begin{aligned}
h(x)[f(x) + g(x)] &= \left[\sum_{j=0}^{\infty} c_jx^j\right]\left[\sum_{i=0}^{\infty}(a_i + b_i)x^i\right] = \sum_{n=0}^{\infty}\left[\sum_{i=0}^{n} c_i(a_{n-i} + b_{n-i})\right]x^n \\
&= \sum_{n=0}^{\infty}\left[\sum_{i=0}^{n} c_ia_{n-i}\right]x^n + \sum_{n=0}^{\infty}\left[\sum_{i=0}^{n} c_ib_{n-i}\right]x^n \\
&= \left[\sum_{j=0}^{\infty} c_jx^j\right]\left[\sum_{i=0}^{\infty} a_ix^i\right] + \left[\sum_{j=0}^{\infty} c_jx^j\right]\left[\sum_{i=0}^{\infty} b_ix^i\right] \\
&= h(x)f(x) + h(x)g(x)
\end{aligned}
$$

so the left distributive law holds.

27. a. Let $f(x) \sum_{i=0}^{\infty} a_x^i$ and $g(x) = \sum_{i=0}^{\infty}$ be polynomials in $F[x]$. Then

$$
\begin{aligned}
D(f(x) + g(x)) &= D\left(\sum_{i=0}^{\infty} a_ix^i + \sum_{i=0}^{\infty} b_ix^i\right) = D\left(\sum_{i=0}^{\infty}(a_i + b_i)x^i\right) \\
&= \sum_{i=1}^{\infty} i(a_i + b_i)x^{i-1} = \sum_{i=1}^{\infty}(ia_i + ib_i)x^{i-1} \\
&= \sum_{i=1}^{\infty} ia_ix^{i-1} + \sum_{i=1}^{\infty} ib_ix^{i-1} = D(f(x)) + D(g(x)).
\end{aligned}
$$

b. The kernel of D is F. [This would not be true if F had characteristic p, for then $D(x^p) = 0$.]

c. The image of D is $F[x]$ because D is additively a homomorphism with $D(1) = 0$ and $D(\frac{1}{i+1}a_i x^{i+1}) = a_i x^i$.

28. a. $\phi_{\alpha_1,\cdots,\alpha_n}(f(x_1,\cdots,x_n))$ is the element of F obtained by replacing each x_i by α_i in the polynomial and computing in E the resulting sum of products. That is $\phi_{\alpha_1,\cdots,\alpha_n}(f(x_1,\cdots,x_n)) = f(\alpha_1,\cdots,\alpha_n)$. This is a map $\phi_{\alpha_1,\cdots,\alpha_n} : F[x_1,\cdots,x_n] \to E$ which is a homomorphism and maps F isomorphically by the identity map, that is, $\phi_{\alpha_1,\cdots,\alpha_n}(a) = a$ for $a \in F$.

b. $\phi_{-3,2}(x_1^2 x_2^3 + 3x_1^4 x_2) = (9)(8) + 3(81)(2) = 72 + 486 = 558$.

c. Let F be a subfield of E. Then $(\alpha_1,\cdots,\alpha_n)$ in $\underbrace{E \times E \times \cdots E}_{n \text{ factors}}$ is a **zero** of $f(x_1,\cdots,x_n) \in F[x_1,\cdots x_n]$ if $\phi_{\alpha_1,\cdots,\alpha_n}(f(x_1,\cdots,x_n)) = 0$.

29. *Addition associative:* Let $\phi, \psi, \mu \in R^R$. Then $[(\phi+\psi)+\mu](r) = (\phi+\psi)(r)+\mu(r) = \phi(r)+\psi(r)+\mu(r) = \phi(r) + (\psi + \mu)(r) = [\phi + (\psi + \mu)](r)$ because addition in R is associative. Because $(\phi + \psi) + \mu$ and $\phi + (\psi + \mu)$ have the same value on each $r \in R$, they are the same function.

Identity for $+$: The function ϕ_0 such that $\phi_0(r) = 0$ for all $r \in R$ acts as additive identity, for $(\phi_0 + \psi)(r) = \phi_0(r) + \psi(r) = 0 + \psi(r) = \psi(r)$. Because ϕ_0 and $\phi_0 + \psi$ have the same value on each $r \in R$, we see that they are the same function. A similar argument shows that $\psi + \phi_0 = \psi$.

Additive inverse: Given $\phi \in R^R$, the function $-\phi$ defined by $(-\phi)(r) = -(\phi(r))$ for $r \in R$ is the additive inverse of ϕ, for $(\phi+(-\phi))(r) = \phi(r)+(-\phi)(r) = \phi(r)+(-\phi(r)) = 0 = \phi_0(r)$, so $\phi+(-\phi) = \phi_0$. A similar argument shows that $(-\phi) + \phi = \phi_0$.

Addition commutative: We have $(\phi+\psi)(r) = \phi(r)+\psi(r) = \psi(r)+\phi(r) = (\psi+\phi)(r)$ because addition in R is commutative. Thus $\phi + \psi$ and $\psi + \phi$ are the same function, so $\phi + \psi = \psi + \phi$.

Multiplication associative: Now $[(\phi \cdot \psi) \cdot \mu](r) = [(\phi \cdot \psi)(r)]\mu(r) = [\phi(r)\psi(r)]\mu(r) = \phi(r)[\psi(r)\mu(r)] = \phi(r)[(\psi \cdot \mu)(r)] = [\phi \cdot (\psi \cdot \mu)](r)$ because multiplication in R is associative. Thus $\phi \cdot (\psi \cdot \mu) = (\phi \cdot \psi) \cdot \mu$ because the functions have the same value on each $r \in R$.

Left distributive law: We have $[\phi \cdot (\psi + \mu)](r) = \phi(r)[(\psi + \mu)(r)] = \phi(r)[\psi(r) + \mu(r)] = \phi(r)\psi(r) + \phi(r)\mu(r) = (\phi \cdot \psi)(r) + (\phi \cdot \mu)(r) = [\phi \cdot \psi + \phi \cdot \mu](r)$ because the left distributive law holds in R. Thus $\phi \cdot (\psi + \mu) = \phi \cdot \psi + \phi \cdot \mu$ because these functions have the same value at each $r \in R$.

Right distributive law: The proof is analogous to that for the left distributive law.

30. a. The map $\mu : F[x] \to F^F$ where $\mu(f(x))$ is the function $\phi \in F^F$ such that $\phi(a) = f(a)$ for all $a \in F$ is easily seen to be a homomorphism of $F[x]$ into the ring F^F, and by definition, $P_F = \mu[F[x]]$. Thus P_F is the homomorphic image of a ring under a ring homomorphism. Theorem 13.12 then shows that $\langle P_F, + \rangle$ is a group. Let $\phi, \psi \in P_F$, and let $f(x), g(x) \in F[x]$ be such that $\mu(f(x)) = \phi$ and $\mu(g(x)) = \psi$. Then $(\phi \cdot \psi)(a) = \phi(a)\psi(a) = f(a)g(a)$ for all $a \in F$, so $\mu(f(x)g(x)) = \phi \cdot \psi$. Thus $\phi \cdot \psi \in P_F$ so P_F is closed under multiplication. By Exercise 48 of Section 18, P_F is a subring of F_F.

b. Let F be the finite field $\mathbb{Z}_2$. A function in $\mathbb{Z}_2^{\mathbb{Z}_2}$ has just two elements in both its domain and range. Thus there are only $2^2 = 4$ such functions in all. However, $\mathbb{Z}_2[x]$ is an infinite set, so it isn't isomorphic to $P_{\mathbb{Z}_2}$.

31. a. There are $2^2 = 4$ elements in $\mathbb{Z}_2^{\mathbb{Z}_2}$ and $3^3 = 27$ in $\mathbb{Z}_3^{\mathbb{Z}_3}$.

b. Because $(\phi + \phi)(a) = \phi(a) + \phi(a) = 2 \cdot \phi(a) = 0$ in $\mathbb{Z}_2$, we see that every element of $\mathbb{Z}_2^{\mathbb{Z}_2}$ is its own additive inverse, so this additive group of order 4 must be isomorphic to $\mathbb{Z}_2 \times \mathbb{Z}_2$. Similarly, if

$\phi \in \mathbb{Z}_3^{\mathbb{Z}_3}$ then $3 \cdot \phi(a) = 0$ for all $a \in \mathbb{Z}_3$. Because this group is abelian, we see that it is isomorphic to $\mathbb{Z}_3 \times \mathbb{Z}_3 \times \mathbb{Z}_3$.

c. Let $g_i(x) = (1/c)f_i(x)$ for $i = 1, \cdots, n$. Note that $g_i(a_j) = 0$ for $j \neq i$. Let $d_i = g_i(a_i)$. Note that d_i appears as a product of nonzero factors $(a_i - a_k)$ for $k \neq i$, and consequently $d_i \neq 0$ for $i = 1, \cdots, n$. Let $\phi \in F^F$ and suppose that $\phi(a_i) = c_i$. Let $f(x) \in F$ be defined by

$$f(x) = \sum_{i=1}^{n} \frac{c_i}{d_i} g_i(x).$$

Because $g_j(a_i) = 0$ if $j \neq i$, we see that only the ith term in the sum defining $f(x)$ contributes a nonzero summand to $f(a_i)$. Because $d_i = g_i(a_i)$, we obtain

$$f(a_i) = \frac{c_i}{d_i} g_i(a_i) = \frac{c_i}{d_i} d_i = c_i \text{ for } i = 1, \cdots, n.$$

We have shown that each function ϕ in F^F is a polynomial function $f(x)$ in P_F.

23. Factorization of Polynomials over a Field

1. We perform the desired division.

$$
\begin{array}{r}
x^4 + x^3 + x^2 + x - 2 = q(x) \\
x^2 + 2x - 3 \overline{\smash{)}\; x^6 + 3x^5 + 4x^2 - 3x + 2} \\
\underline{x^6 + 2x^5 - 3x^4} \\
x^5 + 3x^4 \\
\underline{x^5 + 2x^4 - 3x^3} \\
x^4 + 3x^3 + 4x^2 \\
\underline{x^4 + 2x^3 - 3x^2} \\
x^3 - 3x \\
\underline{x^3 + 2x^2 - 3x} \\
-2x^2 + 2 \\
\underline{-2x^2 - 4x + 6} \\
4x + 3 = r(x)
\end{array}
$$

2. We perform the desired division.

$$
\begin{array}{r}
5x^4 + 5x^2 - x = q(x) \\
3x^2 + 2x - 3 \overline{\smash{)}\; x^6 + 3x^5 + 4x^2 - 3x + 2} \\
\underline{x^6 + 3x^5 + 6x^4} \\
x^4 + 4x^2 \\
\underline{x^4 + 3x^3 + 6x^2} \\
-3x^3 - 2x^2 - 3x \\
\underline{-3x^3 - 2x^2 + 3x} \\
x + 2 = r(x)
\end{array}
$$

3. We perform the desired division.

$$
\begin{array}{r}
6x^4 + 7x^3 + 2x^2 - x + 2 = q(x) \\
\underline{} \\
\end{array}
$$

$$
2x+1 \overline{\smash{\big)}\, x^5 - 2x^4 + + 3x - 5}
$$

$$
\underline{x^5 + 6x^4}
$$
$$
3x^4
$$
$$
\underline{3x^4 + 7x^3}
$$
$$
4x^3
$$
$$
\underline{4x^3 + 2x^2}
$$
$$
-2x^2 + 3x
$$
$$
\underline{-2x^2 - x}
$$
$$
4x - 5
$$
$$
\underline{4x + 2}
$$
$$
4 = r(x)
$$

4. We perform the desired division.

$$
9x^2 + 5x + 10 = q(x)
$$
$$
5x^2 - x + 2 \overline{\smash{\big)}\, x^4 + 5x^3 - 3x^2}
$$
$$
\underline{x^4 + 2x^3 + 7x^2}
$$
$$
3x^3 + x^2
$$
$$
\underline{3x^3 - 5x^2 + 10x}
$$
$$
6x^2 + x
$$
$$
\underline{6x^2 - 10x + 9}
$$
$$
2 = r(x)
$$

5. Trying $2 \in \mathbb{Z}_5$, we find that $2^2 = 4, 2^3 = 3, 2^4 = 1$, so 2 generates the multiplicative subgroup $\{1, 2, 3, 4\}$ of all units in $\mathbb{Z}^5$. By Corollary 6.16, the only generators are $2^1 = 2$ and $2^3 = 3$.

6. Trying $2 \in \mathbb{Z}_7$, we find that $2^3 = 1$, so 2 does not generate. Trying 3, we find that $3^2 = 2, 3^3 = 6, 3^4 = 4, 3^5 = 5$, and $3^6 = 1$, so 3 generates the six units 1, 2, 3, 4, 5, 6 in $\mathbb{Z}_7$. By Corollary 6.16, the only generators are $3^1 = 3$ and $3^5 = 5$.

7. Trying $2 \in \mathbb{Z}_{17}$, we find that $2^4 = -1$, so $2^8 = 1$ and 2 does not generate. Trying 3, we find that $3^2 = 9, 3^3 = 10, 3^4 = 13, 3^5 = 5, 3^6 = 15, 3^7 = 11, 3^8 = 16 = -1$. Because the order of 3 must divide 16, we see that 3 must be of order 16, so 3 generates the units in $\mathbb{Z}_{17}$. By Corollary 6.16, the only generators are $3^1 = 3, 3^3 = 10, 3^5 = 5, 3^7 = 11, 3^9 = 14, 3^{11} = 7, 3^{13} = 12$, and $3^{15} = 6$.

8. Trying $2 \in \mathbb{Z}_{23}$, we find that $2^2 = 4, 2^3 = 8, 2^4 = 16, 2^5 = 9, 2^6 = 18, 2^7 = 13, 2^8 = 3, 2^9 = 6, 2^{10} = 12$, and $2^{11} = 1$, so 2 does not generate. However, this computation shows that $(-2)^{11} = -1$. Because the order of -2 must divide 22, we see that $21 = -2$ must be of order 22, so 21 generates the units of $\mathbb{Z}_{23}$. By Corollary 6.16, the only generators are $(-2)^1 = 21, (-2)^3 = 15, (-2)^5 = 14, (-2)^7 = 10, (-2)^9 = 17, (-2)^{13} = 19, (-2)^{15} = 7, (-2)^{17} = 5, (-2)^{19} = 20$, and $(-2)^{21} = 11$.

9. In $\mathbb{Z}_5$, we have $x^4 + 4 = x^4 - 1 = (x^2 + 1)(x^2 - 1)$. Replacing 1 by - 4 again, we continue and discover that $(x^2 - 4)(x^2 - 1) = (x - 2)(x + 2)(x - 1)(x + 1)$.

10. By inspection, -1 is a zero of $x^3 + 2x^2 + 2x + 1$ in $\mathbb{Z}_7[x]$. Executing the division algorithm as illustrated in our answers to Exercises 1 through 3, we compute $x^3 + 2x^2 + 2x + 1$ divided by $x - (-1) = x + 1$, and find that

$$x^3 + 2x^2 + 2x + 1 = (x+1)(x^2 + x + 1).$$

By inspection, 2 and 4 are zeros of $x^2 + x + 1$. Thus the factorization is

$$x^3 + 2x^2 + 2x + 1 = (x+1)(x-4)(x-2).$$

11. By inspection, 3 is a zero of $2x^3 + 3x^2 - 7x - 5$ in $\mathbb{Z}_{11}[x]$. Dividing by $x - 3$ using the technique illustrated in our answers to Exercises 1 through 3, we find that

$$2x^3 + 3x^2 - 7x - 5 = (x-3)(2)(x^2 - x - 1).$$

By inspection, -3 and 4 are zeros of $x^2 - x - 1$, so the factorization is

$$2x^3 + 3x^2 - 7x - 5 = (x-3)(x+3)(2x-8).$$

12. By inspection, -1 is a zero of $x^3 + 2x + 3$ in $\mathbb{Z}_5[x]$, so the polynomial is not irreducible. We divide by $x + 1$, using the technique of Exercises 1 through 3, and obtain

$$x^3 + 2x + 3 = (x+1)(x^2 - x + 3).$$

By inspection, -1 and 2 are zeros of $x^2 - x + 3$, so the factorization is

$$x^3 + 2x + 3 = (x+1)(x+1)(x-2).$$

13. Let $f(x) = 2x^3 + x^2 + 2x + 2$ in $\mathbb{Z}_5[x]$. Then $f(0) = 2, f(1) = 2, f(-1) = -1, f(2) = 1$, and $f(-2) = 1$, so $f(x)$ has no zeros in $\mathbb{Z}_5$. Because $f(x)$ is of degree 3, Theorem 23.10 shows that $f(x)$ is irreducible over $\mathbb{Z}_5$.

14. $f(x) = x^2 + 8x - 2$ satisfies the Eisenstein condition for irreducibility over $\mathbb{Q}$ with $p = 2$. It is not irreducible over $\mathbb{R}$ because the quadriatic formula shows that is has the real zeros $(-8 \pm \sqrt{72})/2$. Of course it is not irreducible over $\mathbb{C}$ also.

15. The polynomial $g(x) = x^2 + 6x + 12$ is irreducible over $\mathbb{Q}$ because it satisfies the Eisenstein condition with $p = 3$. It is also irreducible over $\mathbb{R}$ because the quadratic formula shows that its zeros are $(-6 \pm \sqrt{-12})/2$, which are not in $\mathbb{R}$. It is not irreducible over $\mathbb{C}$, because its zeros lie in $\mathbb{C}$.

16. If $x^3 + 3x^2 - 8$ is reducible over $\mathbb{Q}$, then by Theorem 23.11, it factors in $\mathbb{Z}[x]$, and must therefore have a linear factor of the form $x - a$ in $\mathbb{Z}[x]$. Then a must be a zero of the polynomial and must divide -8, so the possibilities are $a = \pm 1, \pm 2, \pm 4, \pm 8$. Computing the polyomial at these eight values, we find none of them is a zero of the polynomial, which is therefore irreducible over $\mathbb{Q}$.

17. If $x^4 - 22x^2 + 1$ is reducible over $\mathbb{Z}$, then by Theorem 23.11, it factors in $\mathbb{Z}[x]$, and must therefore either have a linear factor in $\mathbb{Z}[x]$ or factor into two quadratics in $\mathbb{Z}[x]$. The only possibilites for a linear factor are $x \pm 1$, and clearly neither 1 nor -1 is a zero of the polynomial, so a linear factor is impossible. Suppose

$$x^4 - 22x^2 + 1 = (x^2 + ax + b)(x^2 + cx + d).$$

Equating coefficients, we see that

$x^3 \ coefficient : 0 = a + c$

$x^2 \ coefficient : -22 = ac + b + d$

x coefficient $: 0 = bc + ad$

constant term $: 1 = bd$ so either $b = d = 1$ or $b = d = -1$.

Suppose $b = d = 1$. Then $-22 = ac + 1 + 1$ so $ac = -24$. Because $a + c = 0$, we have $a = -c$, so $-c^2 = -24$ which is impossible for an integer c. Similarly, if $b = d = -1$, we deduce that $-c^2 = -20$, which is also impossible. Thus the polynomial is irreducible.

18. Yes, with $p = 3$.   **19.** Yes, with $p = 3$.

20. No, for 2 divides the coefficient 4 of x^{10} and 3^2 divides the constant term -18.

21. Yes, with $p = 5$.

22. Let this polynomial be $f(x)$. If $f(x)$ has a rational zero, then this zero can be expressed as a fraction with numerator dividing 10 and denominator dividing 6. The possibilities are $\pm 10, \pm 5, \pm 10/3, \pm 5/2, \pm 2, \pm 5/3, \pm 1, \pm 5/6, \pm 2/3, \pm 1/2, \pm 1/3$, and $\pm 1/6$.

Experimentation with a calculator shows that there is a negative real zero between -2 and -3 because $f(-2) < 0$ and $f(-3) > 0$. (Recall the intermediate value theorem.) The only possible rational candidate is -5/2. We reach for our calculator and find that $f(-2.5) = 0$, so -5/2 is a zero and $(2x + 5)$ is a linear factor.

Because $f(0) < 0$ and $f(1) > 0$, the intermediate value theorem shows that there is a real zero a satisfying $0 < a < 1$. The rational possibilies are 5/6, 2/3, 1/2, 1/3, and 1/6. Because $2x + 5$ is a factor, accounting for the factor 2 of 6 and the factor 5 of 10, we can discard 5/6, 1/2, and 1/6, leaving 2/3 and 1/3 to try. We reach for our calculator and compute $f(2/3) = 0$, so $3x - 2$ is also a factor. Because we have accounted for the 6 and the 10 with these linear factors, the only other possible rational zeros would have to be 1 or -1, and we easily find that these are not zeros. Thus the rational zeros are 2/3 and -5/2.

23. The definition is incorrect. We must require that $g(x)$ and $h(x)$ have degree less than the degree of $f(x)$, and that the polynomial is nonconstant.

A nonconstant polynomial $f(x) \in F[x]$ is **irreducible over the field** F if $f(x) \neq g(x)h(x)$ for any polynomials $g(x), h(x) \in F[x]$ both of degree less than the degree of $f(x)$.

24. The definition is correct.

25. T T T F T F T T T T

26. Considering $f(x) = x^4 + x^3 + x^2 - x + 1$ in $\mathbb{Z}[x]$, we find that $f(-2) = 16 - 8 + 4 + 2 + 1 = 15$. Thus $p = 3$ and $p = 5$ are primes such that -2 is a zero of $f(x)$ in $\mathbb{Z}_p$, that is, such that $x + 2$ is a factor of $f(x)$ in $\mathbb{Z}_p[x]$.

27. The polynomials of degree 2 in $\mathbb{Z}_2[x]$ are

x^2: not irreducible because 0 is a zero,
$x^2 + 1$: not irreducible because 1 is a zero,
$x^2 + x$: not irreducible because 0 is a zero,
$x^2 + x + 1$: irreducible because neither 0 nor 1 are zeros.

Thus our answer is $x^2 + x + 1$.

28. The Polynomials of degree 3 in $\mathbb{Z}_2[x]$ are

x^3: not irreducible because 0 is a zero,
$x^3 + 1$: not irreducible because 1 is a zero,
$x^3 + x$: not irreducible because 0 is a zero,
$x^3 + x^2$: not irreducible because 0 is a zero,
$x^3 + x + 1$: irreducible, neither 0 nor 1 is a zero,
$x^3 + x^2 + 1$: irreducible, neither 0 nor 1 is a zero,
$x^3 + x^2 + x$: not irreducible because 0 is a zero,
$x^3 + x^2 + x + 1$: not irreducible, 1 is a zero.

Thus the irreducible cubics are $x^3 + x + 1$ and $x^3 + x^2 + 1$.

29. The 18 polynomials of degree 2 in $\mathbb{Z}_3$ are

$x^2, x^2 + x, x^2 + 2x, 2x^2, 2x^2 + x, 2x^2 + 2x$ all reducible because 0 is a zero,

$x^2 + 2, x^2 + x + 1, 2x^2 + 1, 2x^2 + 2x + 2$ all reducible because 1 is a zero,

$x^2 + 2x + 1, 2x^2 + x + 2$ both reducible because 2 is a zero, and

$x^2 + 1, x^2 + x + 2, x^2 + 2x + 2, 2x^2 + 2, 2x^2 + x + 1, 2x^2 + 2x + 1$ which are all irreducible because none has 0, 1, or 2 as a zero.

30. An irreducible polynomial must have a nonzero constant term or 0 is a zero; this eliminates 18 of the 54 cubic polynomials in $\mathbb{Z}_3[x]$. Now a is a zero of $f(x)$ if and only if a is a zero of $2f(x)$, so we can consider just the 18 cubics with leading coefficient 1 and constant term nonzero.

$x^3 + 2, x^3 + x^2 + 1, x^3 + x + 1, x^3 + 2x^2 + 2x + 1, x^3 + x^2 + 2x + 2, x^3 + 2x^2 + x + 2$ have 1 as a zero, so they are reducible.

$x^3 + 1, x^3 + 2x^2 + 2, x^3 + x + 2, x^3 + x^2 + x + 1$ have -1 as a zero, so they are reducible.

The remaining eight cubics with leading coefficient 1 and nonzero constant term, namely:

$x^3 + 2x + 1, \quad x^3 + 2x + 2, \quad x^3 + x^2 + 2, \quad x^3 + 2x^2 + 1, \quad x^3 + x^2 + x + 2,$
$x^3 + x^2 + 2x + 1, \quad x^3 + 2x^2 + x + 1, \quad \text{and} \quad x^3 + 2x^2 + 2x + 2$

and their doubles

$2x^3 + x + 2, \quad 2x^3 + x + 1, \quad 2x^3 + 2x^2 + 1, \quad 2x^3 + x^2 + 2, \quad 2x^3 + 2x^2 + 2x + 1,$
$2x^3 + 2x^2 + x + 2, \quad 2x^3 + x^2 + 2x + 2, \quad \text{and} \quad 2x^3 + x^2 + x + 1$

are irreducible.

31. Following the hint, each reducible quadratic that is of the form $x^2 + ax + b$ is a product $(x + c)(x + d)$ for $c, d \in \mathbb{Z}_p$. There are $\binom{p}{2} = p(p-1)/2$ such products (neglecting order of factors) where $c \neq d$. There are p such products where $c = d$. Thus there are $p(p-1)/2 + p = p^2/2 + p/2 = p(p+1)/2$ reducible quadratics with leading coefficient 1. Because the leading coefficient (upon multiplication) can be any one of $p - 1$ nonzero elements, there are $(p-1)p(p+1)/2$ reducible quadratics altogether. The total number of quadratic polynomials in $\mathbb{Z}_p[x]$ is $(p-1)p^2$. Thus the number of irreducible quadratics is $(p-1)p^2 - (p-1)p(p+1)/2 = p(p-1)[p - (p+1)/2] = p(p-1)^2/2$.

32. Each zero of a polynomial leads to a linear factor, and the number of linear factors in the factorization of a polynomial cannot exceed the degree of the polynomial.

33. If the group were not cyclic, then the Fundamental Theorem for finitely generated abelian groups shows that the least common multiple m of the orders of the elements would be less than the number n of elements, leading to a polynomial $x^m - 1$ with $n > m$ zeros, which is impossible in a field.

34. Note that $x^2 = xx$ and $x^2 + 1 = (x+1)^2$ are reducible in $\mathbb{Z}_p$. For an odd prime p and $a \in \mathbb{Z}_p$, we know that $(-a)^p + a = -a^p + a = -a + a = 0$ by Corollary 20.2. Thus $x^p + a$ has $-a$ as a zero, so it is reducible over $\mathbb{Z}_p$ for every prime p. [Actually, the binomial theorem and Corollary 20.2 show that $x^p + a = x^p + a^p = (x+a)^p$.]

35. We are given that $f(a) = a_0 + a_1 a + \cdots + a_n a^n = 0$ and $a \neq 0$. Dividing by a^n, we find that

$$a_0 \left(\frac{1}{a}\right)^n + a_1 \left(\frac{1}{a}\right)^{n-1} + \cdots + a_n = 0$$

which is just what we wanted to show.

36. By Theorem 23.1, we know that $f(x) = q(x)(x - a) + c$ for some constant $c \in F$. Applying the evaluation homomorphism ϕ_a to both sides of this equation, we find that

$$f(a) = q(a)(a - a) + c = q(a)0 + c = c,$$

so the remainder $r(x) = c$ is actually $f(a)$.

37. a. Let $f(x) = \sum_{i=0}^{\infty} a_i x^i$ and $g(x) = \sum_{i=0}^{\infty} b_i x^i$. Then

$$\overline{\sigma_m}(f(x) + g(x)) = \overline{\sigma_m}\left(\sum_{i=0}^{\infty}(a_i + b_i)x^i\right) = \sum_{i=0}^{\infty}\overline{\sigma_m}(a_i + b_i)x^i$$

$$= \sum_{i=0}^{\infty}[\overline{\sigma_m}(a_i) + \overline{\sigma_m}(b_i)]x^i = \overline{\sigma_m}(f(x)) + \overline{\sigma_m}(g(x))$$

and

$$\overline{\sigma_m}(f(x)g(x)) = \overline{\sigma_m}\left(\sum_{n=0}^{\infty}\left(\sum_{i=0}^{n}a_i b_{n-i}\right)x^n\right) = \sum_{n=0}^{\infty}\overline{\sigma_m}\left(\sum_{i=0}^{n}a_i b_{n-i}\right)x^n$$

$$= \sum_{n=0}^{\infty}\left(\sum_{i=0}^{n}\overline{\sigma_m}(a_i b_{n-i})\right)x^n = \sum_{n=0}^{\infty}\left(\sum_{i=0}^{n}\overline{\sigma_m}(a_i)\overline{\sigma_m}(b_{n-i})\right)x^n$$

$$= \overline{\sigma_m}(f(x))\overline{\sigma_m}(g(x)),$$

so $\overline{\sigma_m}$ is a homomorphism. If $h(x) \in \mathbb{Z}_m[x]$, then if $k(x)$ is the polynomial in $\mathbb{Z}[x]$ obtained from $h(x)$ by just viewing the coefficients as elements of $\mathbb{Z}$ rather than of $\mathbb{Z}_m$, we see that $\overline{\sigma_m}(k(x)) = h(x)$, so the homomorphism $\overline{\sigma_m}$ is onto $\mathbb{Z}_m[x]$.

b. Let $f(x) = g(x)h(x)$ for $g(x), h(x) \in \mathbb{Z}[x]$ with the degrees of both $g(x)$ and $h(x)$ less than the degree n of $f(x)$. Applying the homomorphism $\overline{\sigma_m}$, we see that $\overline{\sigma_m}(f(x)) = \overline{\sigma_m}(g(x))\overline{\sigma_m}(h(x))$ is a factorization of $\overline{\sigma_m}(f(x))$ into two polynomials of degree less than the degree n of $\overline{\sigma_m}(f(x))$, contrary to hypothesis. Thus $f(x)$ is irreducible in $\mathbb{Z}[x]$, and hence in $\mathbb{Q}[x]$ by Theorem 23.11.

c. Taking $m = 5$, we see that $\overline{\sigma_5}(x^3 + 17x + 36) = x^3 + 2x + 1$ which does not have any of the five elements 0, 1, -1, 2, -2 of $\mathbb{Z}_5$ as a zero, and is thus irreducible over $\mathbb{Z}_5$ by Theorem 23.10. By Part(**b**), we conclude that $x^3 + 17x + 36$ is irreducible over $\mathbb{Q}$.

24. Noncommutative Examples

1. $(2e + 3a + 0b) + (4e + 2a + 3b) = e + 0a + 3b$, where coefficients are added in $\mathbb{Z}_5$.

2. Because $\{e, a, b\}$ is a cyclic multiplicative group, we have $aa = b, bb = a, ab = ba = e$, and of course e acts as identity element. With the coefficients of e, a, and b from $\mathbb{Z}_5$, we obtain

$$(2e + 3a + 0b)(4e + 2a + 3b) =$$
$$2e(4e + 2a + 3b) + 3a(4e + 2a + 3b) + 0b(4e + 2a + 3b) =$$
$$(3e + 4a + 1b) + (4e + 2a + 1b) + (0e + 0a + 0b) = 2e + a + 2b.$$

3. Because $\{e, a, b\}$ is a cyclic multiplicative group, we have $aa = b, bb = a, ab = ba = e$, and of course e acts as identity element. With the coefficients of e, a, and b from $\mathbb{Z}_5$, we obtain

$$(3e + 3a + 3b)^2 = 3e(3e + 3a + 3b) + 3a(3e + 3a + 3b) + 3b(3e + 3a + 3b)$$
$$= (4e + 4a + 4b) + (4e + 4a + 4b) + (4e + 4a + 4b) = 2e + 2a + 2b.$$

Having seen how this could have been simplified in view of the equal coefficients of e, a, and b, we can now proceed more quickly and find that $(3e + 3a + 3b)^4 = (2e + 2a + 2b)^2 = 4 \cdot (1e + 1a + 1b)^2 = 4 \cdot (3e + 3a + 3b) = 2e + 2a + 2b$.

4. $(i + 3j)(4 + 2j - k) = 4i + 2ij - ik + 12j + 6jj - 3jk = 4i + 2k + j + 12j - 6 - 3i = -6 + i + 13j + 2k$.

5. $i^2 j^3 kji^5 = (-1)(-j)kji = (jk)(ji) = i(-k) = j$.

6. $(i + j)^{-1} = \frac{1}{i+j} \cdot \frac{-i-j}{-i-j} = \frac{-i-j}{2} = -\frac{1}{2}i - \frac{1}{2}j$.

7. $[(1 + 3i)(4j + 3k)]^{-1} = (4j + 3k + 12k - 9j)^{-1} = (-5j + 15k)^{-1} = \frac{1}{-5j+15k} \cdot \frac{5j-15k}{5j-15k} = \frac{5j-15k}{25+225} = \frac{j-3k}{50} = \frac{1}{50}j - \frac{3}{50}k$.

8. $(0\rho_0 + 1\rho_1 + 0\rho_2 + 0\mu_1 + 1\mu_2 + 1\mu_3)(1\rho_0 + 1\rho_1 + 0\rho_2 + 1\mu_1 + 0\mu_2 + 1\mu_3)$
$= (1\rho_1 + 1\rho_2 + 1\mu_3 + 1\mu_2) + (1\mu_2 + 1\mu_3 + 1\rho_2 + 1\rho_1) + (1\mu_3 + 1\mu_1 + 1\rho_1 + 1\rho_0)$
$= (1\rho_0 + 1\rho_1 + 0\rho_2 + 1\mu_1 + 0\mu_2 + 1\mu_3)$

9. The center is $\{r + 0i + 0j + 0k \mid r \in R, r \neq 0\}$ because nonzero coefficients of i, j, or k lead to an element that does not commute with j, k, or i respectively.

10. Clearly $\{a + bi \mid a, b \in \mathbb{R}\}$ and $\{a + bj \mid a, b \in \mathbb{R}\}$ as well as $\{a + bk \mid a, b \in \mathbb{R}\}$ are subrings of the quaternions that are actually fields isomorphic to $\mathbb{C}$.

11. F F F F F F T F T F

12. a. The polynomial $x^2 + 1 \in \mathbb{H}[x]$ has $i, -i, j, -j, k$, and $-k$ as zeros in $\mathbb{H}$.

   b. The subset $\{1, -1, i, -i, j, -j, k, -k, \}$ of $\mathbb{H}$ is a group under quaternion multiplication and is not cyclic because no element has order greater than 2.

13. Let $\psi(m, n) = (m, -m)$. It is easily seen that ψ is an endomorphism of $\mathbb{Z}$. Then $(\phi\psi)(m, n) = \phi(\psi(m, n)) = \phi(m, -m) = (0, 0)$.

14. The matrices

$$\begin{bmatrix} 1 & 0 \\ 0 & 1 \end{bmatrix}, \begin{bmatrix} 0 & 1 \\ 1 & 0 \end{bmatrix}, \begin{bmatrix} 1 & 1 \\ 0 & 1 \end{bmatrix}, \begin{bmatrix} 1 & 0 \\ 1 & 1 \end{bmatrix}, \begin{bmatrix} 0 & 1 \\ 1 & 1 \end{bmatrix}, \begin{bmatrix} 1 & 1 \\ 1 & 0 \end{bmatrix},$$

have inverses

$$\begin{bmatrix} 1 & 0 \\ 0 & 1 \end{bmatrix}, \begin{bmatrix} 0 & 1 \\ 1 & 0 \end{bmatrix}, \begin{bmatrix} 1 & -1 \\ 0 & 1 \end{bmatrix}, \begin{bmatrix} 1 & 0 \\ -1 & 1 \end{bmatrix}, \begin{bmatrix} -1 & 1 \\ 1 & 0 \end{bmatrix}, \text{ and } \begin{bmatrix} 0 & 1 \\ 1 & -1 \end{bmatrix},$$ respectively in every

field. (Note that if the field has characteristic 2, we have -1 = 1 so the last four matrices in the second row may be the same as four in the top row.)

15. Let $m \in \mathbb{Z}$ [or $m \in \mathbb{Z}_n$ as the case may be]. Let ϕ_m be the endomorphism of the additive abelian group of the ring such that $\phi_m(1) = m$. Then $\{\phi_m \mid m \in \mathbb{Z} \text{ [or } m \in \mathbb{Z}_n]\}$ is the entire homomorphism ring, because a homomorphism of each of these cyclic groups is determined by its value on the generator 1 of the group. Define $\psi : \text{End}(\mathbb{Z}) \to \mathbb{Z}$ [or $\text{End}(\mathbb{Z}_n) \to \mathbb{Z}_n$] by $\psi(\phi_m) = m$. Now $(\phi_i + \phi_j)(1) = \phi_i(1) + \phi_j(1) = i + j = \phi_{i+j}(1)$, so $\phi_i + \phi_j = \phi_{i+j}$ because these homormophisms agree on the generator 1. Hence $\psi(\phi_i + \phi_j) = \psi(\phi_{i+j}) = i + j = \psi(\phi_i) + \psi(\phi_j)$, so ψ is an additive homomorphism. Also, $(\phi_i\phi_j)(1) = \phi_i(\phi_j(1)) = \phi_i(j) = ij = \phi_{ij}(1)$, so $\phi_i\phi_j = \phi_{ij}$. Therefore $\psi(\phi_i\phi_j) = \psi(\phi_{ij}) = ij = \psi(\phi_i)\psi(\phi_j)$. Hence ψ is a ring homomorphism. By definition, the image under ψ is the entire ring $\mathbb{Z}$ [or $\mathbb{Z}_n$]. If $\psi(\phi_i) = \psi(\phi_j)$, then $i = j$ in $\mathbb{Z}$ [or $\mathbb{Z}_n$] so ϕ_i and ϕ_j map the generator 1 into the same element and thus are the same homomorphism. Thus ψ is an isomorphism.

16. A homomorphism of $\mathbb{Z}_2 \times \mathbb{Z}_2$ can map the generators (1, 0) and (0, 1) onto any elements of the ring. Thus there are a total of $4 \cdot 4 = 16$ homomorphisms of $\mathbb{Z}_2 \times \mathbb{Z}_2$ into itself, while the ring itself has only 4 elements . Thus the ring of all homomorphisms cannot be isomorphic to the ring itself, for they have different cardinality.

17. Because we are dealing with homomorphisms, it suffices to show that $(YX - XY)(ax^n) = 1(ax^n)$ for a monomial $ax^n \in F[x]$. We have

$$\begin{aligned} (YX - XY)(ax^n) &= (YX)(ax^n) - (XY)(ax^n) \\ &= Y(ax^{n+1}) - X(nax^{n-1}) \\ &= (n+1)ax^n - nax^n = 1(ax^n). \end{aligned}$$

18. Let $\phi : RG \to R$ be defined by $\phi(re) = r$. Then $\phi(re + se) = \phi((r + s)e) = r + s = \phi(re) + \phi(se)$, and $\phi((re)(se)) = \phi((rs)e) = rs = \phi(re)\phi(se)$, so ϕ is a homomorphism. Clearly, the image of RG under ϕ is all of R. If $\phi(re) = \phi(se)$, then $r = s$, so ϕ is one to one. Thus ϕ is an isomorphism.

19. **a.** From the statement of the problem, we expect the identity matrix I_2 to play the roll of 1, the matrix B with coefficient b to play the role i, and the matrix C with coefficient c to play the roll of j in a quaternion $a + bi + cj + dk$. Note that $B^2 = C^2 = -I_2$. Thus we let

$$K = BC = \begin{bmatrix} 0 & 1 \\ -1 & 0 \end{bmatrix} \begin{bmatrix} 0 & i \\ i & 0 \end{bmatrix} = \begin{bmatrix} i & 0 \\ 0 & -i \end{bmatrix}.$$

b. (See the answer in the text.)

c. We should check that ϕ is one to one.

25. Ordered Rings and Fields

1. Under $P_{\text{high}}, a - 0$ is positive, $x - a$ is positive, $x^2 - x$ is positive, etc. Thus we have the ordering $a < x < x^2 < x^3 < \cdots < x^n < \cdots$ for any $a \in R$ and $n \in \mathbb{Z}^+$.

2. The ordering is P_{low} so $(x^i - x^j) \in P$ and $x^i > x^j$ if $i < j$. Thus we have the ordering $\cdots < x^3 < x^2 < x < x^0 = 1 < x^{-1} < x^{-2} < x^{-3} < \cdots$.

3. Because $\sqrt{2}$ is negative, we must have $n < 0$ for $n\sqrt{2}$ to be positive. We see that $m + n\sqrt{2}$ is positive if $m > 0$ and $n < 0$, or if $m > 0$ and $m^2 > 2n^2$, or if $n < 0$ and $2n^2 > m^2$.

4. (i) $a\,c\,d\,e\,b$ (ii) $d\,b\,a\,e\,c$ 5. (i) $a\,c\,e\,d\,b$ (ii) $e\,c\,b\,a\,d$

6. (i) $c\,a\,b\,e\,d$ (ii) $e\,c\,a\,b\,d$ 7. (i) $d\,a\,b\,c\,e$ (ii) $d\,c\,e\,a\,b$

8. (i) $e\,a\,c\,b\,d$ (ii) $c\,d\,a\,e\,b$ 9. (i) $c\,a\,e\,d\,b$ (ii) $e\,c\,b\,a\,d$

10. $b\,d\,e\,a\,c$ 11. $d\,b\,a\,e\,c$ 12. $b\,e\,c\,d\,a$ 13. $d\,e\,b\,a\,c$

14. The smallest subfield of $\mathbb{C}$ containing $\sqrt[3]{2}$ is the intersection F of all subfields of $\mathbb{C}$ that contain $\sqrt[3]{2}$. Because $\mathbb{R}$ is one such subfield of $\mathbb{C}$, we see that $F \le \mathbb{R}$ and thus can be ordered using the induced ordering from $\mathbb{R}$. Because the smallest subfield K of $\mathbb{C}$ containing $\alpha = \sqrt[3]{2}(\frac{-1+i\sqrt{3}}{2})$ is isomorphic to F, Theorem 25.10 shows that K can be ordered, and K contains α which is not a real number.

15. T T F T T F T F F T

16. Let $\mathbb{Q}[x]$ be ordered using the ordering P_{high}, in which x is greater than every element of $\mathbb{Q}$. Let $\mathbb{Q}[\pi]$ have the ordering provided by the ordering P_{high} of $\mathbb{Q}[x]$ and the isomorphism of $\mathbb{Q}[x]$ with $\mathbb{Q}[\pi]$ provided by the evaluation homomorphism $\phi_\pi : \mathbb{Q}[x] \to \mathbb{R}$, as described in Example 25.11. Because $\phi_\pi(x) = \pi$ and $\phi_\pi(q) = q$ for all $q \in \mathbb{Q}$, the inequality $q < x$ in $\mathbb{Q}[x]$ is carried into $q < \pi$ in $\mathbb{Q}[\pi]$ for all $q \in \mathbb{Q}$.

17. Let $m + n\sqrt{2}$ and $r + s\sqrt{2}$ be any elements of $\mathbb{Z}[\sqrt{2}]$, so that $m, n, r, s \in \mathbb{Z}$. Then

$$
\begin{aligned}
\phi((m + n\sqrt{2}) + (r + s\sqrt{2})) &= \phi((m + r) + (n + s)\sqrt{2}) \\
&= (m + r) - (n + s)\sqrt{2} \\
&= (m - n\sqrt{2}) + (r - s\sqrt{2}) \\
&= \phi(m + n\sqrt{2}) + \phi(r + s\sqrt{2}),
\end{aligned}
$$

showing that ϕ is an additive homomorphism. Turning to the multiplication, we have

$$
\begin{aligned}
\phi((m + n\sqrt{2})(r + s\sqrt{2})) &= \phi((mr + 2ns) + (ms + nr)\sqrt{2}) \\
&= (mr + 2ns) - (ms + nr)\sqrt{2} \\
&= (m - n\sqrt{2})(r - s\sqrt{2}) \\
&= \phi(m + n\sqrt{2})\phi(r + s\sqrt{2}).
\end{aligned}
$$

Thus we have a homomorphism. It is clear that ϕ is a one-to-one map of $\mathbb{Z}(\sqrt{2})$ onto $\mathbb{Z}(\sqrt{2})$, so ϕ is an isomorphism.

18. If $a \in P$, then $a - 0 = a \in P$. By definition of $<$ in Theorem 25.5, $(a - 0) \in P$ means that $0 < a$.

19. If $c = 0$, then $ac = bd = 0$. Because $b \in P$ and R can have can no divisors of zero, we conclude that $d = 0$. By similar argument, $d = 0$ implies that $c = 0$.

Suppose now that c and d are nonzero. From $ac = bd$, we obtain $acd = bd^2$. Now $b \in P$, and $d^2 \in P$ implies $bd^2 \in P$ so $acd \in P$. Then $-acd = a(-cd) \notin P$, so $-cd \notin P$ and thus $cd \in P$.

20. If $a < b$, then $(b - a) \in P$. Now $b - a = (-a) - (-b)$ so $((-a) - (-b)) \in P$. Thus $-b < -a$.

21. If $a < 0$ then $(0 - a) \in P$ so $-a \in P$. If $0 < b$ then $(b - 0) \in P$ so $b \in P$. Consequently $(-a)b = -(ab) \in P$. Thus $(0 - ab) \in P$ so $ab < 0$.

22. Either $a/b \in P$ or $-(a/b) \in P$. If $-(a/b) \in P$, then $b(-a/b) = -a \in P$, contradicting the hypothesis that $a \in P$, so $a/b \in P$.

23. Now $0 < a$ implies that $a = (a - 0) \in P$, and $a < 1$ implies that $(1 - a) \in P$. Either $(\frac{1}{a} - 1) \in P$ or $(1 - \frac{1}{a}) \in P$. If $(1 - \frac{1}{a}) \in P$, then $(a - 1) = a(1 - \frac{1}{a}) \in P$, contradicting $(1 - a) \in P$. Thus $(\frac{1}{a} - 1) \in P$ so $1 < \frac{1}{a}$.

24. Now $-1 < a$ implies that $(a + 1) \in P$ and $a < 0$ implies that $(0 - a) = -a \in P$. If $(\frac{1}{a} + 1) \in P$, then $(-1 - a) = -a(\frac{1}{a} + 1)$ is in P, contradicting $(a + 1) \in P$. Therefore $-(\frac{1}{a} + 1) = (-1 - \frac{1}{a}) \in P$, so $\frac{1}{a} < -1$.

25. *Closure:* Let $a', b' \in P'$ and let $\phi(a) = a'$ and $\phi(b) = b'$. Because ϕ is one to one and $P' = \phi[P]$, we must have $a \in P$ and $b \in P$. Therefore $ab \in P$, so $\phi(ab) = \phi(a)\phi(b) = a'b' \in P'$. Likewise $(a + b) \in P$ so $\phi(a + b) = \phi(a) + \phi(b) = (a' + b') \in P$.

Trichotomy: Let $c' \in R'$ and let c be the unique element of R such that $\phi(c) = c'$. If $c \in P$, then $\phi(c) = c' \in P$. If $c = 0$, then $\phi(c) = c' = 0'$. If $-c \in P$, then $\phi(-c) = -\phi(c) = -c' \in P'$. The fact that only one of $c \in P, c = 0, -c \in P$ holds shows that only one of $c' \in P', c' = 0', -c' \in P'$ holds.

Furthermore, $a < b$ in R if and only if $(b - a) \in P$, which is true if and only if $\phi(b - a) = (\phi(b) - \phi(a)) \in P'$, which is true if and only if $\phi(a) <' \phi(b)$.

26. *Closure:* Let $a, b \in P \cap S$. Then $ab \in P$ by closure of P and $ab \in S$ by closure of S as a subring. Thus $ab \in P \cap S$. Likewise, $(a + b) \in P$ and $(a + b) \in S$ so $(a + b) \in P \cap S$.

Trichotomy: Let $s \in S$. Then $s \in R$ so either $s = 0, s \in P$, or $-s \in P$, and only one of these holds. Thus in S, either $s = 0, s \in S \cap P$, or $-s \in S \cap P$, and only one of these holds.

27. Let $<$ be a relation on R satisfying trichotomy, transitivity, and isotonicity as stated in Theorem 25.5. Let $P = \{x \in R \mid 0 < x\}$.

Closure: Let $x, y \in P$. Then $0 < x$ and $0 < y$. By the second condition in isotonicity, we have $0y < xy$ so $0 < xy$ and $xy \in P$. Also, from $0 < x$ and the first condition of isotonicity, we obtain $0 + y < x + y$, so $y < x + y$. From $0 < y$ and $y < x + y$, we obtain $0 < x + y$ by transitivity.

Trichotomy for P: Let $x \in R$. By trichotomy for $<$, precisely one of $0 < x, 0 = x$, or $x < 0$ holds. Now $0 < x$ if and only if $x \in P$. By isotonicity, $x < 0$ implies $(-x + x) < (-x + 0)$. Again by isotonicity, $(-x + x) < (-x + 0)$ implies $x + (-x + x) < x + (-x + 0)$, that is, it implies $x < 0$. Thus $x < 0$ if and only if $0 < -x$, which is true if and only if $-x \in P$.

By isotonicity, $a < b$ for the given $<$ on R if and only if $-a + a < -a + b$, that is, if and only if $0 < b - a$. By definition of $P, 0 < b - a$ if and only if $(b - a) \in P$, which is true if and only if $a <_P b$. Thus $<$ and $<_P$ are the same relation on R.

28. Note that if $a < 0$ and $b < 0$, then $-a \in P$ and $-b \in P$ so $ab = (-a)(-b) \in P$, so $0 < ab$. It follows at once that a product of an even number of elements x_i where every $x_i < 0$ is an element of P. Thus any product of an even number of elements, all of which are greater than zero or all of which are less than zero, is sure to be positive. From $a^{2n+1} = b^{2n+1}$, we obtain $(a^2)^n a = (b^2)^n b$ and we see the either $a < 0$ and $b < 0$, or $0 < a$ and $0 < b$. Consider the factorization

$$0 = a^{2n+1} - b^{2n+1} = (a - b)(a^{2n} + a^{2n-1}b + a^{2n-2}b^2 + \cdots + b^{2n}).$$

Every summand in parentheses is the product of an even number of factors that are either a or b. Because a and b are either both greater than zero or both less than zero, every summand in parentheses is positive, and thus their sum is positive, and hence nonzero. Because R has no zero divisors, we must have $a - b = 0$, so $a = b$.

29. In the chart following, the order in the left column indicates the order in which the indeterminants were adjoined to R, and whether the order when they were adjoined was P_{high} or P_{low}. The other columns indicate whether the inequality the top of the column is true (T) or false (F). Because no two rows have the same sequence of T's and F's, the orderings are all different.

Ordering	$x < y$	$1 < x$	$1 < y$	$xy < 1$
x-high, y-high	T	T	T	F
x-high, y-low	F	T	F	T
x-low, y-high	T	F	T	F
x-low, y-low	F	F	F	T
y-high, x-high	F	T	T	F
y-high, x-low	T	F	T	T
y-low, x-high	F	T	F	F
y-low, x-low	T	F	F	T

26. Homomorphisms and Factor Rings

1. Let ϕ be a homomorphism of $\mathbb{Z} \times \mathbb{Z}$ into $\mathbb{Z} \times \mathbb{Z}$. Suppose that $\phi(1,0) = (m,n)$. From $\phi(1,0) = \phi[(1,0)(1,0)]$, we see that $m^2 = m$ and $n^2 = n$, so $\phi(1,0)$ must be one of the elements $(0, 0)$, $(1, 0)$, $(0, 1)$, or $(1, 1)$. By a similar argument, $\phi(0,1)$ must be one of these same four elements. We also must have $\phi(1,0)\phi(0,1) = \phi(0,0) = (0,0)$. This gives just 9 possibilities.

$\phi(1,0) = (1,0)$ while $\phi(0,1) = (0,0)$ or $(0, 1)$,
$\phi(1,0) = (0,1)$ while $\phi(0,1) = (0, 0)$ or $(1, 0)$,
$\phi(1,0) = (1, 1)$ while $\phi(0,1) = (0, 0)$, and
$\phi(1,0) = (0,0)$ while $\phi(0,1) = (0, 0)$, $(1, 0)$, $(0, 1)$ or $(1, 1)$.

It is easily checked that each of these does give rise to a homomorphism.

2. In order for $\mathbb{Z}_n$ to contain a subring isomorphic to $\mathbb{Z}_2$, we see that Z_n must contain a nonzero element s such that $s + s = 0$ and $s^2 = s$, so that s can play the role of 1 in $\mathbb{Z}_2$. From $s + s = 0$, we see that n must be even. Let $n = 2m$, so that the group $\langle\{0, m\}, +_n\rangle \simeq \langle\mathbb{Z}_2, +_2\rangle$. In order to have $\langle\{0, m\}, \cdot_n\rangle \simeq \langle\mathbb{Z}_2, \cdot_2\rangle$, we must have $mm = m$. In $\mathbb{Z}_n$, we have $2 \cdot m = 0, 3 \cdot m = (2 \cdot m) + m = m, 4 \cdot m = 0, 5 \cdot m = m$, etc. Thus we have $mm = m$ in $\mathbb{Z}_n$ if and only if m is an odd integer. Hence $\mathbb{Z}_n$ contains a subring isomorphic to $\mathbb{Z}_2$ if and only if $n = 2m$ for an odd integer m.

3. Because the ideals must be additive subgroups, by group theory we see that the possibilities are restricted to the cyclic additive subgroups

$$\langle 0 \rangle = \{0\},$$
$$\langle 1 \rangle = \{0, 1, 2, 3, 4, 5, 6, 7, 8, 9, 0, 10, 11\},$$
$$\langle 2 \rangle = \{0, 2, 4, 6, 8, 10\},$$
$$\langle 3 \rangle = \{0, 3, 6, 9\},$$
$$\langle 4 \rangle = \{0, 4, 8\}, \text{ and}$$
$$\langle 6 \rangle = \{0, 6\}.$$

It is easily checked that each of these is closed under multiplication by any element of $\mathbb{Z}_{12}$, so they are ideals. We have $\mathbb{Z}_{12}/\langle 0 \rangle \simeq \mathbb{Z}_{12}$, $\mathbb{Z}_{12}/\langle 1 \rangle \simeq \{0\}$, $\mathbb{Z}_{12}/\langle 2 \rangle \simeq \mathbb{Z}_2$, $\mathbb{Z}_{12}/\langle 3 \rangle \simeq \mathbb{Z}_3$, $\mathbb{Z}_{12}/\langle 4 \rangle \simeq \mathbb{Z}_4$, and $\mathbb{Z}_{12}/\langle 6 \rangle \simeq \mathbb{Z}_6$.

4. Here are the tables for addition and multiplication in $2\mathbb{Z}/8\mathbb{Z}$.

+	$8\mathbb{Z}$	$2+8\mathbb{Z}$	$4+8\mathbb{Z}$	$6+8\mathbb{Z}$
$8\mathbb{Z}$	$8\mathbb{Z}$	$2+8\mathbb{Z}$	$4+8\mathbb{Z}$	$6+8\mathbb{Z}$
$2+8\mathbb{Z}$	$2+8\mathbb{Z}$	$4+8\mathbb{Z}$	$6+8\mathbb{Z}$	$8\mathbb{Z}$
$4+8\mathbb{Z}$	$4+8\mathbb{Z}$	$6+8\mathbb{Z}$	$8\mathbb{Z}$	$2+8\mathbb{Z}$
$6+8\mathbb{Z}$	$6+8\mathbb{Z}$	$8\mathbb{Z}$	$2+8\mathbb{Z}$	$4+8\mathbb{Z}$

$\cdot$	$8\mathbb{Z}$	$2+8\mathbb{Z}$	$4+8\mathbb{Z}$	$6+8\mathbb{Z}$
$8\mathbb{Z}$	$8\mathbb{Z}$	$8\mathbb{Z}$	$8\mathbb{Z}$	$8\mathbb{Z}$
$2+8\mathbb{Z}$	$8\mathbb{Z}$	$4+8\mathbb{Z}$	$8\mathbb{Z}$	$4+8\mathbb{Z}$
$4+8\mathbb{Z}$	$8\mathbb{Z}$	$8\mathbb{Z}$	$8\mathbb{Z}$	$8\mathbb{Z}$
$6+8\mathbb{Z}$	$8\mathbb{Z}$	$4+8\mathbb{Z}$	$8\mathbb{Z}$	$4+8\mathbb{Z}$

The rings $2\mathbb{Z}/8\mathbb{Z}$ and $\mathbb{Z}_4$ are not isomorphic, for $2\mathbb{Z}/8\mathbb{Z}$ has no unity while $\mathbb{Z}_4$ does.

5. The definition is incorrect; ϕ must map R onto R'.

An **isomorphism** of a ring R with a ring R' is a homomorphism $\phi : R \to R'$ mapping R onto R' such that $\text{Ker}(\phi) = \{0\}$.

6. The definition is correct.

7. The definition is incorrect. The set description is nonsense.

The **kernel of a homomorphism** ϕ mapping a ring R into a ring R' is $\{r \in R \mid \phi(r) = 0'\}$.

8. The differentiation map δ is not a homomorphism; $\delta(f(x)g(x)) = f'(x)g(x) + f(x)g'(x) \neq f'(x)g'(x) = \delta(f(x))\delta(g(x))$. To connect this with Example 26.12, we note that the kernel of δ as an additive group homomorphism is the set of all constant functions. Example 26.12 shows that this is not an ideal, so δ cannot be a ring homomorphism.

9. Let $\phi : \mathbb{Z} \to \mathbb{Z} \times \mathbb{Z}$ be defined by $\phi(n) = (n, 0)$. Then $\mathbb{Z}$ has unity 1, but $\phi(1) = (1, 0)$ is not the unity of $\mathbb{Z} \times \mathbb{Z}$; the unity of $\mathbb{Z} \times \mathbb{Z}$ is $(1, 1)$.

10. T F T F T F T T T T

11. (See the text answer.)

12. We know that $\mathbb{Z}/2\mathbb{Z} \simeq \mathbb{Z}_2$, which is a field.

13. $\mathbb{Z}$ is an integral domain. $\mathbb{Z}/4\mathbb{Z} \simeq \mathbb{Z}_4$, where 2 is a divisor of 0.

14. $\mathbb{Z} \times \mathbb{Z}$ has divisors of zero, but $(\mathbb{Z} \times \mathbb{Z})/(\mathbb{Z} \times \{0\}) \simeq \mathbb{Z}$ which has no divisors of zero.

15. $\{(n, n) \mid n \in \mathbb{Z}\}$ is a subring of $\mathbb{Z} \times \mathbb{Z}$, but it is not an ideal because $(2, 1)(n, n) = (2n, n)$.

16. a. The notations r and s would be used to denote elements of R, not of R/N. The student probably does not understand the structure of a factor ring.

b. Assume that R/N is commutative. Then

$$(r + N)(s + N) = (s + N)(r + N) \text{ for all } r, s \in R.$$

c. Let $r, s \in R$. Then

$(r + N)(s + N) = (s + N)(r + N)$ for all $r, s \in R$
if and only if $rs + N = sr + N$ for all $r, s \in R$, so
if and only if $(rs + N) - (sr + N) = N$ for all $r, s \in R$, so
if and only if $(rs - sr) + N = N$ for all $r, s \in R$, so
if and only $(rs - sr) \in N$ for all $r, s \in R$.

17. Because $(a + b\sqrt{2}) + (c + d\sqrt{2}) = (a + c) + (b + d)\sqrt{2}$ and $0 = 0 + 0\sqrt{2}$ and $-(a + b\sqrt{2}) = (-a) + (-b)\sqrt{2}$, we see that R is closed under addition, has an additive identity, and contains additive inverses. Thus $\langle R, + \rangle$ is a group. Now $(a + b\sqrt{2})(c + d\sqrt{2}) = (ac + 2bd) + (ad + bc)\sqrt{2}$, so R is closed under multiplication and is thus a ring. We will show that R' is a ring by showing that it is the image of R under a homomorphism $\phi : R \to M_2$. Let

$$\phi(a + b\sqrt{2}) = \begin{bmatrix} a & 2b \\ b & a \end{bmatrix}.$$

Then

$$\begin{aligned}
\phi((a + b\sqrt{2}) + (c + d\sqrt{2})) &= \phi((a + c) + (b + d)\sqrt{2} \\
&= \begin{bmatrix} a + c & 2(b + d) \\ b + d & a + c \end{bmatrix} = \begin{bmatrix} a & 2b \\ b & a \end{bmatrix} + \begin{bmatrix} c & 2d \\ d & c \end{bmatrix} \\
&= \phi(a + b\sqrt{2}) + \phi(c + d\sqrt{2})
\end{aligned}$$

and

$$\begin{aligned}
\phi((a + b\sqrt{2})(c + d\sqrt{2})) &= \phi((ac + 2bd) + (ad + bc)\sqrt{2}) \\
&= \begin{bmatrix} ac + 2bd & 2(ad + bc)) \\ ad + bc & ac + 2bd \end{bmatrix} = \begin{bmatrix} a & 2b \\ b & a \end{bmatrix} \begin{bmatrix} c & 2d \\ d & c \end{bmatrix} \\
&= \phi(a + b\sqrt{2})\phi(c + d\sqrt{2}).
\end{aligned}$$

This shows that ϕ is a homomorphism. Because $R' = \phi[R]$, we see that R' is a ring. If $\phi(a + b\sqrt{2})$ is the matrix with all entries zero, then we must have $a = b = 0$, so $\mathrm{Ker}(\phi) = 0$ and ϕ is one to one. Thus ϕ is an isomorphism of R onto R'.

18. Let $\phi : F \to R$ be a homomorphism of a field F into a ring R, and let $N = \mathrm{Ker}(\phi)$. If $N \neq \{0\}$, then N contains a nonzero element u of F which is a unit. Because N is an ideal, we see that $u^{-1}u = 1$ is in N, and then N contains $a1 = a$ for all $a \in F$. Thus N is either $\{0\}$, in which case ϕ is one to one by group theory, or $N = F$, so that ϕ maps every element of F onto 0.

19. By Exercise 49 of Section 13, $\psi\phi(r + s) = \psi\phi(r) + \psi\phi(s)$ for all $r, s \in R$. For multiplication, we note that $\psi\phi(rs) = \psi(\phi(rs)) = \psi(\phi(r)\phi(s)) = [\psi\phi(r)][\psi\phi(s)]$ because both ϕ and ψ are homomorphisms. Thus $\psi\phi$ is also a homomorphism.

20. In a commutative ring R, the binomial expansion $(a + b)^n = \sum_{i=0}^{n} \binom{n}{i} a^i b^{n-i}$ is valid. If p is a prime and $n = p$, then all the binomial coefficients $\binom{p}{i}$ for $1 \le i \le p - 1$ are divisible by p, and thus the term $\binom{p}{i} a^i b^{p-i} = 0$ for a and b in a commutative ring of characteristic p. This shows at once that $\phi_p(a + b) = (a + b)^p = a^p + b^p = \phi_p(a) + \phi_p(b)$. Also $\phi_p(ab) = (ab)^p = a^p b^p$ because R is commutative. But $a^p b^p = \phi_p(a)\phi_p(b)$, so ϕ is a homomorphism.

21. By Theorem 26.3, we know $\phi(1)$ is unity for $\phi[R]$. Suppose that R' has unity $1'$. Then $\phi(1) = \phi(1)1' = \phi(1)\phi(1)$ so that we have $\phi(1)1' - \phi(1)\phi(1) = 0'$. Consequently, $\phi(1)(1' - \phi(1)) = 0'$. Now if $\phi(1) = 0'$, then $\phi(a) = \phi(1a) = \phi(1)\phi(a) = 0'\phi(a) = 0'$ for all $a \in R$, so $\phi[R] = \{0'\}$ contrary to hypothesis. Thus $\phi(1) \ne 0'$. Because R' has no divisors of zero, we conclude from $\phi(1)(1' - \phi(1)) = 0'$ that $1' - \phi(1) = 0'$, so $\phi(1)$ is the unity $1'$ of R'.

22. a. Because the ideal N is also a subring of R, Theorem 26.3 shows that $\phi[N]$ is a subring of R'. To show that is is an ideal of $\phi[R]$, we show that $\phi(r)\phi[N] \subseteq \phi[N]$ and $\phi[N]\phi(r) \subseteq \phi[N]$ for all $r \in R$. Let $r \in R$ and let $s \in N$. Then $rs \in N$ and $sr \in N$ because N is an ideal. Applying ϕ, we see that $\phi(r)\phi(s) = \phi(rs) \in \phi[N]$ and $\phi(s)\phi(r) = \phi(sr) \in \phi[N]$.

b. Let $\phi : \mathbb{Z} \to \mathbb{Q}$ be the injection map given by $\phi(n) = n$ for all $n \in \mathbb{Z}$. Now $2\mathbb{Z}$ is an ideal of $\mathbb{Z}$, but $2\mathbb{Z}$ is not an ideal of $\mathbb{Q}$ because $(1/2)2 = 1$ and 1 is not in $2\mathbb{Z}$.

c. Let N' be an ideal of R' or of $\phi[R]$. We know that $\phi^{-1}[N']$ is at least a subring of R by Theorem 26.3. We must show that $r\phi^{-1}[N'] \subseteq \phi^{-1}[N']$ and that $\phi^{-1}[N']r \subseteq \phi^{-1}[N']$ for all $r \in R$. Let $s \in \phi^{-1}[N']$, so that $\phi(s) \in N'$. Then $\phi(rs) = \phi(r)\phi(s)$ and $\phi(r)\phi(s) \in N'$ because N' is an ideal. This shows that $rs \in \phi^{-1}[N']$, so $r\phi^{-1}[N'] \subseteq \phi^{-1}[N']$. Also $\phi(sr) = \phi(s)\phi(r)$ and $\phi(s)\phi(r) \in N'$ because N' is an ideal. This shows that $sr \in \phi^{-1}[N']$, so $\phi^{-1}[N']r \subseteq \phi^{-1}[N']$.

23. If $f(x_1, \cdots, x_n)$ and $g(x_1, \cdots, x_n)$ both have every element of S as a zero, then so do their sum, product, and any multiple of one of them by any element $h(x_1, \cdots, x_n)$ in $F[x_1, \cdots, x_n]$. Because the possible multipliers from $F[x_1, \cdots, x_n]$ include 0 and -1, we see that the set N_S is indeed a subring closed under multiplication by elements of $F[x_1, \cdots, x_n]$, and thus is an ideal of this polynomial ring.

24. Let N be an ideal of a field F. If N contains a nonzero element a, then N contains $(1/a)a = 1$, because N is an ideal. But then N contains $s1 = s$ for every $s \in F$, so $N = F$. Thus N is either $\{0\}$ or F. If $N = F$, then $F/N = F/F$ is the trivial ring of one element. If $N = \{0\}$, then $F/N = F/\{0\}$ is isomorphic to F, because each element $s + \{0\}$ of $F/\{0\}$ can be renamed s.

25. If $N \ne R$, then the unity 1 of R is not an element of N, for if $1 \in N$, then so is $r1 = r$ for all $r \in R$. Thus $1 + N \ne N$, that is, $1 + N$ is not the zero element of R/N. Clearly $(1 + N)(r + N) = r + N = (r + N)(1 + N)$ in R/N, which shows that $1 + N$ is unity for R/N.

26. Let $x, y \in I_a$ so $ax = ay = 0$. Then $a(x + y) = ax + ay = 0 + 0 = 0$ so $(x + y) \in I_a$. Also, $a(xy) = (ax)y = 0y = 0$ so $xy \in I_a$. Because $a0 = 0$ and $a(-x) = -(ax) = -0 = 0$, we see that I_a contains 0 and additive inverses of each of its elements x, so I_a is a subring of R. (Note that thus far, we have not used commutativity in R.) Let $r \in R$. Then $a(xr) = (ax)r = 0r = 0$ so $xr \in I_a$, and because R is commutative, we see that $a(rx) = r(ax) = r0 = 0$, so $rx \in I_a$. Thus I_a is an ideal of R.

27. Let $\{N_i \mid i \in I\}$ be a collection of ideals in R. Each of these ideals is a subring of R, and Exercise 49 of Section 18 shows that $N = \bigcap_{i \in I} N_i$ is also a subring of R. We need only show that N is closed under multiplication by elements of R. Let $r \in R$ and let $s \in N$. Then $s \in N_i$ for all $i \in I$. Because each N_i is an ideal of R, we see that $rs \in N_i$ and $sr \in N_i$ for all $i \in I$. Thus $rs \in N$.

28. By Exercise 39 of Section 14, the map $\phi_* : R/N \to R'/N'$ defined by $\phi_*(r + N) = \phi(r) + N'$ is well defined and satisfies the additive requirements for a homomorphism. Now we have $\phi_*((r+N)(s+N)) = \phi_*(rs + N) = \phi(rs) + N' = [\phi(r)\phi(s)] + N' = [\phi(r) + N'][\phi(s) + N'] = [\phi_*(r + N)][\phi_*(s + N)]$ so ϕ_* also satisfies the multiplicative condition, and is a ring homomorphism.

29. The condition that ϕ maps R onto a *nonzero* ring R' shows that no unit of R is in $\text{Ker}(\phi)$, for if $\text{Ker}(\phi)$ contains a unit u, then it contains $(ru^{-1})u = r$ for all $r \in R$, which would mean that $\text{Ker}(\phi) = R$ and R' would be the zero ring.

Let u be a unit in R. Because $\phi[R] = R'$, we know that $\phi(1)$ is unity $1'$ in R'. From $uu^{-1} = u^{-1}u = 1$, we obtain $\phi(uu^{-1}) = \phi(u)\phi(u^{-1}) = 1'$ and $\phi(u^{-1}u) = \phi(u^{-1})\phi(u) = 1'$. Thus $\phi(u)$ is a unit of R', and its inverse is $\phi(u^{-1})$.

30. Let $\sqrt{\{0\}}$ be the collection of all nilpotent elements of R. Let $a, b \in \sqrt{\{0\}}$. Then there exist positive integers m and n such that $a^m = b^n = 0$. In a *commutative* ring, the binomial expansion is valid. Consider $(a + b)^{m+n}$. In the binomial expansion, each summand contains a term $a^i b^{m+n-i}$. Now either $i \geq m$ so that $a^i = 0$ or $m + n - i \geq n$ so that $b^{m+n-i} = 0$. Thus each summand of $(a + b)^{m+n}$ is zero, so $(a + b)^{m+n} = 0$ and $\sqrt{\{0\}}$ is closed under addition. For multiplication, we note that because R is commutative, $(ab)^{mn} = (a^m)^n(b^n)^m = (0)(0) = 0$, so $ab \in \sqrt{\{0\}}$. If $s \in R$, then $(sa)^m = a^m s^m = 0 s^m = 0$ so $\sqrt{\{0\}}$ is also closed under left and right multiplication by elements of R. Taking $x = 0$, we see that $0 \in \sqrt{\{0\}}$. Also $(-a)^m$ is either a^m or $-a^m$, so $(-a)^m = 0$ and $-a \in \sqrt{\{0\}}$. Thus $\sqrt{\{0\}}$ is an ideal of R.

31. The nilradical of $\mathbb{Z}_{12}$ is $\{0, 6\}$. The nilradical of $\mathbb{Z}$ is $\{0\}$ and the nilradical of $\mathbb{Z}_{32}$ is $\{0, 2, 4, 6, 8, \cdots, 30\}$.

32. Suppose $(a + N)^m = N$ in R/N. Then $a^m \in N$. Because N is the nilradical of R, there exists $n \in \mathbb{Z}^+$ such that $(a^m)^n = 0$. But then $a^{mn} = 0$ so $a \in N$. Thus $a + N = N$ so $\{N\}$ is the nilradical of R/N.

33. Let $a \in R$. Because the nilradical of R/N is R/N, there is some positive integer m such that $(a + N)^m = N$. Then $a^m \in N$. Because every element of N is nilpotent, there exists a positive integer n such that $(a^m)^n = 0$ in R. But then $a^{mn} = 0$, so a is an element of the nilradical of R. Thus the nilradical of R is R.

34. Let $a, b \in \sqrt{N}$. Then $a^m \in N$ and $b^n \in N$ for some positive integers m and n. Precisely as in the answer to Exercise 30, we argue that $(a + b) \in N$, $ab \in N$, and also that $sa \in N$ and $as \in N$ for any $s \in R$. Because $0^1 \in N$, we see that $0 \in \sqrt{N}$. Also $(-a)^m$ is either a^m or $-(a^m)$, and both a^m and $-(a^m)$ are in N. Thus $-a \in \sqrt{N}$. This shows that $\sqrt{N}$ is an ideal of R.

35. **a.** Let $R = \mathbb{Z}$ and let $N = 4\mathbb{Z}$. Then $\sqrt{N} = 2\mathbb{Z} \neq 4\mathbb{Z}$.

   b. Let $R = \mathbb{Z}$ and let $N = 2\mathbb{Z}$. Then $\sqrt{N} = 2\mathbb{Z}$.

36. If $\sqrt{N}/N$ is viewed as a subring of R/N, then it is the nilradical of R/N, in the sense of the definition in Exercise 30.

37. We have

$$\phi[(a + bi) + (c + di)] = \phi[(a + c) + (d + b)i] = \begin{bmatrix} a + c & b + d \\ -b - d & a + c \end{bmatrix}$$

$$= \begin{bmatrix} a & b \\ -b & a \end{bmatrix} + \begin{bmatrix} c & d \\ -d & c \end{bmatrix} = \phi(a + bi) + \phi(c + di).$$

Also

$$\phi[(a+bi)(c+di)] \; = \; \phi[(ac-bd)+(ad+bc)i] = \begin{bmatrix} ac-bd & ad+bc \\ -ad-bc & ac-bd \end{bmatrix}$$

$$= \begin{bmatrix} a & b \\ -b & a \end{bmatrix}\begin{bmatrix} c & d \\ -d & c \end{bmatrix} = \phi(a+bi)\phi(c+di).$$

Thus ϕ is a homomorphism. It is obvious that ϕ is one to one. Hence ϕ exhibits an isomorpism of $\mathbb{C}$ with the subring $\phi[\mathbb{C}]$, which therefore must be a field.

38. a. For $x, y \in R$, we have

$$\lambda_a(x+y) = a(x+y) = ax + ay = \lambda_a(x) + \lambda_a(y).$$

Thus λ_a is a homomorphism of $\langle R, + \rangle$ with itself, that is, an element of $\text{End}(\langle R, + \rangle)$

b. Note that for $a, b \in R$, we have $(\lambda_a \lambda_b)(x) = \lambda_a(\lambda_b(x)) = \lambda_a(bx) = a(bx) = (ab)x = \lambda_{ab}(x)$. Thus $\lambda_a \lambda_b = \lambda_{ab}$ and R' is closed under multiplication. We also have $(\lambda_a + \lambda_b)(x) = \lambda_a(x) + \lambda_b(x) = ax + bx = (a+b)x = \lambda_{a+b}(x)$, so $\lambda_a + \lambda_b = \lambda_{a+b}$. Thus R' is closed under addition. From what we have shown, it follows that $\lambda_0 + \lambda_a = \lambda_{0+a} = \lambda_a$ and $\lambda_a + \lambda_0 = \lambda_{a+0} = \lambda_a$ so λ_0 acts as additive identity. Finally, $\lambda_{-a} + \lambda_a = \lambda_{-a+a} = \lambda_0$ and $\lambda_a + \lambda_{-a} = \lambda_{a-a} = \lambda_0$ so R' contains an additive inverse of each element. Thus R' is a ring.

c. Let $\phi : R \to R'$ be defined by $\phi(a) = \lambda_a$. By our work in Part(**b**), we see that $\phi(a+b) = \lambda_{a+b} = \lambda_a + \lambda_b = \phi(a) + \phi(b)$, and $\phi(ab) = \lambda_{ab} = \lambda_a \lambda_b = \phi(a)\phi(b)$. Thus ϕ is a homomorphism, and is clearly onto R'. Suppose that $\phi(a) = \phi(b)$. Then $ax = bx$ for all $x \in R$. Because R has unity (and this is the only place where that hypothesis is needed), we have in particular $a1 = b1$ so $a = b$. Thus ϕ is one to one and onto R', so it is an isomorphism.

27. Prime and Maximal Ideals

1. Because a finite integral domain is a field, the prime and the maximal ideals coincide. The ideals $\{0, 2, 4\}$ and $\{0, 3\}$ are both prime and maximal because the factor rings are isomorphic to the fields $\mathbb{Z}_2$ and $\mathbb{Z}_3$ respectively.

2. Because a finite integral domain is a field, the prime and the maximal ideals coincide. The prime and maximal ideals are $\{0, 2, 4, 6, 8, 10\}$ and $\{0, 3, 6, 9\}$ because the factor rings are isomorphic to the fields $\mathbb{Z}_2$ and $\mathbb{Z}_3$ respectively.

3. Because a finite integral domain is a field, the prime and the maximal ideals coincide. The prime and maximal ideals are $\{(0,0), (1,0)\}$ and $\{(0,0), (0,1)\}$ because the factor rings are isomorphic to the field $\mathbb{Z}_2$.

4. A finite integral domain is a field, so prime and maximal ideals coincide. The prime and maximal ideals are $\{(0,0), (0,1), (0,2), (0,3)\}$ and $\{(0,0), (1,0), (0,2), (1,2))\}$ leading to factor rings isomorphic to the field $\mathbb{Z}_2$.

5. By Theorem 27.25, we need only find all values c such that $x^2 + c$ is irreducible over $\mathbb{Z}_3$. Let $f(x) = x^2$. Then $f(0) = 0, f(1) = 1$, and $f(2) = 1$. We must find $c \in \mathbb{Z}_3$ such that $0 + c$ and $1 + c$ are both nonzero. Clearly $c = 1$ is the only choice.

6. By Theorem 27.25, we need only find all values c such that $x^3 + x^2 + c$ is irreducible over $\mathbb{Z}_3$. Let $f(x) = x^3 + x^2$. Then $f(0) = 0, f(1) = 2$, and $f(2) = 0$. We must find $c \in \mathbb{Z}_3$ such that $0 + c$ and $2 + c$ are both nonzero. Clearly $c = 2$ is the only choice.

7. By Theorem 27.25, we need only find all values c such that $g(x) = x^3 + cx^2 + 1$ is irreducible over $\mathbb{Z}_3$. When $c = 0, g(2) = 0$ and when $c = 1, g(1) = 0$, butwhen $c = 2, g(x)$ has no zeros. Thus $c = 2$ is the only choice.

8. By Theorem 27.25, we need only find all values c such that $x^2 + x + c$ is irreducible over $\mathbb{Z}_5$. Let $f(x) = x^2 + x$. Then $f(0) = 0, f(1) = 2, f(2) = 1, f(3) = 2$, and $f(4) = 0$. We must find $c \in \mathbb{Z}_5$ such that $0 + c, 1 + c$,and $2 + c$ are all nonzero. Clearly $c = 1$ and $c = 2$ both work.

9. By Theorem 27.25, we need only find all values c such that $g(x) = x^2 + cx + 1$ is irreducible over $\mathbb{Z}_5$. We compute that when $c = 0, g(2) = 0$, when $c = 1, g(x)$ has no zeros, when $c = 2, g(-1) = 0$, when $c = 3, g(1) = 0$, and when $c = 4, g(x)$ has no zeros. Thus c can be either 1 or 4.

10. The definition is incorrect. We need to specify that the ideals are not R.

      A **maximal ideal** of a ring R is an ideal M such that there is no ideal N of R such that $M \subset N \subset R$.

11. The definition is incorrect nonsense.

      A **prime ideal** of a commutative ring R is an ideal N such that if $a, b \in R$ and $ab \in N$, then either $a \in N$ or $b \in N$.

12. The definition is correct, although this is not the way it is phrased in the text.

13. The definition is correct, although this is not the way it is phrased in the text.

14. F T T F T T T F T F

15. $\mathbb{Z} \times 2\mathbb{Z}$ is a maximal ideal of $\mathbb{Z} \times \mathbb{Z}$, for the factor ring is isomorphic to $\mathbb{Z}_2$, which is a field.

16. $\mathbb{Z} \times \{0\}$ is a prime ideal of $\mathbb{Z} \times \mathbb{Z}$ that is not maximal, for the factor ring is isomorphic to $\mathbb{Z}$ which is an integral domain, but not a field.

17. $\mathbb{Z} \times 4\mathbb{Z}$ is a proper ideal of $\mathbb{Z} \times \mathbb{Z}$ that is not prime, for the factor ring is isomorphic to $\mathbb{Z}_4$ which has divisors of zero.

18. $\mathbb{Q}/\langle x^2 - 5x + 6 \rangle$ is not a field, because $x^2 - 5x + 6 = (x - 2)(x - 3)$ is not an irreducible polynomial, so the ideal $\langle x^2 - 5x + 6 \rangle$ is not maximal.

19. $\mathbb{Q}/\langle x^2 - 6x + 6 \rangle$ is a field, because the polynomial $x^2 - 6x + 6$ is irreducible by the Eisenstein condition with $p = 2$ or $p = 3$, so $\langle x^2 - 6x + 6 \rangle$ is a maximal ideal.

20. If $a + M$ has no multiplicative inverse in R/M, then the principal ideal generated by $a + M$ does not contain $1 + M$, so it is not R/M. Then the inverse image of this ideal under the canonical homomorphism of R into R/M would be an ideal strictly between M and R.

21. If there were an ideal N strictly between the ideal M and the ring R, then its image under the canonical homomorphism of R into R/M would be an ideal of R/M strictly between $\{0 + M\}$ and R/M. This is impossible because there are no nontrivial proper ideals in a field.

22. If F is a field, then the division algorithm can be used to show that every ideal N in $F[x]$ is principal, generated by any element of N of minimum possible degee N.

23. Because a maximal ideal in $F[x]$ is a prime ideal, a factorization of $p(x)$ into two polymials both having degree less than $\deg(p(x))$ would mean that $\langle p(x) \rangle$ would contain a polynomial of degree less than the degree of $p(x)$, which is impossible.

24. Theorem 19.11 shows that every finite integral domain is a field. Let N be a prime ideal in a finite commutative ring R with unity. Then R/N is a finite integral domain, and therefore a field, and therefore N is a maximal ideal.

25. Yes, it is possible; $\mathbb{Z}_2 \times \mathbb{Z}_3$ contains a subring isomorphic to $\mathbb{Z}_2$ and one isomorphic to $\mathbb{Z}_3$.

26. Yes, it is possible; $\mathbb{Z}_2 \times \mathbb{Z}_3$ contains a subring isomorphic to $\mathbb{Z}_2$ and one isomorphic to $\mathbb{Z}_3$.

27. No, it is not possible. Enlarging the integal domain to a field of quotients, we would then have a field containing (up to isomorphism) two different prime fields $\mathbb{Z}_p$ and $\mathbb{Z}_q$. The unity of each of these fields would be a zero of $x^2 - x$, but this polynomial has only one nonzero zero in a field, namely the unity of the field.

28. Let M be a maximal ideal of R and suppose that $ab \in M$ but a is not in M. Let $N = \{ra + m \mid r \in R, m \in M\}$. From $(r_1 a + m_1) + (r_2 a + m_2) = (r_1 + r_2)a + (m_1 + m_2)$, we see that N is closed under addition. From $r(r_1 a + m_1) = (rr_1)a + (rm_1)$ and the fact that M is an ideal, we see that N is closed under multiplication by elements of R, and is of course closed itself under multiplication. Also $0 = 0a + 0$ is in N and furthermore $(-r)a + (-m) = -(ra) - m = -(ra + m)$ is in N. Thus N is an ideal. Clearly N contains M, but $N \neq M$ because $1a + 0 = a$ is in N but a is not in M. Because M is maximal, we must have $N = R$. Therefore $1 \in N$, so $1 = ra + m$ for some $r \in R$ and $m \in M$. Multiplying by b, we find that $b = rab + mb$. But ab and mb are both in M, so $b \in M$. We have shown that if $ab \in M$ and a is not in M, then $b \in M$. This is the definition of a prime ideal.

29. We use the addendum to Theorem 26.3 stated in the final paragraph of Section 26 and proved in Exercise 22 of that section. Suppose that N is any ideal of R. By the addendum mentioned and using the canonical homomorphism $\gamma : R \to R/N$, if M is a proper ideal of R properly containing N, then $\gamma[M]$ is a proper nontrivial ideal of R/N. This shows that if M is not maximal, then R/N is not a simple ring. On the other hand, suppose that R/N is not a simple ring, and let N' be a proper nontrivial ideal of R/N. By the addendum mentioned, $\gamma^{-1}[N']$ is an ideal of R, and of course $\gamma^{-1}[N'] \neq R$ because N' is a proper ideal of R/N, and also $\gamma^{-1}[N']$ properly contains N because N' is nontrival in R/N. Thus $\gamma^{-1}[N']$ is a proper ideal of R that properly contains N, so N is not maximal. We have proved p if and only if q by proving not p if and only if not q.

This exercise is the straightforward analogue of Theorem 15.18 for groups, that is, a maximal ideal of a ring is analogous to a maximal normal subgroup of a group.

30. Every ideal of $F[x]$ is principal by Theorem 26.24. Suppose $\langle f(x) \rangle \neq \{0\}$ is a proper prime ideal of $F[x]$. Then every polynomial in $\langle f(x) \rangle$ has degree greater than or equal to the degree of $f(x)$. Thus if $f(x) = g(x)h(x)$ in $F[x]$ where the degrees of both $g(x)$ and $h(x)$ are less than the degree of $f(x)$, neither $g(x)$ nor $h(x)$ can be in $\langle f(x) \rangle$. This would contradict the fact that $\langle f(x) \rangle$ is a prime ideal, so no such factorization of $f(x)$ in $F[x]$ can exist, that is, $f(x)$ is irreducible in $F[x]$. By Theorem 26.25, $\langle f(x) \rangle$ is therefore a maximal ideal of $F[x]$.

31. If $f(x)$ divides $g(x)$, then $g(x) = f(x)q(x)$ for some $q(x) \in F[x]$, so $g(x) \in \langle f(x) \rangle$ because this ideal consists of all multiples of $f(x)$. Conversely, if $g(x) \in \langle f(x) \rangle$, then $g(x)$ is some multiple $h(x)f(x)$ of $f(x)$ for $h(x) \in F[x]$. The equation $g(x) = h(x)f(x)$ is the definition of $f(x)$ dividing $g(x)$.

32. The equation

$$[r_1(x)f(x) + s_1(x)g(x)] + [r_2(x)f(x) + s_2(x)g(x)] = [r_1(x) + r_2(x)]f(x) + [s_1(x) + s_2(x)]g(x)$$

shows that N is closed under addition. The equation

$$[r(x)f(x) + s(x)(g(x)]h(x) = h(x)[r(x)f(x) + s(x)g(x)] = [h(x)r(x)]f(x) + [h(x)s(x)]g(x)$$

shows that N is closed under multiplication by any $h(x) \in F[x]$; in particular N is closed under multiplication. Now $0 = 0f(x) + 0g(x)$ and $-[r(x)f(x) + s(x)g(x)] = [-r(x)]f(x) + [-s(x)]g(x)$ are in N, so we see that N is an ideal.

Suppose now that $f(x)$ and $g(x)$ have different degrees and that $N \neq F[x]$. Suppose that $f(x)$ is irreducible. By Theorem 26.25, we know that then $\langle f(x) \rangle$ is a maximal ideal of $F[x]$. But clearly $\langle f(x) \rangle \subseteq N$. Because $N \neq F[x]$, we must have $\langle f(x) \rangle = N$. In particular $g(x) \in N$ so $g(x) = f(x)q(x)$. Because $f(x)$ and $g(x)$ have different degrees, we see that $g(x) = f(x)q(x)$ must be a factorization of $g(x)$ into polynomials of smaller degree than the degree of $g(x)$. Hence $g(x)$ is not irreducible.

33. Given that the Fundamental Theorem of Algebra holds, let N be the smallest ideal of $\mathbb{C}[x]$ containing r polynomials $f_1(x), f_2(x), \cdots, f_r(x)$. Because every ideal in $\mathbb{C}[x]$ is a principal ideal, we have $N = \langle h(x) \rangle$ for some polynomial $h(x) \in \mathbb{C}[x]$. Let $\alpha_1, \alpha_2, \cdots, \alpha_s$ be all the zeros in $\mathbb{C}$ of $h(x)$, and let α_i be a zero of multiplicity m_i. By the Fundamental Theorem of Algebra, $h(x)$ must factor into linear factors in $\mathbb{C}[x]$, so that

$$h(x) = c(x - \alpha_1)^{m_1}(x - \alpha_2)^{m_2} \cdots (x - \alpha_s)^{m_s}.$$

Note each α_i is a zero of every $f_j(x)$ because each $f_j(x)$ is a multiple of the generator $h(x)$ of N. Thus by hypothesis, each α_i is a zero of $g(x)$. The Fundamental Theorem of Algebra shows that $g(x) = k(x)(x - \alpha_1)(x - \alpha_2) \cdots (x - \alpha_s)$ for some polynomial $k(x) \in \mathbb{C}[x]$. Let m be the maximum of $m_1, m_2, \cdots, m_s$. Then $g(x)^m$ has each $(x - \alpha_i)^{m_i}$ as a factor, and thus has $h(x)$ as a factor, so $g(x)^m \in \langle h(x) \rangle = N$.

Conversely, let the Nullstellensatz for $\mathbb{C}[x]$ hold. Suppose that the Fundamental Theorem of Algebra does not hold, so that there exists a nonconstant polynomial $f_1(x)$ in $\mathbb{C}[x]$ having no zero in $\mathbb{C}$. Then every zero of $f_1(x)$ is also a zero of every polynomial in $\mathbb{C}[x]$, because there are no zeros of $f_1(x)$. By the Nullstellensatz for $\mathbb{C}[x]$, every element of $\mathbb{C}[x]$ has the property that some power of it is in $\langle f_1(x) \rangle$, so that some power of every polynomial in $\mathbb{C}[x]$ has $f_1(x)$ as a factor. This is certainly impossible, because $1 \in \mathbb{C}[x]$ and $f_1(x)$ is a nonconstant polynomial and thus is not a factor of $1^n = 1$ for any positive integer n. Thus there can be no such polynomial $f_1(x)$ in $\mathbb{C}[x]$, and the Fundamental Theorem of Algebra holds.

34. a. Let $a_1, a_2 \in A$ and $b_1, b_2 \in B$. Then $(a_1 + b_1) + (a_2 + b_2) = (a_1 + a_2) + (b_1 + b_2)$ because addition is commutative. This shows that $A + B$ is closed under addition. For $r \in R$, we know that $ra_1 \in A$ and $rb_1 \in B$, so $r(a_1 + b_1) = ra_1 + rb_1$ is in $A + B$. A similar argument with multiplication on the right shows that $(a_1 + b_1)r = a_1r + b_1r$ is in $A + B$. Thus $A + B$ is closed under multiplication on the left or right by elements of R, in particular, multiplication is closed on $A + B$. Because $0 = 0 + 0 \in A + B$ and $-(a_1 + b_1) = (-a_1) + (-b_1)$ is in $A + B$, we see that $A + B$ is an ideal.

b. Because $a + 0 = a$ is in $A + B$ and $0 + b = b$ is in $A + B$ for all $a \in A$ and $b \in B$, we see that $A \subseteq (A + B)$ and $B \subseteq (A + B)$.

35. a. It is clear that AB is closed under addition; [a sum of m products of the form a_ib_i] + [a sum of n products of the form a_jb_k] is a sum of $m + n$ prodcts of this form, and hence is in AB. Because A

and B are ideals, we see that $r(a_ib_i) = (ra_i)b_i$ and $(a_ib_i)r = a_i(b_ir)$ are again of the form a_jb_j. The distributive laws then show that each sum of products a_ib_i when multiplied on the left or right by $r \in R$ produces again a sum of such products. Thus AB is closed under multiplication by elements of R, and hence is closed itself under multiplication. Because $0 = 00$ and $-(a_ib_i) = (-a_i)(b_i)$ are in AB, we see that AB is indeed an ideal.

b. Regarding a_ib_i as a_i in A multiplied on the right by an element b_i of R, we see that a_ib_i is in the ideal A. Regarding a_ib_i as b_i in B multiplied on the left by an element a_i of R, we see that a_ib_i is in B, so $a_ib_i \in A \cap B$. Because A and B are closed under addition, we see that any element of AB is contained in both A and B, so $AB \subseteq A \cap B$.

36. Let $x, y \in A : B$, and let $b \in B$. Then $xb \in A$ and $yb \in A$ for all $b \in B$, so $(x + y)b = xb + yb$ is in A for all $b \in B$, because A is closed under addition. Thus $A : B$ is closed under addition.

Turning to multiplication, let $r \in R$. We want to show that xr and rx are in $A : B$, that is, that $(xr)b$ and $(rx)b$ are in A for all $b \in B$. Because multiplication is commutative by hypothesis, it suffices to show that xbr is in A for all $b \in B$. But because $x \in A : B$, we know that $xb \in A$ and A is an ideal, so $(xb)r \in A$. Thus $A : B$ is closed under multiplication by elements in R; in particular, it is itself closed under multipliction.

Because $0b = 0$ and $0 \in A$, we see that $0 \in A : B$. Because $(-x)b = -(xb)$ and $xb \in A$ implies $(-xb) \in A$, we see that $A : B$ contains the additive identity and additive inverse of each of its elements.

37. Clearly S is closed under addition, contains the zero matrix, and contains the additive inverse of each of its elements. The computation

$$\begin{bmatrix} a & b \\ 0 & 0 \end{bmatrix}\begin{bmatrix} c & d \\ 0 & 0 \end{bmatrix} = \begin{bmatrix} ac & ad \\ 0 & 0 \end{bmatrix}$$

shows that S is closed under multiplication, so it is a subring of $M_2(F)$. The computations

$$\begin{bmatrix} 0 & 1 \\ 1 & 0 \end{bmatrix}\begin{bmatrix} a & b \\ 0 & 0 \end{bmatrix} = \begin{bmatrix} 0 & 0 \\ a & b \end{bmatrix}$$

and

$$\begin{bmatrix} a & b \\ 0 & 0 \end{bmatrix}\begin{bmatrix} c & d \\ e & f \end{bmatrix} = \begin{bmatrix} ac + be & ad + bf \\ 0 & 0 \end{bmatrix}$$

show that S is not closed under left multiplication by elements of $M_2(F)$, but is closed under right multiplication by those elements. Thus S is a right deal, but not a left ideal, of $M_2(F)$.

38. Let $R = M_2(\mathbb{Z}_2)$. The computations

$$\begin{bmatrix} 0 & 1 \\ 1 & 0 \end{bmatrix}\begin{bmatrix} a & b \\ c & d \end{bmatrix} = \begin{bmatrix} c & d \\ a & b \end{bmatrix} \quad \text{and} \quad \begin{bmatrix} a & b \\ c & d \end{bmatrix}\begin{bmatrix} 0 & 1 \\ 1 & 0 \end{bmatrix} = \begin{bmatrix} b & a \\ d & c \end{bmatrix}$$

show that for every matrix in an ideal N of R, the matrix obtained by interchanging its rows and the matrix obtained by interchanging its columns are again in N. Thus if N contains any one of the four matrices having 1 for one entry and 0 for all the others, then N contains all four such matrices, and hence all nonzero matrices because any matrix in R is a sum of such matrices and N is closed under addition. By interchanging rows and columns, every nonzero matix with at least two nonzero entries can be brought to one of the following forms:

two zero entries: $\begin{bmatrix} 1 & 1 \\ 0 & 0 \end{bmatrix}$, $\begin{bmatrix} 1 & 0 \\ 1 & 0 \end{bmatrix}$, $\begin{bmatrix} 1 & 0 \\ 0 & 1 \end{bmatrix}$

one zero entry: $\begin{bmatrix} 0 & 1 \\ 1 & 1 \end{bmatrix}$ no zero entry: $\begin{bmatrix} 1 & 1 \\ 1 & 1 \end{bmatrix}$.

The following computations then show that every nontrivial ideal of R must contain one of the four matrices with only one nonzero entry, and hence must be all of R:

$$\begin{bmatrix} 1 & 1 \\ 0 & 0 \end{bmatrix}\begin{bmatrix} 1 & 0 \\ 0 & 0 \end{bmatrix} = \begin{bmatrix} 1 & 0 \\ 0 & 0 \end{bmatrix}, \quad \begin{bmatrix} 1 & 0 \\ 0 & 0 \end{bmatrix}\begin{bmatrix} 1 & 0 \\ 1 & 0 \end{bmatrix} = \begin{bmatrix} 1 & 0 \\ 0 & 0 \end{bmatrix},$$

$$\begin{bmatrix} 1 & 0 \\ 0 & 0 \end{bmatrix}\begin{bmatrix} 1 & 0 \\ 0 & 1 \end{bmatrix} = \begin{bmatrix} 1 & 0 \\ 0 & 0 \end{bmatrix}, \quad \begin{bmatrix} 1 & 0 \\ 0 & 0 \end{bmatrix}\begin{bmatrix} 0 & 1 \\ 1 & 1 \end{bmatrix} = \begin{bmatrix} 0 & 1 \\ 0 & 0 \end{bmatrix},$$

$$\begin{bmatrix} 1 & 0 \\ 0 & 0 \end{bmatrix}\begin{bmatrix} 1 & 1 \\ 1 & 1 \end{bmatrix}\begin{bmatrix} 1 & 0 \\ 0 & 0 \end{bmatrix} = \begin{bmatrix} 1 & 0 \\ 0 & 0 \end{bmatrix}.$$

28. Gröbner Bases for Ideals

1. $-3x^3 + 7x^2y^2z - 5x^2yz^3 + 2xy^3z^5$

2. $-4x + 5y^3z^3 + 3y^2z^5 - 8z^7$

3. $2x^2yz^2 - 2xy^2z^2 - 7x + 3y + 10z^3$

4. $-8xy - 4xz + 3yz^3 + 2yz + 38$

5. $2z^5y^3x - 5z^3yx^2 + 7zy^2x^2 - 3x^3$

6. $-8z^7 + 3z^5y^2 + 5z^3y^3 - 4x$

7. $10z^3 - 2z^2y^2x + 2z^2yx^2 + 3y - 7x$

8. $3z^3y + 2zy - 4zx - 8yx + 38$

9. (See the answer in the text.)

10. $2xy^3z^5 - 5x^2yz^3 + 7x^2y^2z - 3x^3$

11. $3y^2z^5 - 8z^7 + 5y^3z^3 - 4x$

12. $2x^2yz^2 - 2xy^2z^2 + 10z^3 - 7x + 3y$

13. $3yz^3 - 8xy - 4xz + 2yz + 38$

14. We write the given ideal as $\langle x^2y + 4xy, xy^2 - 2x, xy - y^2 \rangle$ so that the polynomials are listed in decreasing order; the maximum order term is x^2y. Multiplying the third by $-x$ and adding to the first (or dividing the first by the third), we write the ideal as $\langle xy^2 + 4xy, xy^2 - 2x, xy - y^2 \rangle$ with maximum order term $xy^2 < x^2y$.

15. We write the given ideal as $\langle xy + y^3, x - y^4, y^3 + z \rangle$ so that the polynomials are listed in decreasing order; the maximum order term is xy. Multiplying the second by $-y$ and adding to the first (or dividing the first by the second), we write the ideal as $\langle y^5 + y^3, x - y^4, y^3 + z \rangle = \langle x - y^4, y^5 + y^3, y^3 + z \rangle$ with maximum order term $x < xy$.

16. We write the given ideal as $\langle x^3 + y^2z^3, x^2yz^3 + 4, xyz - 3z^3 \rangle$ so that the polynomials are listed in decreasing order; the maximum order term is x^3. Because the leading terms of the second and third polynomials do not divide x^3, we cannot perform a single-step division algorithm reduction that gives a basis with a smaller maximum term order.

17. We write the given ideal as $\langle y^3z^2 - 2z, y^2z^3 + 3, y^2z^2 + 3 \rangle$ so that the polynomials are listed in decreasing order; the maximum order term is y^3z^2. Multiplying the third by $-y$ and adding to the first (or dividing the first by the third), we write the ideal as $\langle -3y - 2z, y^2z^3 + 3, x^2z^2 + 3 \rangle = \langle y^2z^3 + 3, y^2z^2 + 3, -3y - 2z \rangle$ with maximum order term $y^2z^3 < y^3z^2$.

18. Starting with the given basis and adding (-2)(1st) to the 2nd, and adding (-1)(1st) to the 3rd yields

$$\langle w + x - y + 4z - 3, -x + 3y - 10z + 10, 2x - 2y - 3z - 2 \rangle.$$

Adding (2)(2nd) to the 3rd yields

$$\langle w + x - y + 4z - 3, -x + 3y - 10z + 10, 4y - 23z + 18 \rangle$$

Thus $\{w + x - y + 4z - 3, -x + 3y - 10z + 10, 4y - 23z + 18\}$ is a Gröbner basis.

19. Starting with the given basis and adding (-2)(1st) to the 2nd and adding (-1)(1st) to the 3rd yields

$$\langle w - 4x + 3y - z + 2, 6x - 5y + 1, -6x + 5y - 3 \rangle.$$

Adding the 2nd to the 3rd yields

$$\langle w - 4x + 3y - z + 2, 6x - 5y + 1, -2 \rangle.$$

Because the ideal contains a unit, -2, we know that the ideal is equal to $\mathbb{R}[w, x, y, z] = \langle -2 \rangle = \langle 1 \rangle$. Thus any set $\{a\}$, where $a \neq 0$ and $a \in \mathbb{R}$, is a Gröbner basis. (The corresponding algebraic variety is $\varnothing$.)

20. Because every ideal in $\mathbb{R}[x]$ is principal, a Gröbner basis will consist of a polynomial of minimum degree that has the form $f(x)$(1st) $+ g(x)$(2nd). Now when we add $(-x)$(2nd) to the 1st, we obtain $\langle x^2 - 4, x^3 + x^2 - 4x - 4 \rangle$. Adding $(-x)$(1st) to the 2nd, we obtain $\langle x^2 - 4, x^2 - 4 \rangle = \langle x^2 - 4 \rangle$. Thus $\{x^2 - 4\}$ is a Gröbner basis.

21. Because every ideal in $\mathbb{R}[x]$ is principal, a Gröbner basis will consist of a polynomial of minimum degree that has the form $f(x)$(1st) $+ g(x)$(2nd) $+ h(x)$(3rd). Adding $(-x)$(2nd) to the 1st and adding (-1)(2nd) to the 3rd, we obtain

$$\langle -3x^3 + 9x^2 - 6x, x^3 - x^2 - 4x + 4, x^2 + x - 2 \rangle.$$

Adding $(3x)$(3rd) to the 1st and adding $(-x)$(3rd) to the 2nd yields

$$\langle 12x^2 - 12x, -2x^2 - 2x + 4, x^2 + x - 2 \rangle.$$

Adding (-12)(3rd) to the 1st and adding (2)(3rd) to the 2nd yields

$$\langle -24x + 24, 0, x^2 + x - 2 \rangle = \langle x - 1, (x - 1)(x + 2) \rangle = \langle x - 1 \rangle.$$

Thus $\{x - 1\}$ is a Gröbner basis.

22. Because every ideal in $\mathbb{R}[x]$ is principal, a Gröbner basis will consist of a polynomial of minimum degree that has the form $f(x)$(1st) $+ g(x)$(2nd). Now when we add $(-x^2)$(2nd) to the 1st, we obtain $\langle x^4 - x^3 + 2x^2 + 2x - 5, x^3 - x^2 + x - 1 \rangle$. Adding $(-x)$(2nd) to the 1st yields $\langle x^2 + 3x - 5, x^3 - x^2 + x - 1 \rangle$. Adding $(-x)$(1st) to the 2nd yields $\langle x^2 + 3x - 5, -4x^2 + 6x - 1 \rangle$. Adding (4)(1st) to the 2nd yields $\langle x^2 + 3x - 5, 18x - 21 \rangle$. Adding $(-x/18)$(2nd) to the 1st, we obtain $\langle \frac{25}{6}x - 5, 18x - 21 \rangle = \langle \frac{5}{6}x - 1, 6x - 7 \rangle$. Adding $(-36/5)$(1st) to the 2nd yields $\langle \frac{5}{6}x - 1, \frac{1}{5} \rangle = \langle 1 \rangle$. Thus $\{1\}$ is a Gröbner basis. (The corresponding algebraic variety is $\varnothing$.)

23. Adding $(-x)$(2nd) to the 1st, we get $\langle -2xy + 8x - 2, xy + 2y - 9 \rangle$. Adding (2)(2nd) to the 1st yields $\langle 8x + 4y - 20, xy + 2y - 9 \rangle = \langle 2x + y - 5, xy + 2y - 9 \rangle$. Adding $(-y/2)$(1st) to the 2nd yields $\langle 2x + y - 5, -\frac{1}{2}y^2 + \frac{9}{2}y - 9 \rangle = \langle 2x + y - 5, y^2 - 9y + 18 \rangle = \langle g_1, g_2 \rangle$.

We now proceed to test for a Gröbner basis according to Theorem 28.12. Maximum term degree can't be reduced by the division algorithm. Form $S(g_1, g_2) = (y^2)(1st) - (2x)(2nd) = 18xy - 36x + y^3 - 5y^2$. This is of greater term order than either g_2 or g_1. We see if this can be reduced to zero using g_1 and g_2, that is, repeatedly using the division algorithm on remainders with just g_1 or g_2 as divisors. We have

$$\begin{aligned} S(g_1, g_2) - 9y(g_1) &= 18xy - 36x + y^3 - 5y^2 - 9y(2x + y - 5) \\ &= -36x + y^3 - 14y^2 + 45y. \end{aligned}$$

We add $18g_1 = 36x + 18y - 90$ to this and obtain

$$y^3 - 14y^2 + 63y - 90.$$

We add $(-y)g_2 = -y^3 + 9y^2 - 18y$ to this and obtain

$$-5y^2 + 45y - 90.$$

Finally, adding $5g_2 = 5y^2 - 45y + 90$ to this we obtain 0. Thus by Theorem 18.12, we see that $\{g_1, g_2\} = \{2x + y - 5, y^2 - 9y + 18\}$ is a Gröbner basis.

Because $y^2 - 9y + 18 = (y - 6)(y - 3)$, any point on the corresponding variety has y-coordinate 3 or 6. Requiring that the point be a zero of $2x + y - 5$, we find that the variety is $\{(1,3), (-\frac{1}{2},6)\}$.

24. Let $N = \langle x^2y + x, xy^2 - y \rangle = \langle g_1, g_2 \rangle$. Maximum term degree in the basis $\{g_1, g_2\}$ cannot be reduced further by using the division algorithm. We compute $S(g_1, g_2) = yg_1 - xg_2 = y(x^2y + x) - x(xy^2 - y) = 2xy$, and proceed to reduce the basis $\{x^2y + x, xy^2 - y, xy\}$ using the division algorithm. Adding $(-x)$(3rd) to the 1st and adding $(-y)$(3rd) to the 2nd, we obtain $N = \langle xy, x, y \rangle = \langle x, y \rangle$. Thus $\{x, y\}$ is a Gröbner basis and the corresponding algebraic variety is the origin, $\{(0,0)\}$.

25. Let $N = \langle x^2y + x + 1, xy^2 + y - 1 \rangle = \langle g_1, g_2 \rangle$. Maximum term degree in the basis $\{g_1, g_2\}$ cannot be reduced further by using the division algorithm. We therefore compute $S(g_1, g_2) = y(x^2y + x + 1) - x(xy^2 + y - 1) = x + y$, and proceed to reduce the basis $\{x^2y + x + 1, xy^2 + y - 1, x + y\}$ using the division algorithm. Adding $(-xy)$(3rd) to the 1st, we obtain $N = \langle -xy^2 + x + 1, xy^2 + y - 1, x + y \rangle$. Adding the 2nd to the 1st, we have $N = \langle xy^2 + y - 1, x + y \rangle$. Adding $(-y^2)$(2nd) to the 1st, we obtain $N = \langle x + y, -y^3 + y - 1 \rangle = \langle h_1, h_2 \rangle$ where maximum term degree cannot be reduced further using the division algorithm.

While we can determine the algebraic variety easily now, for illustration, we verify that we do have a Gröbner basis using Theorem 28.12. Now $S(h_1, h_2) = y^3h_1 + xh_2 - xy - x + y^4 - x(y - 1) + y^4$, and we test if it can be reduced to zero using the division algorithm with h_1 and h_2 as divisors. We obtain $(1 - y)h_1 + x(y - 1) + y^4 = (1 - y)(x + y) + x(y - 1) + y^4 = y^4 - y^2 + y$. Adding yh_2 to this yields 0, so by Theorem 28.12, we see that $\{h_1, h_2\} = \{x + y, -y^3 + y - 1\}$ is a Gröbner basis for N.

Using our calculator, we find that $y^3 - y + 1$ has one real zero which is approximately -1.3247, so $V(N) = \{a, -a\}$ for $a \approx 1.3247$.

26. Let $N = \langle x^2y + xy^2, xy - x \rangle$. Adding $(-x)$(2nd) to the 1st, we discover that $N = \langle x^2 + xy^2, xy - x \rangle = \langle g_1, g_2 \rangle$ where maximum term degree cannot be reduced further using the division algorithm. We compute $S(g_1, g_2) = y(x^2 + xy^2) - x(xy - x) = x^2 + xy^3$, and test if it can be reduced to zero using the division algorithm repeatedly with g_1 and g_2 as divisors. We find that $(x^2 + xy^3) + (-1)g_1 = (x^2 + xy^3) - (x^2 + xy^2) = xy^3 - xy^2$. Adding $(-y^2)g_2 = (-y^2)(xy - x)$ to $xy^3 - xy^2$ yields 0, so $\{x^2 + xy^2, xy - x\}$ is a Gröbner basis for N by Theorem 28.12. We can obtain a slightly simpler Gröbner basis by adding $(-y)$(2nd) to the 1st, which yields $\{x^2 + xy, xy - x\}$. Then adding (-1)(2nd) to the 1st gives us $\{x^2 + x, xy - x\}$ as Gröbner basis. Note that the initial power products x^2 and xy of the basis polynomials remain the same.

Now $x^2 + x$ has 0 and -1 as zeros. We see that $(0, a)$ is a zero of $xy - x$ for all $a \in \mathbb{R}$, but $(-1, b)$ is a zero of $xy - x$ only if $b = 1$. Thus $V(N) = \{(-1, 1), (0, a) \mid a \in \mathbb{R}\}$.

27. T F T T T T T T F F (The answer T to Part(**h**) assumes that the student has had a course in linear algebra where matrix reduction was used to solve linear systems.)

28. Let $P_i = xy$ and $P_j = x^2$ in lex with $y < x$. Then $xy < x^2$ but xy does not divide x^2.

29. *Additive closure:* With $c_i, d_i \in R$ for $i = 1, 2, \cdots, r$, we have

$$(c_1 f_1 + c_2 f_2 + \cdots + c_r f_r) + (d_1 f_1 + d_2 f_2 + \cdots, + d_r f_r) = (c_1 + d_1)f_1 + (c_2 + d_2)f_2 + \cdots + (c_r f_r + d_r f_r)$$

and $(c_i + d_i)$ is in R for $i = 1, 2, \cdots, r$.

Additive identity: Set all $c_i = 0$

Additive inverses: Replace all c_i by $-c_i$.

Multiplicative property: For $a \in R$, $a(c_1 f_1 + c_2 f_2 + \cdots + c_r f_r) = (ac_1)f_1 + (ac_2)f_2 + \cdots + (ac_r)f_r$ and $ac_i \in R$ for $i = 1, 2, \cdots, r$.

30. Let $h(\mathbf{x})$ be a common divisor of $f(\mathbf{x})$ and $g(\mathbf{x})$, so that $f(\mathbf{x}) = q_1(\mathbf{x})h(\mathbf{x})$ and $g(\mathbf{x}) = q_2(\mathbf{x})h(\mathbf{x})$ for $q_1(\mathbf{x}), q_2(\mathbf{x})$ in $F[\mathbf{x}]$. Then

$$
\begin{aligned}
r(\mathbf{x}) = f(\mathbf{x}) - g(\mathbf{x})q(\mathbf{x}) &= q_1(\mathbf{x})h(\mathbf{x}) - q_2(\mathbf{x})h(\mathbf{x})q(\mathbf{x}) \\
&= [q_1(\mathbf{x}) - q_2(\mathbf{x})q(\mathbf{x})]h(\mathbf{x}),
\end{aligned}
$$

so $h(\mathbf{x})$ divides $r(\mathbf{x})$ as well as $g(\mathbf{x})$. Thus divisors of both $f(\mathbf{x})$ and $g(\mathbf{x})$ also divide both $g(\mathbf{x})$ and $r(\mathbf{x})$.

Going the other way, suppose $k(\mathbf{x})$ divides both $g(\mathbf{x})$ and $r(\mathbf{x})$, so that $g(\mathbf{x}) = q_3(\mathbf{x})k(\mathbf{x})$ and $r(\mathbf{x}) = q_4(\mathbf{x})k(\mathbf{x})$ for $q_3(\mathbf{x}), q_4(\mathbf{x}) \in F[\mathbf{x}]$. Then

$$
\begin{aligned}
f(\mathbf{x}) = g(\mathbf{x})q(\mathbf{x}) + r(\mathbf{x}) &= q_3(\mathbf{x})k(\mathbf{x})q(\mathbf{x}) + q_4(\mathbf{x})k(\mathbf{x}) \\
&= [q_3(\mathbf{x})q(\mathbf{x}) + q_4(\mathbf{x})]k(\mathbf{x}),
\end{aligned}
$$

so $k(\mathbf{x})$ divides $f(\mathbf{x})$ as well as $g(\mathbf{x})$. Thus divisors of both $g(\mathbf{x})$ and $r(\mathbf{x})$ also divide both $g(\mathbf{x})$ and $f(\mathbf{x})$.

Thus the set of divisors of $f(\mathbf{x})$ and $g(\mathbf{x})$ is the same as the set of divisors of $g(\mathbf{x})$ and $r(\mathbf{x})$.

31. Let $N = \langle xy, y^2 - y \rangle = \langle g_1, g_2 \rangle$. By Theorem 28.12, we need to show that $S(g_1, g_2) = y(xy) - x(y^2 - y) = xy$ can be reduced to 0 using the division algorithm with just xy and $y^2 - y$ as divisors. Adding $(-1)g_1$ to xy, we immediately obtain 0, and we are done.

32. *Additive closure:* Let $f(\mathbf{x}), g(\mathbf{x}) \in I(S)$, so that $f(\mathbf{s}) = 0$ and $g(\mathbf{s}) = 0$ for all $\mathbf{s} \in S$. Applying the evaluation homomorphism $\phi_\mathbf{s}$, we get $\phi_\mathbf{s}(f(\mathbf{x}) + g(\mathbf{x})) = \phi_\mathbf{s}(f(\mathbf{x})) + \phi_\mathbf{s}(g(\mathbf{x})) = f(\mathbf{s}) + g(\mathbf{s}) = 0 + 0 = 0$ for all $\mathbf{s} \in S$, so $(f + g) \in I(S)$.

Additive identity: $\phi_\mathbf{s}(0) = 0$ for all $\mathbf{s} \in S$, so $0 \in I(S)$.

Additive inverses: $f(\mathbf{x}) \in I(S)$ implies $\phi_\mathbf{s}(-f(\mathbf{x})) = -\phi_\mathbf{s}(f(\mathbf{x})) = -f(\mathbf{s}) = -0 = 0$ for all $\mathbf{s} \in S$, so $-f(\mathbf{x}) \in I(S)$.

Multiplicative property: Let $f(\mathbf{x}) \in I(S)$ and $h(\mathbf{x}) \in F[\mathbf{x}]$. Then $\phi_\mathbf{s}(h(\mathbf{x})f(\mathbf{x})) = \phi_\mathbf{s}(h(\mathbf{x}))\phi_\mathbf{s}(f(\mathbf{x})) = h(\mathbf{s})f(\mathbf{s}) = h(\mathbf{s})(0) = 0$ for all $\mathbf{s} \in S$, so $h(\mathbf{x})f(\mathbf{x}) \in I(S)$.

33. By Definition 28.1, $V(I(S))$ consists of all common zeros of elements of $I(S)$. By definition of $I(S)$ in Exercise 32, every $f(\mathbf{x}) \in I(S)$ has every $\mathbf{s} \in S$ as a zero, so $\mathbf{s} \in V(I(S))$ for all $\mathbf{s} \in S$. Thus $S \subseteq V(I(S))$.

34. Let $n = 1, F = \mathbb{Q}$ and $S = \mathbb{Z} \subseteq \mathbb{Q}$ in Exercise 32. Then $I(S) = \{0\}$, for a nonzero element of $\mathbb{Q}[x]$ can have only a finite number of zeros in $\mathbb{Q}$. However, $V(I(S)) = V(\{0\}) = \mathbb{Q} \neq \mathbb{Z}$.

35. Let $f(\mathbf{x}) \in N$. Now $V(N)$ consists of all $\mathbf{s} \in F^n$ such that $h(\mathbf{s}) = 0$ for all $h(\mathbf{x}) \in N$. Thus $\mathbf{s} \in V(N)$ implies that $f(\mathbf{s}) = 0$ because $f(\mathbf{x})$ is an example of an $h(\mathbf{x}) \in N$. Thus $f(\mathbf{s}) = 0$ for all $\mathbf{s} \in V(N)$. By definition of $I(S)$ in Exercise 32, this means that $f(\mathbf{x}) \in I(V(N))$. Thus $N \subseteq I(V(N))$.

36. Let $N = \langle x^2, y^2 \rangle$. Then $V(N) = \{(0,0)\}$, and $I(V(N)) = \langle x, y \rangle \neq N$.

29. Introduction to Extension Fields

1. Let $\alpha = 1 + \sqrt{2}$. Then $(\alpha - 1)^2 = 2$ so $\alpha^2 - 2\alpha - 1 = 0$. Thus α is a zero of $x^2 - 2x - 1$ in $\mathbb{Q}[x]$.

2. Let $\alpha = \sqrt{2} + \sqrt{3}$. Then $\alpha^2 = 2 + 2\sqrt{6} + 3$ so $\alpha^2 - 5 = 2\sqrt{6}$. Squaring again, we obtain $\alpha^4 - 10\alpha^2 + 1 = 0$, so α is a zero of $x^4 - 10x^2 + 1$ in $\mathbb{Q}[x]$.

3. Let $\alpha = 1 + i$. Then $(\alpha - 1)^2 = -1$, so $\alpha^2 - 2\alpha + 2 = 0$. Thus α is a zero of $x^2 - 2x + 2$ in $\mathbb{Q}[x]$

4. Let $\alpha = \sqrt{1 + \sqrt[3]{2}}$. Then $\alpha^2 = 1 + \sqrt[3]{2}$ so $\alpha^2 - 1 = \sqrt[3]{2}$. Cubing, we obtain $\alpha^6 - 3\alpha^4 + 3\alpha^2 - 3 = 0$, so α is a zero of $x^6 - 3x^4 + 3x^2 - 3$ in $\mathbb{Q}[x]$.

5. Let $\alpha = \sqrt{\sqrt[3]{2} - i}$. Then $\alpha^2 + i = \sqrt[3]{2}$. Cubing, we obtain $\alpha^6 + 3\alpha^4 i - 3\alpha^2 - i = 2$, so $\alpha^6 - 3\alpha^2 - 2 = (1 - 3\alpha^4)i$. Squaring, we obtain $\alpha^{12} - 6\alpha^8 - 4\alpha^6 + 9\alpha^4 + 12\alpha^2 + 4 = -1 + 6\alpha^4 - 9\alpha^8$. Thus $\alpha^{12} + 3\alpha^8 - 4\alpha^6 + 3\alpha^4 + 12\alpha^2 + 5 = 0$, so α is a zero of $x^{12} + 3x^8 - 4x^6 + 3x^4 + 12x^2 + 5$ in $\mathbb{Q}[x]$.

6. Let $\alpha = \sqrt{3 - \sqrt{6}}$. Then $\alpha^2 - 3 = -\sqrt{6}$. Squaring again, we obtain $\alpha^4 - 6\alpha^2 + 3 = 0$, so α is a zero of $f(x) = x^4 - 6x^2 + 3$ in $\mathbb{Q}[x]$. Now $f(x)$ is monic and is irreducible by the Eisenstein condition with $p = 3$. Thus $\deg(\alpha, \mathbb{Q}) = 4$ and $\text{irr}(\alpha, \mathbb{Q}) = f(x)$.

7. Let $\alpha = \sqrt{\frac{1}{3} + \sqrt{7}}$. Then $\alpha^2 - \frac{1}{3} = \sqrt{7}$. Squaring again, we obtain $\alpha^4 - \frac{2}{3}\alpha^2 - \frac{62}{9} = 0$, or $9\alpha^4 - 6\alpha^2 - 62 = 0$. Let $f(x) = 9x^4 - 6x^2 - 62$. Then $f(x)$ is irreducible by the Eisenstein condition with $p = 2$. Thus $\deg(\alpha, \mathbb{Q}) = 4$ and $\text{irr}(\alpha, \mathbb{Q}) = \frac{1}{9}f(x)$.

8. Let $\alpha = \sqrt{2} + i$. Then $\alpha^2 = 2 + 2\sqrt{2}i - 1$ so $\alpha^2 - 1 = 2\sqrt{2}i$. Squaring again, we obtain $\alpha^4 - 2\alpha^2 + 1 = -8$, so $\alpha^4 - 2\alpha^2 + 9 = 0$. Let $f(x) = x^4 - 2x^2 + 9$. One can show that $f(x)$ is irreducible by the technique of Example 23.14. Thus $\deg(\alpha, \mathbb{Q}) = 4$ and $\text{irr}(\alpha, \mathbb{Q}) = f(x)$.

9. We see that i is algebraic over $\mathbb{Q}$ because it is a zero of $x^2 + 1$ in $\mathbb{Q}[x]$; $\deg(i, \mathbb{Q}) = 2$.

10. Let $\alpha = 1 + i$. Then $\alpha - 1 = i$ so $\alpha^2 - 2\alpha + 2 = 0$. Because α is not in $\mathbb{R}$, we see that α is algebraic of degree 2 over $\mathbb{R}$.

11. The text told us that π is transcendental over $\mathbb{Q}$, behaving just like an indeterminant. Thus $\sqrt{\pi}$ is also transcendental over $\mathbb{Q}$. [It is easy to see that if a polynomial expression in $\sqrt{\pi}$ is zero, then a polynomial in π is zero. Namely, starting with $f(\sqrt{\pi}) = 0$, move all odd-degree terms to the right-hand side, factor $\sqrt{\pi}$ out from them, and then square both sides.]

12. Because $\sqrt{\pi} \in \mathbb{R}$, it is algebraic over $\mathbb{R}$ of degree 1. It is a zero of $x - \sqrt{\pi}$ in $\mathbb{R}[x]$.

13. Now $\sqrt{\pi}$ is algebraic over $\mathbb{Q}(\pi)$ of degree 2. It is not in $\mathbb{Q}(\pi)$. Remember that π behaves just like an indeterminant x over $\mathbb{Q}$. Note that $\sqrt{x}$ is not in $\mathbb{Q}(x)$, but it is a zero of $y^2 - x$ in $(\mathbb{Q}(x))[y]$.

14. Now π^2 is transcendental over $\mathbb{Q}$ for the text told us that π is transcendental over $\mathbb{Q}$, and a polynomial expression in π^2 equal to zero and having rational coefficients can be viewed as a polynomial expression in π equal to zero with coefficients in $\mathbb{Q}$ and having all terms of even degree.

15. Now $\pi^2 \in \mathbb{Q}(\pi)$ so it is algebraic over $\mathbb{Q}(\pi)$ of degree 1. It is a zero of $x - \pi^2$ in $(\mathbb{Q}(\pi))[x]$.

16. Now π^2 is algebraic over $\mathbb{Q}(\pi^3)$ of degree 3. It is not in $\mathbb{Q}(\pi^3)$, (note that x^2 is not a polynomial in x^3.) but it is a zero of $x^3 - (\pi^3)^2 = x^3 - \pi^6$ in $(\mathbb{Q}(\pi^3))[x]$.

17. We perform a division.

$$
\begin{array}{r}
x + (1 + \alpha) \\
\hline
x - \alpha \, {\big|} \, x^2 + x + 1 \\
\end{array}
$$

$$
\begin{array}{l}
x^2 - \alpha x \\
\hline
\quad (1+\alpha)x \\
\quad (1+\alpha)x - \alpha^2 - \alpha \\
\hline
\qquad\qquad \alpha^2 + \alpha + 1 = 2 \cdot (\alpha + 1) = 0.
\end{array}
$$

We have $x^2 + x + 1 = (x - \alpha)(x + \alpha + 1)$.

18. a. Let $f(x) = x^2 + 1$. Then $f(0) = 1$, $f(1) = 2$, and $f(-1) = 2$ so $f(x)$ is a cubic with no zeros in $\mathbb{Z}_3$ and thus is irreducible in $\mathbb{Z}_3[x]$.

b.

+	0	1	2	α	2α	$1+\alpha$	$1+2\alpha$	$2+\alpha$	$2+2\alpha$
0	0	1	2	α	2α	$1+\alpha$	$1+2\alpha$	$2+\alpha$	$2+2\alpha$
1	1	2	0	$1+\alpha$	$1+2\alpha$	$2+\alpha$	$2+2\alpha$	α	2α
2	2	0	1	$2+\alpha$	$2+2\alpha$	α	2α	$1+\alpha$	$1+2\alpha$
α	α	$1+\alpha$	$2+\alpha$	2α	0	$1+2\alpha$	1	$2+2\alpha$	2
2α	2α	$1+2\alpha$	$2+2\alpha$	0	α	1	$1+\alpha$	2	$2+\alpha$
$1+\alpha$	$1+\alpha$	$2+\alpha$	α	$1+2\alpha$	1	$2+2\alpha$	2	2α	0
$1+2\alpha$	$1+2\alpha$	$2+2\alpha$	2α	1	$1+\alpha$	2	$2+\alpha$	0	α
$2+\alpha$	$2+\alpha$	α	$1+\alpha$	$2+2\alpha$	2	2α	0	$1+2\alpha$	1
$2+2\alpha$	$2+2\alpha$	2α	$1+2\alpha$	2	$2+\alpha$	0	α	1	$1+\alpha$

·	0	1	2	α	2α	$1+\alpha$	$1+2\alpha$	$2+\alpha$	$2+2\alpha$
0	0	0	0	0	0	0	0	0	0
1	0	1	2	α	2α	$1+\alpha$	$1+2\alpha$	$2+\alpha$	$2+2\alpha$
2	0	2	1	2α	α	$2+2\alpha$	$2+\alpha$	$1+2\alpha$	$1+\alpha$
α	0	α	2α	2	1	$2+\alpha$	$1+\alpha$	$2+2\alpha$	$1+2\alpha$
2α	0	2α	α	1	2	$1+2\alpha$	$2+2\alpha$	$1+\alpha$	$2+\alpha$
$1+\alpha$	0	$1+\alpha$	$2+2\alpha$	$2+\alpha$	$1+2\alpha$	2α	2	1	α
$1+2\alpha$	0	$1+2\alpha$	$2+\alpha$	$1+\alpha$	$2+2\alpha$	2	α	2α	1
$2+\alpha$	0	$2+\alpha$	$1+2\alpha$	$2+2\alpha$	$1+\alpha$	1	2α	α	2
$2+2\alpha$	0	$2+2\alpha$	$1+\alpha$	$1+2\alpha$	$2+\alpha$	α	1	2	2α

19. The definition is incorrect. The polynomial must be nonzero and in $F[x]$.

An element α of an extension field E of a field F is **algebraic over** F if and only if α is a zero of some nonzero polynomial in $F[x]$.

20. The definition is incorrect. The polynomial must be nonzero.

An element β of an extension field E of a field F is **transcendental over** F if and only if β is not a zero of any nonzero polynomial in $F[x]$.

21. The definition is incorrect. Only the coefficient of the leading term need be 1.

A **monic polynomial** in $F[x]$ is a nonzero polynomial having 1 as the coefficient in the term of highest degree.

22. The definition is incorrect. The subfields examined must contain F as well as α.

A field E is a **simple extension** of a subfield F if and only if there exists some $\alpha \in E$ such that no proper subfield of E contains both F and α.

23. T T T T F T F T F T

24. a. One such field is $\mathbb{Q}(\pi^3)$.

b. One such field is $\mathbb{Q}(e^{10})$.

25. a. Let $f(x) = x^3 + x^2 + 1$. Then $f(0) = 1$ and $f(1) = 1$ so $f(x)$ has no zeros in $\mathbb{Z}_2$ and is thus irreducible.

b. The long division uses the relations $\alpha^3 = \alpha^2 + 1$ and $-\alpha^3 - \alpha^2 = -\alpha^2 - 1 - \alpha^2 = -1$.

$$
\begin{array}{r}
x^2 + (1+\alpha) + (\alpha^2 + \alpha) \\
\hline
x - \alpha \,\big)\, x^3 \;+\; x^2 \qquad\quad +\quad 1 \\
\underline{x^3 \;-\; \alpha x^2} \\
(1+\alpha)x^2 \\
\underline{(1+\alpha)x^2 - (\alpha^2 + \alpha)x} \\
(\alpha^2 + \alpha)x + 1 \\
\underline{(\alpha^2 + \alpha)x - 1} \\
0
\end{array}
$$

Continuing, we try α^2 as a zero of $q(x) = x^2 + (1+\alpha)x + (\alpha^2 + \alpha)$. Substituting, we obtain

$$
\begin{aligned}
\alpha^4 + (1+\alpha)\alpha^2 + (\alpha^2 + \alpha) &= \alpha(\alpha^2 + 1) + \alpha^2 + (\alpha^2 + 1) + (\alpha^2 + \alpha) \\
&= (\alpha^2 + 1) + \alpha + \alpha^2 + (\alpha^2 + 1) + (\alpha^2 + \alpha) \\
&= 2 \cdot (\alpha^2 + 1) + 2 \cdot \alpha^2 + 2 \cdot \alpha = 0
\end{aligned}
$$

so α^2 is a zero of $q(x)$. We do another long division. This one involves the computation

$$
\begin{aligned}
\alpha^2(\alpha^2 + \alpha + 1) &= \alpha\alpha^3 + \alpha^3 + \alpha^2 \\
&= \alpha(\alpha^2 + 1) + (\alpha^2 + 1) + \alpha^2 \\
&= (\alpha^2 + 1) + \alpha + (\alpha^2 + 1) + \alpha^2 \\
&= \alpha^2 + \alpha.
\end{aligned}
$$

$$
\begin{array}{r}
x + (\alpha^2 + \alpha + 1) \\
x - \alpha^2 \overline{\smash{\big)}\, x^2 \ + \ (1 + \alpha)x + (\alpha^2 + \alpha)} \\
\underline{x^2 \ - \quad\ \alpha^2 x} \\
(\alpha^2 + \alpha + 1)x + (\alpha^2 + \alpha) \\
\underline{(\alpha^2 + \alpha + 1)x + (\alpha^2 + \alpha)} \\
0
\end{array}
$$

Thus in $(\mathbb{Z}_2)[x]$,

$$x^3 + x^2 + 1 = (x - \alpha)(x - \alpha^2)[x - (\alpha^2 + \alpha + 1)].$$

26. The group $\langle \mathbb{Z}_2(\alpha), + \rangle$ is abelian of order 8 with the property that $a + a = 0$ for all elements a in the group. Thus the group must be isomorphic to $\mathbb{Z}_2 \times \mathbb{Z}_2 \times \mathbb{Z}_2$. The group $\langle \mathbb{Z}_2(\alpha)^*, \cdot \rangle$ is abelian of order 7, and must be cyclic (both because it has prime order and because it is the multiplicative group of nonzero elements of a finite field) and is isomorphic to $\mathbb{Z}_7$.

27. It is the monic polynomial in $F[x]$ of *minimal* degree having α as a zero.

28. Take an irreducible factor $p(x)$ of $f(x)$, and form the field $E = F[x]/\langle p(x) \rangle$. If we identify each $a \in F$ with the coset $a + \langle p(x) \rangle$ in $F[x]/\langle p(x) \rangle$, then we can view E as an extension field of F. The coset $\alpha = x + \langle p(x) \rangle$ can be viewed as a zero in E of $p(x)$, and hence as a zero in E of $f(x)$.

29. Every element of $F(\beta)$ can be expressed as a quotient of polynomials in β with coefficients in F. Because α is algebraic over $F(\beta)$, there is a polynomial expression in α with coefficients in $F(\beta)$ which is equal to zero. By multiplying this equation by the polynomial in β which is the product of the denominators of the coefficients in this equation, we obtain a polynomial in α equal to zero and having as coefficients polynomials in β. Now a polynomial in α with coefficients that are polynomials in β can be formally rewritten as a polynomial in β with coefficients that are polynomials in α. [Recall that $(F[x])[y] \simeq (F[y])[x]$.] This polynomial expression is still zero, which shows that β is algebraic over $F(\alpha)$.

30. Theorem 29.18 shows that every element of $F(\alpha)$ can be uniquely expressed in the form

$$b_0 + b_1\alpha + b_2\alpha^2 + \cdots + b_{n-1}\alpha^{n-1}.$$

Because F has q elements, there are q choices for b_0, then q choices for b_1, etc. Thus there are q^n such expressions altogether. The *uniqueness* property shows that different expressions correspond to distinct elements of $F(\alpha)$, which must therefore have q^n elements.

31. **a.** Let $f(x) = x^3 + x^2 + 2$. Then $f(0) = 2, f(1) = 1$, and $f(-1) = 2$ so $f(x)$ has no zeros in $\mathbb{Z}_3$ and thus is irreducible over $\mathbb{Z}_3[x]$.

   b. Exercise 30 shows that the field $\mathbb{Z}_3[x]/\langle f(x) \rangle$, which can be viewed as an extension field of $\mathbb{Z}_3$ of degree 3, has $3^3 = 27$ elements.

32. a. If $p \neq 2$, then $1 \neq p - 1$ in $\mathbb{Z}_p$, but $1^2 = (p-1)^2$. Thus the squaring function mapping $\mathbb{Z}_p \to \mathbb{Z}_p$ is not one to one; in fact, its image can have at most $p - 1$ elements. Thus some element of $\mathbb{Z}_p$ is not a square if $p \neq 2$.

b. We saw in Example 29.19 that there is a finite field of four elements. Let p be an odd prime. By Part(**a**), there exists $a \in \mathbb{Z}_p$ such that $x^2 - a$ has no zeros in $\mathbb{Z}_p$. This means that $x^2 - a$ is irreducible in $\mathbb{Z}_p[x]$. Let α be a zero of $x^2 - a$ in an extension field of $\mathbb{Z}_p$. By Exercise 30, $\mathbb{Z}_p(\alpha)$ has p^2 elements.

33. Let $\beta \in F(\alpha)$. Then β is equal to a quotient $r(\alpha)/s(\alpha)$ of polynomials in α with coefficients in F. Suppose that $f(\beta) = 0$ where $f(x) \in F[x]$ and is of degree n. Multiplying the equation $f(\beta) = 0$ by $s(\alpha)^n$, we obtain a polynomial in α with coefficients in F which is equal to zero. But then, α is algebraic over F, which is contrary to hypothesis. Therefore there is no such nonzero polynomial expression $f(\beta) = 0$, that is, β is transcendantal over F.

34. We know that $x^3 - 2$ is irreducible in $\mathbb{Q}[x]$ by the Eisenstein condition with $p = 2$. Therefore $\sqrt[3]{2}$ is algebraic of degree 3 over $\mathbb{Q}$. By Theorem 29.18, the field $\mathbb{Q}(\sqrt[3]{2})$ consists of all elements of $\mathbb{R}$ of the form $a + b(\sqrt[3]{2}) + c(\sqrt[3]{2})^2$ for $a, b, c \in \mathbb{Q}$, and distinct values of $a, b,$ and c give distinct elements of $\mathbb{R}$. The set given in the problem consists of precisely these elements of $\mathbb{R}$, so the given set is the field $\mathbb{Q}(\sqrt[3]{2})$.

35. We keep using Theorem 29.18 and Exercise 30. Now the polynomial $x^3 + x + 1$ in $\mathbb{Z}_2[x]$ has no zeros in $\mathbb{Z}_2$ and is therefore irreducible in $\mathbb{Z}_2[x]$. If α is a zero of this polynomial in an extension field, then $\mathbb{Z}_2(\alpha)$ has $2^3 = 8$ elements by Exercise 30.

Similarly, let α be a zero of the irreducible polynomial $x^4 + x + 1$ in $\mathbb{Z}_2[x]$. Then $\mathbb{Z}_2(\alpha)$ has $2^4 = 16$ elements.

Finally, let α be a zero of the irreducible polynomial $x^2 - 2$ in $\mathbb{Z}_5[x]$. Then $\mathbb{Z}_5(\alpha)$ has $5^2 = 25$ elements.

36. Following the hint, we let F^* be the multiplicative group of nonzero elements of F. We are given that F is finite; suppose that F has m elements. Then F^* has $m - 1$ elements. Because the order of an element of a finite group divides the order of the group, we see that for all $a \in F^*$ we have $a^{m-1} = 1$. Thus every $a \in F^*$ is a zero of the polynomial $x^{m-1} - 1$. Of course, 0 is a zero of x. Thus every $\alpha \in F$ is algebraic over the prime field $\mathbb{Z}_p$ of F, for the polynomial $x^{m-1} - 1$ is in $\mathbb{Z}_p$ for all primes p.

37. Let E be a finite field with prime subfield $\mathbb{Z}_p$. If $E = \mathbb{Z}_p$, then the order of E is p and we are done. Otherwise, let $\alpha_1 \in E$ where $\alpha_1 \notin \mathbb{Z}_p$. Let $F_1 = \mathbb{Z}_p(\alpha_1)$. By Exercise 30, the field F_1 has order p^{n_1} where n_1 is the degree of α_1 over $\mathbb{Z}_p$. If $F_1 = E$, we are done, Otherwise, we find $\alpha_2 \in E$ where $\alpha_2 \notin F_1$, and form $F_2 = F_1(\alpha_2)$, obtaining a field of order $p^{n_1 n_2}$ where n_2 is the degree of α_2 over F_1. We continue this process, constructing fields F_i of order $p^{n_1 n_2 \cdots n_i}$. Because E is a finite field, this process must eventually terminate with a field $F_r = E$. Then E has order $p^{n_1 n_2 \cdots n_r}$ which is a power of p as asserted.

30. Vector Spaces

1. $\{(0, 1), (1, 0)\}, \quad \{(1, 1), (-1, 1)\}, \quad \text{and} \quad \{(2, 1), (1, 2)\}$

2. Suppose that $a(1, 1, 0) + b(1, 0, 1) + c(0, 1, 1) = (d, e, f)$. Then $a + b = d, a + c = e,$ and $b + c = f$. Subtracting the second equation from the first, we obtain $b - c = d - e$. Adding this to the last equation, we obtain $2b = f + d - e$, so $b = (f + d - e)/2$. Then $a = (d + e - f)/2$ and $c = (e + f - d)/2$.

This shows that the given vectors span $\mathbb{R}^3$. Setting $d = e = f = 0$, we see that we must then have $a = b = c = 0$ so the vectors are also independent, and hence are a basis for $\mathbb{R}^3$.

3. We claim the vectors are dependent, and thus cannot form a basis. If $a(-1, 1, 2) + b(2, -3, 1) + c(10, -14, 0) = (0, 0, 0)$, then

$$
\begin{array}{rrrrl}
-a & + \; 2b & + \; 10c & = & 0 \\
a & - \; 3b & - \; 14c & = & 0 \\
2a & + \; b & & = & 0.
\end{array}
$$

Adding the first two equations, we find that $-b - 4c = 0$. Adding twice the first equation to the last, we find that $5b + 20c = 0$, which is essentially the same equation. Let $c = 1$ so $b = -4$ and $a = 2$. We find that

$$2(-1, 1, 2) + (-4)(2, -3, 1) + 1(10, -14, 0) = (0, 0, 0),$$

so the vectors are indeed independent.

4. Because $\sqrt{2}$ is a zero of irreducible $x^2 - 2$ of degree 2, Theorem 30.23 shows that a basis is $\{1, \sqrt{2}\}$.

5. Because $\sqrt{2}$ is in $\mathbb{R}$ and is a zero of $x - \sqrt{2}$ of degree 1, Theorem 30.23 shows that a basis is $\{1\}$.

6. Because $\sqrt[3]{2}$ is a zero of irreducible $x^3 - 2$ of degree 3, by Theorem 30.23 a basis is $\{1, \sqrt[3]{2}, (\sqrt[3]{2})^2\}$.

7. Because $\mathbb{C} = \mathbb{R}(i)$ where i is a zero of irreducible $x^2 + 1$ of degree 2, Theorem 30.23 shows that a basis is $\{1, i\}$.

8. Because i is a zero of irreducible $x^2 + 1$ of degree 2, Theorem 30.23 shows that a basis is $\{1, i\}$.

9. Since $\sqrt[4]{2}$ is a zero of irreducible $x^4 - 2$ of degree 4, by Theorem 30.23 a basis is $\{1, \sqrt[4]{2}, \sqrt{2}, (\sqrt[4]{2})^3\}$.

10. Recall that α is a zero of $x^2 + x + 1$, so $\mathbb{Z}_2(\alpha)$ is a 2-dimensional vector space over $\mathbb{Z}_2$. Thus the three elements $1, 1 + \alpha$, and $(1 + \alpha)^2 = 1 + 2 \cdot \alpha + \alpha^2 = 1 + 1 + \alpha = \alpha$ must be independent. By inspection, we see that

$$1(1) + 1(1 + \alpha) + 1(1 + \alpha)^2 = 1 + (1 + \alpha) + \alpha = 0,$$

so α is a zero of $x^2 + x + 1$, which is thus not only irr$(\alpha, \mathbb{Z}_2)$ but is also irr$(1 + \alpha, \mathbb{Z}_2)$. Of course, we already knew this because the polynomial's other zero, bedsides α, must lie in $\mathbb{Z}_2(\alpha)$ since $x^2 + x + 1$ has to factor into linear factors there, and $1 + \alpha$ is the only possibility for the other zero. However, we wanted to present the technique given in our first argument.

11. The definition is incorrect. Delete the "uniquely".

The vectors in a subset S of a vector space V over a field F **span** V if and only if each $\beta \in V$ an be expressed as a linear combination of the vectors in S.

12. The definition is incorrect. Some coefficients in the linear combination must be nonzero.

The vectors in a subset S of a vector space V over a field F are **linearly independent over** F if and only if the zero vector cannot be expressed as a linear combination, having some coefficients nonzero, of vectors in S.

13. The definition is correct.

14. The definition is incorrect. Replace "dependent" by "independent".

A **basis** for a vector space V over a field F is a set of vectors in V that span V and are linearly independent.

15. T F T T F F F T T T

16. a. A **subspace** of the vector space V over F is a subset W of V that is closed under vector addition and under multiplication by scalars in F, and is itself a vector space over F under these two operations.

b. Let $\{W_i \mid i \in I\}$ be a collection of subspaces of V. Because $\langle W_i, + \rangle$ is an abelian group, Theorem 7.4 shows that $\bigcap_{i \in I} W_i$ is again an abelian group. Let $a \in F$ and let $\alpha \in \bigcap_{i \in I} W_i$. Then $\alpha \in W_i$ for each $i \in I$, so $a\alpha \in W_i$ for each $i \in I$ because each W_i is a vector space. Hence $a\alpha \in \bigcap_{i \in I} W_i$ so the intersection is closed under scalar multiplication. All the other axioms for a vector space (distributive laws, etc.) certainly hold in this intersection, because they hold for all elements in V.

17. a. Let S be a subset of a vector space V over a field F. The **subspace generated by** S is the intersection of all subspaces of V that contain S.

b. Clearly, the sum of two finite linear combinations of elements of S is again a finite linear combination of elements of S. Also, a scalar times a finite linear combination is again a finite linear combination:

$$a(b_1\alpha_1 + \cdots + b_n\alpha_n) = (ab_1)\alpha_n + \cdots + (ab_n)\alpha_n.$$

Because $0 = 0\alpha$ for $\alpha \in S$, we see that $0 \in V$ is a finite linear combination of elements of S. Multiplying by the scalar -1, we see that an additive inverse of such a linear combination is again a finite linear combination of elements of S. Therefore the set of all finite linear combinations of elements of S is a vector space, and is clearly the smallest vector space that contains S.

This result is analogous to the case of Theorem 7.6 for abelian groups, although we do have to check here that scalar multiplication is closed.

18. The direct sum of vector spaces $V_1, V_2, \cdots, V_n$ over the same field F is

$$\{(\alpha_1, \alpha_2, \cdots, \alpha_n) \mid \alpha_1 \in V_i \text{ for } i = 1, 2, \cdots, n\},$$

with addition and scalar multiplication defined by

$$(\alpha_1, \alpha_2, \cdots, \alpha_n) + (\beta_1, \beta_2, \cdots, \beta_n) = (\alpha_1 + \beta_1, \alpha_2 + \beta_2, \cdots, \alpha_n + \beta_n).$$

and

$$a(\alpha_1, \alpha_2, \cdots, \alpha_n) = (a\alpha_1, a\alpha_2, \cdots, a\alpha_n) \text{ for all } a \in F.$$

Because addition and multiplication are defined by performing the operations in each component, and because the vectors appearing in each component form a vector space over F, it is clear that this direct sum is again a vector space.

19. Let F be any field and let $F^n = \{(a_1, a_2, \cdots, a_n) \mid a_i \in F\}$. Then F^n is a vector space with addition and multiplication of n-tuples defined by performing those operations in each component. (It is the direct sum of F with itself n times, as defined in Exercise 18.) A basis for F^n is

$$\{(1, 0, 0, \cdots, 0), (0, 1, 0, \cdots, 0), \cdots, (0, 0, 0, \cdots, 1)\}.$$

20. Let V and V' be vector spaces over the same field F. A map $\phi : V \to V'$ is an **isomorphism** if ϕ is a one-to-one map, $\phi[V] = V'$, and furthermore

$$\phi(\alpha + \beta) = \phi(\alpha) + \phi(\beta) \quad \text{and} \quad \phi(a\alpha) = a\phi(\alpha)$$

for all $\alpha, \beta \in V$ and all $a \in F$.

21. Because each vector in V can be expressed as a linear combination of the β_i, we see that $\{\beta_1, \beta_2, \cdots, \beta_n\}$ generates V. Now $0 = 0\beta_1 + 0\beta_2 + \cdots + 0\beta_n$. By hypothesis, this is the *unique* linear combination of the β_i that yields 0, so the vectors β_i are independent. Therefore, $\{\beta_1, \beta_2, \cdots, \beta_n\}$ is a basis for V.

   For the other direction, suppose that $\{\beta_1, \beta_2, \cdots, \beta_n\}$ is a basis for V. Then every vector α is a linear combination of the β_i. Suppose that $\alpha = \sum_{i=1}^n c_i \beta_i$ and also $\alpha = \sum_{i=1}^n d_i \beta_i$. Subtracting, we obtain $0 = \alpha - \alpha = \sum_{i=1}^n (c_i - d_i)\beta_i$. Because the β_i are linearly independent, we must have $c_i - d_i = 0$ so $c_i = d_i$ for $i = 1, 2, \cdots n$, and the expression for α as a linear combination of the β_i is unique.

22. a. The system can be rewritten as

$$X_1\alpha_1 + X_2\alpha_2 + \cdots + X_n\alpha_n = \beta \tag{1}$$

because the ith component of the vector on the left side of the equation (1) is $a_{i1}X_1 + a_{i2}X_2 + \cdots + a_{in}X_n$ and the ith component of β is β_i. Equation (1) shows that the system has a solution if and only if β is a finite linear combination of the vectors α_j for $j = 1, 2, \cdots, n$. By Exercise 17, this means that the system has a solution if and only if β lies in the subspace generated by the vectors $\alpha_1, \alpha_2, \cdots, \alpha_n$.

b. Note that if $\{\alpha_j \mid j = 1, \cdots, n\}$ is a basis for a vector space, then Exercise 21 shows that the linear combination of the α_j that equals a vector β is unique. In terms of Part(**a**), this means that the system (1) has a unique solution.

23. Let $\{\gamma_1, \gamma_2, \cdots, \gamma_n\}$ be a basis for V. Let $\phi : F^n \to V$ be defined by $\phi(a_1, a_2, \cdots, a_n) = a_1\gamma_1 + a_2\gamma_2 + \cdots + a_n\gamma_n$. Because addition and multiplication in F^n is by components, it is obvious that $\phi(\alpha + \beta) = \phi(\alpha) + \phi(\beta)$ and $\phi(a\alpha) = a\phi(\alpha)$ for all $\alpha, \beta \in V$ and all $a \in F$. Because $\{\gamma_1, \gamma_2, \cdots, \gamma_n\}$ is a basis for V, every vector in V can be expressed as a linear combination of these vectors, so ϕ maps F^n *onto* V. By Exercise 21, the expression for a vector in V as a linear combination of the vectors γ_i is *unique*, so ϕ is one to one. Thus ϕ is an isomorphism.

24. a. Let $\alpha \in V, \alpha \neq 0$. Because $\{\beta_i \mid i \in I\}$ is a basis for V, we know by Exercise 21 that there are *unique* vectors $\beta_{i_1}, \beta_{i_2}, \cdots, \beta_{i_n}$ and nonzero scalars $a_1, a_2, \cdots a_n$ such that $\alpha = a_1\beta_{i_1} + a_2\beta_{i_2} + \cdots + a_n\beta_{i_n}$. By the conditions for a linear transformation, we then have

$$\phi(\alpha) = a_1\phi(\beta_{i_1}) + a_2\phi(\beta_{i_2}) + \cdots + a_n\phi(\beta_{i_n}).$$

This shows that the map ϕ is completely determined by the values $\phi(\beta_i)$ for $i \in I$.

b. Let $\phi : V \to V'$ be defined as follows: For nonzero $\alpha \in V$, express α as a linear combination

$$\alpha = a_1\beta_{i_1} + a_2\beta_{i_2} + \cdots + a_n\beta_{i_n} \tag{2}$$

with nonzero scalars $a_1, a_2, \cdots a_n$. This can be done because $\{\beta_i \mid i \in I\}$ is a basis for V. Define $\phi(\alpha) = a_1\beta_{i_1}' + a_2\beta_{i_2}' + \cdots + a_n\beta_{i_n}'$. Because the expression (2) for α with nonzero scalars is unique by Exercise 21, we see that ϕ is well defined, and of course, $\phi(\beta_i) = \beta_i'$ for $i \in I$. Because addition and scalar multiplication of linear combinations of the β_i and the β_i' are both achieved by adding and scalar multiplying respectively the coefficients in the linear combinations, we see at once that ϕ satisfies the required properties for a linear transformation. Part(**a**) shows that this transformation is completely determined by the vectors β_i', that is, the linear transformation is unique.

25. a. A linear transformation of vector spaces is analogous to a homomorphism of groups.

b. The **kernel** or **nullspace** of ϕ is $\text{Ker}(\phi) = \phi^{-1}[\{0'\}] = \{\alpha \in V \mid \phi(\alpha) = 0'\}$. Considering just the additive groups of V and V', group theory shows that $\text{Ker}(\phi)$ is an additive group. Let $\alpha \in \text{Ker}(\phi)$. Then $\phi(a\alpha) = a\phi(\alpha) = a0' = 0'$, so $\text{Ker}(\phi)$ is closed under scalar multiplication by scalars $a \in F$. Hence $\text{Ker}(\phi)$ is a subspace of V.

c. ϕ is an isomorphism of V with V' if ϕ is one to one (equivalently, if $\text{Ker}(\phi) = \{0\}$) and if ϕ maps V onto V'.

26. Let V/S be the factor group $\langle V, + \rangle / \langle S, + \rangle$, which is abelian because V is abelian. Define scalar multiplication on V/S by $a(\alpha + S) = a\alpha + S$ for $a \in F, (\alpha + S) \in V/S$. For $\sigma \in S$, we have $a(\alpha + \sigma) = a\alpha + a\sigma$ and $a\sigma \in S$ for all σ in the subspace S, so this scalar multiplication is well defined, independent of the choice of representative in the coset $\alpha + S$. Because addition and scalar multiplication in V/S are computed in terms of representatives in V and because V is a vector space, we see that addition and scalar multiplication in V/S satisfy the axioms for a vector space.

27. a. By group theory, we know that $\langle \phi[V], + \rangle$ is a subgroup of $\langle V', + \rangle$. Let $\alpha \in V$ and $a \in F$. Then $a\phi(\alpha) = \phi(a\alpha)$ shows that $\phi[V]$ is closed under multiplication by scalars in F. Thus $\phi[V]$ is a subspace of V'.

b. Let $\{\alpha_1, \alpha_2, \cdots, \alpha_r\}$ be a basis for $\text{Ker}(\phi)$. By Theorem 30.19, this set can be enlarged to a basis

$$\{\alpha_1, \alpha_2, \cdots, \alpha_r \, \beta_1, \beta_2, \cdots, \beta_m\}$$

for V. Let $\gamma \in V$. Then $\gamma = a_1\alpha_1 + \cdots + a_r\alpha_r + b_1\beta_1 + \cdots + b_m\beta_m$ for scalars $a_i, b_j \in F$. Because $\phi(\alpha_i) = 0$ for $i = 1, \cdots, r$, we see that $\phi(\gamma) = b_1\phi(\beta_1) + \cdots + b_m\phi(\beta_m)$. Thus $\{\phi(\beta_1), \cdots, \phi(\beta_m)\}$ spans $\phi[V]$. We claim that this set is independent, and hence is actually a basis for $\phi[V]$. Suppose that $c_1\phi(\beta_1) + \cdots + c_m\phi(\beta_m) = 0$ for scalars $c_j \in F$. Then $\phi(c_1\beta_1 + \cdots + c_m\beta_m) = 0$, so $(c_1\beta_1 + \cdots + c_m\beta_m) \in \text{Ker}(\phi)$, and thus

$$c_1\beta_1 + \cdots + c_m\beta_m = d_1\alpha_1 + \cdots + d_r\alpha_r$$

for some scalars d_i. Moving everything to the lefthand side of this equation, we obtain a linear combination of the vectors α_i and β_j which is equal to 0. Because the α_i and β_j form a basis for V, they are independent so all the coefficients d_i and c_j must be zero. The fact that the c_j must be zero shows that $\{\phi(\beta_1), \cdots, \phi(\beta_m)\}$ is independent, and thus is a basis for $\phi[V]$. By our construction, $\dim(\text{Ker}(\phi)) = r$, $\dim(V) = r + m$, and we have shown that $\dim(\phi[V]) = m$. Thus $\dim(\phi[V]) = m = (r + m) - r = \dim(V) - \dim(\text{Ker}(\phi))$.

31. Algebraic Extensions

1. Because $\text{irr}(\sqrt{2}, \mathbb{Q}) = x^2 - 2$, the degree is 2 and a basis is $\{1, \sqrt{2}\}$.

2. By Example 31.9, the degree is 4 and a basis is $\{1, \sqrt{2}, \sqrt{3}, \sqrt{6}\}$.

3. We notice that $\sqrt{18} = \sqrt{2}\sqrt{3}\sqrt{3}$. Thus $\mathbb{Q}(\sqrt{2}, \sqrt{3}, \sqrt{18})$ and $\mathbb{Q}(\sqrt{2}, \sqrt{3})$ are the same field. Thus the degree is 4 and a basis is $\{1, \sqrt{2}, \sqrt{3}, \sqrt{6}\}$ by Example 31.9.

4. Now $\sqrt{3} \notin \mathbb{Q}(\sqrt[3]{2})$ because $\mathbb{Q}(\sqrt{3})$ is of degree 2 over $\mathbb{Q}$ while $\mathbb{Q}(\sqrt[3]{2})$ is of degree 3, and 2 does not divide 3. Thus the degree of $\mathbb{Q}(\sqrt[3]{2}, \sqrt{3})$ over $\mathbb{Q}$ is 6. We form products from the bases $\{1, \sqrt{3}\}$ for $\mathbb{Q}(\sqrt{3})$ over $\mathbb{Q}$ and $\{1, \sqrt[3]{2}, (\sqrt[3]{2})^2\}$ for $\mathbb{Q}(\sqrt[3]{2})$ over $\mathbb{Q}(\sqrt{3})$, obtaining $\{1, \sqrt[3]{2}, (\sqrt[3]{2})^2, \sqrt{3}, \sqrt{3}(\sqrt[3]{2}), \sqrt{3}(\sqrt[3]{2})^2\}$ as a basis.

5. As in the solution to Exercise 4, the extension has degree 6. Taking products from bases $\{1, \sqrt{2}\}$ for $\mathbb{Q}(\sqrt{2})$ over $\mathbb{Q}$ and $\{1, \sqrt[3]{2}, (\sqrt[3]{2})^2\}$ for $\mathbb{Q}(\sqrt[3]{2})$ over $\mathbb{Q}(\sqrt{2})$, we see $\{1, \sqrt[3]{2}, (\sqrt[3]{2})^2, \sqrt{2}, \sqrt{2}(\sqrt[3]{2}), \sqrt{2}(\sqrt[3]{2})^2\}$ is a basis. It is easy to see that $\mathbb{Q}(\sqrt{2}, \sqrt[3]{2}) = \mathbb{Q}(\sqrt[6]{2})$ since $2^{1/6} = 2^{7/6}/2 = 2^{3/6}2^{4/6}/2 = 2^{1/2}(2^{1/3})^2/2$, so another basis is $\{1, 2^{1/6}, 2^{2/6}, 2^{3/6}, 2^{4/6}, 2^{5/6}\}$.

6. As shown in Example 31.9, we have $\deg(\sqrt{2} + \sqrt{3}, \mathbb{Q}) = 4$, so $\mathbb{Q}(\sqrt{2} + \sqrt{3}) = \mathbb{Q}(\sqrt{2}, \sqrt{3})$ and a basis over $\mathbb{Q}$ is $\{1, \sqrt{2}, \sqrt{3}, \sqrt{6}\}$ just as in Example 31.9.

7. Because $\sqrt{2}\sqrt{3} = \sqrt{6}$, we see that the field is $\mathbb{Q}(\sqrt{6})$ which has degree 2 over $\mathbb{Q}$ and a basis $\{1, \sqrt{6}\}$.

8. As in the solution to Exercise 4, we see the extension is of degree 6 because 2 does not divide 3. We form products from the bases $\{1, \sqrt{2}\}$ for $\mathbb{Q}(\sqrt{2})$ over $\mathbb{Q}$ and $\{1, \sqrt[3]{5}, (\sqrt[3]{5})^2\}$ for $\mathbb{Q}(\sqrt[3]{5})$ over $\mathbb{Q}(\sqrt{2})$, yielding $\{1, \sqrt[3]{5}, (\sqrt[3]{5})^2, \sqrt{2}, \sqrt{2}(\sqrt[3]{5}), \sqrt{2}(\sqrt[3]{5})^2\}$ as a basis.

9. Now $\sqrt[3]{6}/\sqrt[3]{2} = \sqrt[3]{3}$ and $\sqrt[3]{24} = 2(\sqrt[3]{3})$, so $\mathbb{Q}(\sqrt[3]{2}, \sqrt[3]{6}, \sqrt[3]{24}) = \mathbb{Q}(\sqrt[3]{2}, \sqrt[3]{3})$. The degree over $\mathbb{Q}$ is 9, and we take products from the bases $\{1, \sqrt[3]{2}, (\sqrt[3]{2})^2\}$ and $\{1, \sqrt[3]{3}, (\sqrt[3]{3})^2\}$ for $\mathbb{Q}(\sqrt[3]{2})$ over $\mathbb{Q}$ and $\mathbb{Q}(\sqrt[3]{2}, \sqrt[3]{3})$ over $\mathbb{Q}(\sqrt[3]{2})$ respectively, obtaining the basis

$$\{1, \sqrt[3]{2}, \sqrt[3]{4}, \sqrt[3]{3}, \sqrt[3]{6}, \sqrt[3]{12}, \sqrt[3]{9}, \sqrt[3]{18}, \sqrt[3]{36}\}.$$

10. Because $\mathbb{Q}(\sqrt{2}, \sqrt{6}) = \mathbb{Q}(\sqrt{2}, \sqrt{3})$, the extension is of degree 2 over $\mathbb{Q}(\sqrt{3})$ and we can take the set $\{1, \sqrt{2}\}$ as a basis.

11. Example 31.9 shows that $\mathbb{Q}(\sqrt{2} + \sqrt{3}) = \mathbb{Q}(\sqrt{2}, \sqrt{3})$, so the extension has degree 2 and we can take as basis over $\mathbb{Q}(\sqrt{3})$ the set $\{1, \sqrt{2}\}$.

12. By Example 31.9, $\mathbb{Q}(\sqrt{2} + \sqrt{3}) = \mathbb{Q}(\sqrt{2}, \sqrt{3})$ so the degree of the extension is 1 and $\{1\}$ is a basis.

13. Now $\sqrt{6} + \sqrt{10} = \sqrt{2}(\sqrt{3} + \sqrt{5})$ so we have $\mathbb{Q}(\sqrt{2}, \sqrt{6} + \sqrt{10}) = \mathbb{Q}(\sqrt{2}, \sqrt{3} + \sqrt{5})$. The degree of the extension is 2 and a basis over $\mathbb{Q}(\sqrt{3} + \sqrt{5})$ is $\{1, \sqrt{2}\}$.

14. The definition is incorrect. An algebraic extension need not be a finite extension.

An **algebraic extension** of a field F is an extension field E of F with the property that each $\alpha \in E$ is a zero of some nonzero polynomial in $F[x]$.

15. The definition is incorrect. If an element adjoined is transcendental over F, the extension is not finite.

A **finite extension field** of a field F is an extension field of F that has finite dimension when regarded as a vector space having F as its field of scalars.

16. The definition is correct.

17. The definition is not quite correct. We should insert "nonconstant" before "polynomial" and "in $F[x]$" after it.

A field F is **algebraically closed** if and only if every nonconstant polnomial in $F[x]$ has a zero in F.

18. Let $E = \mathbb{Q}(\sqrt{2})$ and let $F = \mathbb{Q}$. The algebraic closure of $\mathbb{Q}$ in $\mathbb{Q}(\sqrt{2})$ is $\mathbb{Q}(\sqrt{2})$ because it is an algebraic extension of $\mathbb{Q}$. However, $\mathbb{Q}(\sqrt{2})$ is not algebraically closed, because the polynomial $x^2 + 1$ has no zeros in $\mathbb{Q}(\sqrt{2})$.

19. F T F T F T F F F F

20. Because E is a finite dimensional vector space over F, the set of all powers of an element α in E cannot be independent, so some finite linear combination with nonzero coefficents in F of these powers must be zero.

21. Taking a basis of n elements α_i for E over F and a basis of m elements β_j for K over E, the set of the mn possible products $\alpha_i\beta_j$ is a basis for K over F.

22. If $b \neq 0$, then $a + bi \in \mathbb{C}$ but $a + bi \notin \mathbb{R}$. By Theorem 31.3, $a + bi$ is algebraic over $\mathbb{R}$. Then by Theorem 31.4,
$$[\mathbb{C} : \mathbb{R}] = [\mathbb{C} : \mathbb{R}(a + bi)][\mathbb{R}(a + bi) : \mathbb{R}] = 2,$$
and because $a + bi \notin \mathbb{R}$, we must have $[\mathbb{R}(a + bi) : \mathbb{R}] = 2$, so $[\mathbb{C} : \mathbb{R}(a + bi)] = 1$. Thus $\mathbb{C} = \mathbb{R}(a + bi)$.

23. Let α be any element in E that is not in F. Then $[E : F] = [E : F(\alpha)][F(\alpha) : F] = p$ for some prime p by Theorem 31.4. Because α is not in F, we know that $[F(\alpha) : F] > 1$, so we must have $[F(\alpha) : F] = p$ and therefore $[E : F(\alpha)] = 1$. As we remarked after Definition 31.2, this shows that $E = F(\alpha)$, which is what we wish to show.

24. If $x^2 - 3$ were reducible over $\mathbb{Q}(\sqrt[3]{2})$, then it would factor into linear factors over $\mathbb{Q}(\sqrt[3]{2})$, so $\sqrt{3}$ would lie in the field $\mathbb{Q}(\sqrt[3]{2})$, and we would have $\mathbb{Q}(\sqrt{3}) \leq \mathbb{Q}(\sqrt[3]{2})$. But then by Theorem 31.4,
$$[\mathbb{Q}(\sqrt[3]{2}) : \mathbb{Q}] = [\mathbb{Q}(\sqrt[3]{2}) : \mathbb{Q}(\sqrt{3})][\mathbb{Q}(\sqrt{3}) : \mathbb{Q}].$$
This equation is impossible because $[\mathbb{Q}(\sqrt[3]{2}) : \mathbb{Q}] = 3$ while $[\mathbb{Q}(\sqrt{3}) : \mathbb{Q}] = 2$.

25. Corollary 31.6 shows that the degree of an extension of F by successively adjoining square roots must be 2^n for some $n \in \mathbb{Z}^+$. Because $x^{14} - 3x^2 + 12$ is irreducible over $\mathbb{Q}$ by the Eisenstein condition with $p = 3$, and because $[\mathbb{Q}(\alpha) : \mathbb{Q}] = 14$ for any zero α of this polynomial, and because 14 is not a divisor of 2^n for any $n \in \mathbb{Z}^+$, we see that α cannot lie in any field obtained by adjoining just square roots. Therefore α cannot be expressed as a rational function of square roots, square roots of rational functions of square roots, etc.

26. We need only show that for each $\alpha \in D, \alpha \neq 0$, its multiplicative inverse $1/\alpha$ is in D also. Because E is a finite extension of F, we know that α is algebraic over F. Let $\deg(\alpha, F) = n$. Then by Theorem 30.23, we have
$$F(\alpha) = \{a_0 + a_1\alpha + a_2\alpha^2 + \cdots + a_{n-1}\alpha^{n-1} \mid a_i \in F \text{ for } i = 0, \cdots n - 1\}.$$
In particular, $1/\alpha \in F(\alpha)$, so $1/\alpha$ is equal to polynomial in α with coefficients in F, and is in D.

27. Obviously $\mathbb{Q}(\sqrt{3} + \sqrt{7}) \subseteq \mathbb{Q}(\sqrt{3}, \sqrt{7})$. Now $(\sqrt{3} + \sqrt{7})^2 = 10 + 2\sqrt{21}$ so $\sqrt{21} \in \mathbb{Q}(\sqrt{3} + \sqrt{7})$. Hence $(3\sqrt{7} + 7\sqrt{3}) - 7(\sqrt{3} + \sqrt{7}) = -4\sqrt{7}$ is in $\mathbb{Q}(\sqrt{3} + \sqrt{7})$, so this field contains $\sqrt{7}$ and also $(\sqrt{3} + \sqrt{7}) - \sqrt{7} = \sqrt{3}$. Therefore $\mathbb{Q}(\sqrt{3}, \sqrt{7}) \subseteq \mathbb{Q}(\sqrt{3} + \sqrt{7})$, so $\mathbb{Q}(\sqrt{3}, \sqrt{7}) = \mathbb{Q}(\sqrt{3} + \sqrt{7})$. [One can also make an argument like that in Example 31.9 of the text, finding $\text{irr}(\sqrt{3} + \sqrt{7}, \mathbb{Q})$ and showing that it is of degree 4 over $\mathbb{Q}$. Then we would have $[\mathbb{Q}(\sqrt{3}, \sqrt{7}) : \mathbb{Q}(\sqrt{3} + \sqrt{7})] = 1$, so the fields are equal.]

28. If $a = b$ the result is clear; we assume $a \neq b$. It is obvious that $\mathbb{Q}(\sqrt{a} + \sqrt{b}) \subseteq \mathbb{Q}(\sqrt{a}, \sqrt{b})$.

We now show that $\mathbb{Q}(\sqrt{a}, \sqrt{b}) \subseteq \mathbb{Q}(\sqrt{a} + \sqrt{b})$. Let $\alpha = \frac{a - b}{\sqrt{a} + \sqrt{b}} \in \mathbb{Q}(\sqrt{a} + \sqrt{b})$. Now $\alpha = \sqrt{a} - \sqrt{b}$. Thus $\mathbb{Q}(\sqrt{a} + \sqrt{b})$ contains $\frac{1}{2}[\alpha + (\sqrt{a} + \sqrt{b})] = \frac{1}{2}(2\sqrt{a}) = \sqrt{a}$ and hence also contains $(\sqrt{a} + \sqrt{b}) - \sqrt{a} = \sqrt{b}$. Thus $\mathbb{Q}(\sqrt{a}, \sqrt{b}) \subseteq \mathbb{Q}(\sqrt{a} + \sqrt{b})$.

29. If a zero α of $p(x)$ were in E, then because $p(x)$ is irreducible over F, we would have $[F(\alpha) : F] = \deg(p(x))$, and $[F(\alpha : F]$ would be a divisor of $[E : F]$ by Theorem 31.4. By hypothesis, this is not the case. Therefore $p(x)$ has no zeros in E.

30. Because $F(\alpha)$ is a finite extension of F and $\alpha^2 \in F(\alpha)$, Theorem 31.3 shows that α^2 is algebraic over F. If $F(\alpha^2) \neq F(\alpha)$, then $F(\alpha)$ must be an extension of $F(\alpha^2)$ of degree 2, because α is a zero of $x^2 - \alpha^2$. By Theorem 31.4, we would then have $2 = [F(\alpha) : F(\alpha^2)]$ dividing $[F(\alpha) : F]$, which is impossible because $[F(\alpha) : F]$ is an odd number. Therefore $F(\alpha^2) = F(\alpha)$, so $\deg(\mathrm{irr}(\alpha^2, F)) = \deg(\mathrm{irr}(\alpha, F)) = [F(\alpha) : F]$ which is an odd number.

31. Suppose K is algebraic over F. Then every element of K is a zero of a nonzero polynomial in F[x], and hence in E[x]. This shows that K is algebraic over E. Of course E is algebraic over F, because each element of E is also an element of K.

Conversely, suppose that K is algebraic over E and that E is algebraic over F. Let $\alpha \in K$. We must show that α is algebraic over F. Because K is algebraic over E, α is a zero of some polynomial $a_0 + a_1x + a_2x^2 + \cdots + a_nx^n$ in $E[x]$. Because E is algebraic over F, the a_i are algebraic over F for $i = 0, 1, 2, \cdots, n$. Hence $F(a_0, a_1, a_2, \cdots, a_n)$ is an extension of F of some finite degree m by Theorem 31.11. Because α is algebraic over E of degree $r \leq n$, Theorem 31.4 shows that $F(a_0, a_1, a_2, \cdots, a_n, \alpha)$ is a finite extension of F of degree $\leq mr$. By Theorem 31.3, α is algebraic over F.

32. If α is algebraic over $\overline{F}_E$, then $\overline{F}_E(\alpha)$ is algebraic over $\overline{F}_E$ and by definition, $\overline{F}_E$ is algebraic over F. By Exercise 31, then $\overline{F}_E(\alpha)$ is algebraic over F so, in particular, α is algebraic over F. But then $\alpha \in \overline{F}_E$ contrary to hypothesis. Thus α is transcendental over $\overline{F}_E$.

33. Let $f(x)$ be a nonconstant polynomial in $\overline{F}_E[x]$. We must show that $f(x)$ has a zero in $\overline{F}_E$. Now $f(x) \in E[x]$ and E is agebraically closed by hypothesis, so $f(x)$ has a zero α in E. By Exercise 32, if α is not in $\overline{F}_E$, then α is transcendental over $\overline{F}_E$. But by construction, α is a zero of $f(x) \in \overline{F}_E[x]$, so this is impossible. Hence $\alpha \in \overline{F}_E$, which shows that $\overline{F}_E$ is algebraically closed.

34. Let $\alpha \in E$ and let $p(x) = \mathrm{irr}(\alpha, F)$ have degree n. Now $p(x)$ factors into $(x - \alpha_1)(x - \alpha_2) \cdots (x - \alpha_n)$ in $\overline{F}[x]$. Because by hypotheses all zeros of $p(x)$ in $\overline{F}$ are also in E, we see that this same factorization is also valid in $E[x]$. Hence

$$p(\alpha) = (\alpha - \alpha_1)(\alpha - \alpha_2) \cdots (\alpha - \alpha_n) = 0,$$

so $\alpha = \alpha_i$ for some i. This shows that $F \leq E \leq \overline{F}$. Because by definition $\overline{F}$ contains only elements that are algebraic over F and E contains all of these, we see that $E = \overline{F}$ and is therefore algebraically closed.

35. If F is a finite field of odd characteristic, then $1 \neq -1$ in F. Because $1^2 = (-1)^2 = 1$, the squares of elements of F can run through at most $|F| - 1$ elements of F, so there is some $a \in F$ that is not a square. The polynomial $x^2 - a$ then has no zeros in F, so F is not algebraically closed.

36. For all $n \in \mathbb{Z}, n \geq 2$, the polynomial $x^n - 2$ is irreducible in $\mathbb{Q}[x]$ by the Eisenstein condition with $p = 2$. This shows that $\mathbb{Q}$ has finite extensions contained in $\mathbb{C}$ of arbitrarily high degree. If $\overline{\mathbb{Q}}_\mathbb{C}$ were a finite extension of $\mathbb{Q}$ of degree r, then there would be no algebraic extensions of $\mathbb{Q}$ in $\mathbb{C}$ of degree greater than r. Thus the algebraic closure $\overline{\mathbb{Q}}_\mathbb{C}$ of $\mathbb{Q}$ in $\mathbb{C}$ cannot be a finite extension of $\mathbb{Q}$.

37. Because $[\mathbb{C} : \mathbb{R}] = 2$ and $\mathbb{C}$ is an algebraic closure of $\mathbb{R}$, it must be that every irreducible $p(x)$ in $\mathbb{R}[x]$ of degree > 1 is actually of degree 2. Because $\mathbb{C} = \mathbb{R}$ where $\alpha \in \mathbb{C}$ is any zero of any such polynomial, we know by the construction in Theorem 29.3 that $\mathbb{C} \simeq \mathbb{R}[x]/\langle p(x) \rangle$ for any irreducible polynomial $p(x)$ in $\mathbb{R}[x]$ of degree 2.

Now let E be any finite extension of $\mathbb{R}$. If $E \neq \mathbb{R}$, then let β be in E but not in $\mathbb{R}$. Then $p(x) = \text{irr}(\beta, \mathbb{R})$ has degree 2, because we have seen that there are no irreducible polynomials in $\mathbb{R}[x]$ of greater degree. The construction in Theorem 29.3 shows that $\mathbb{R}(\beta) \simeq \mathbb{R}[x]/\langle p(x) \rangle$ and hence $\mathbb{R}(\beta) \simeq \mathbb{C}$. Because $\mathbb{C}$ is algebraically closed, $\mathbb{R}(\beta)$ is algebraically closed also and admits no proper algebraic extensions. Because E is an algebraic extension of $\mathbb{R}(\beta)$, we must have $E = \mathbb{R}(\beta)$, so $E \simeq \mathbb{C}$.

38. If R contains no nontrivial proper ideals, then $\{0\}$ is the only proper ideal, and is a maximal ideal, and of course it is contained in itself so we are done.

Suppose R contains a nontrivial proper ideal N which of course does not contain the unity 1 of R. The set S of ideals of R that do not contain 1 is partially ordered by inclusion. Let $T = \{N_i \mid i \in I\}$ be a chain of S. We claim that $U = \bigcup_{i \in I} N_i$ is an element of S that is an upper bound of T. Let $x, y \in U$. Then $x \in N_j$ and $y \in N_k$ for some $j, k \in I$. Because T is a chain, one of these ideals is contained in the other, say $N_j \subseteq N_k$. Then $x, y \in N_k$ which is an ideal, so for all $r \in R$, we see that $x \pm y, 0, rx$, and xr are all in N_k and hence in U. This shows that U is an ideal. Clearly $N_i \subseteq U$ for all $i \in I$, and 1 is not in U because 1 is not in N_i for any $i \in I$. Thus $U \in S$ and is an upper bound for T, so the hypotheses of Zorn's lemma are satisfied.

Let M be a maximal element of S; such an element of S exists by Zorn's lemma. Because $M \in S$, we see that M is an ideal of R, and does not contain 1 so $M \neq R$. Suppose that L is an ideal of R such that $M \subseteq L \subseteq R$. If $L \in S$, then $M = L$ because M is a maximal element of S under set inclusion. Otherwise, $1 \in L$ so $L = R$. Thus M is a maximal ideal of R.

32. Geometric Constructions

1. By Euler's formula, $e^{i(3\theta)} = \cos 3\theta + i \sin 3\theta$. On the other hand, $e^{3i\theta} = (e^{i\theta})^3 = (\cos \theta + i \sin \theta)^3$, which we compute using the binomial theorem; the real part of the answer should be equal to $\cos 3\theta$. We have
$$(\cos \theta + i \sin \theta)^3 = (\cos^3 \theta - 3 \cos \theta \sin^2 \theta) + i(\text{don't care}).$$
Thus $\cos 3\theta = \cos^3 \theta - 3 \cos \theta(1 - \cos^2 \theta) = 4 \cos^3 \theta - 3 \cos \theta$.

2. T T T F T F T T T F

3. If a regular 9-gon could be constructed, the angle $(360/9)^\circ = 40^\circ$ could be constructed, and could then be bisected to construct an angle of 20°. The proof of Theorem 32.11 shows, however, that an angle of 20° is not constructible.

4. One can construct an angle of 30° if and only if one can construct the number $\cos 30^\circ = \sqrt{3}/2$. Because $\sqrt{3}$ is constructible and quotients of constructible numbers are constructible, an angle of 30° is constructible.

5. Because $|\overline{OA}| = |\overline{OP}|, \angle OAP = \angle APO = (180^\circ - 36^\circ)/2 = 72^\circ$. Then $\angle QAP = 36^\circ$ so triangle OAP is similar to triangle APQ. Now $|\overline{AP}| = |\overline{AQ}| = |\overline{OQ}| = r$, so $|\overline{QP}| = 1 - r$. Taking ratios of corresponding sides, we obtain $|\overline{AP}|/|\overline{OP}| = |\overline{OA}|/|\overline{AP}|$ so $\frac{r}{1-r} = \frac{1}{r}$. Thus $r^2 = 1 - r$ so $r^2 + r - 1 = 0$. By the quadratic formula, we find that $r = \frac{-1+\sqrt{5}}{2}$ which is a constructible number. Thus we can construct an angle of 36° by taking a line segment $\overline{OP}$ of length 1, drawing a circle of radius 1 at O and one of radius r at P, and finding a point A of intersection of the two circles. Then $\angle AOP$ measures

36°. Thus a regular 10-gon which has central angles of 36° is constructible. A regular pentagon is obtain by starting at vertex 1 of a regular 10-gon and drawing line segments to vertex 3, on to vertex 5, on to vertex 7, on to vertex 9, and then to vertex 1.

6. A regular 20-gon is constructible because we can bisect the constructible angle of 36° (see Exercise 5) to obtain an angle of 18° = (360/20)°.

7. Because we can construct an angle of 72° = 2(36°) by Exercise 5, and because 60° is a constructible angle, we can construct an angle of 72° − 60° = 12° = (360/30)°. Therefore a regular 30-gon can be constructed.

8. Exercise 7 shows that a 12° angle can be constructed, so a 24° = (2 · 12)° = (72/3)° angle can be constructed. Thus an angle of 72° can be trisected.

9. Exercise 7 shows that an angle of 12° can be constructed so an angle of 24° = (2 · 12)° = (360/15)° can be constructed. Hence a regular 15-gon can be constructed.

10. Starting with a line segment of length 1, from 0 to 1 on the usual x-axis, we can construct a line segment of any rational length in a finite number of steps, and thus find any point with rational coordinates in the x, y-plane in a finite number of steps. Thinking analytically, the only other points in the plane we can locate must appear as an intersection of two lines, of a line and a circle, or of two circles, which reduces algebraically to finding the solutions of only linear or quadratic equations. Considering a right triangle with an acute angle θ and hypotenuse of length 1, we see that we can construct the angle θ if and only if we can construct a line segment of length $\cos \theta$. It can be shown that $\cos 20°$ is not a solution of a linear or quadratic equation, but rather of a cubic equation, so while we can construct a 60° angle because $\cos 60° = 1/2$, we cannot trisect it to obtain a 20° angle.

33. Finite Fields

1. Because $4096 = 2^{12}$ is a power of a prime, a finite field of order 4096 does exist.

2. Because $3127 = 53 \cdot 59$ is not a power of a prime, no finite field of order 3127 exists.

3. Because $68921 = 41^3$ is a power of a prime, a finite field of order 68921 does exist.

4. $GF(9)^*$ is a cyclic group under multiplication of order 8 and has $\phi(8) = 4$ generators, so there are 4 primitive 8th roots of unity.

5. $GF(19)^*$ is a cyclic group under multiplication of order 18 and has $\phi(18) = 6$ generators, so there are 6 primitive 18th roots of unity.

6. $GF(31)^*$ is a cyclic group under multiplication of order 30. Its cyclic subgroup of order 15 has $\phi(15) = 8$ generators, so it contains 8 primitive 15th roots of unity.

7. $GF(23)^*$ is a cyclic group under multiplication of order 22. Because 10 is not a divisor of 22, it contains no elements of order 10, so $GF(23)^*$ contains no primitive 10th roots of unity.

8. T F T F T F T T F T

9. Because both the given polynomials are irreducible over $\mathbb{Z}_2$, both $\mathbb{Z}_2(\alpha)$ and $\mathbb{Z}_2(\beta)$ are extension of $\mathbb{Z}_2$ of degree 3 and thus are subfields of $\overline{\mathbb{Z}}_2$ containing $2^3 = 8$ elements. By Theorem 33.3, both of these fields must consist precisely of the zeros in $\overline{\mathbb{Z}}_2$ of the polynomial $x^8 - x$. Thus the fields are the same.

10. Let $p(x)$ be irreducible of degree m in $\mathbb{Z}_p[x]$. Let K be the finite extension of $\mathbb{Z}_p$ obtained by adjoining *all* the zeros of $p(x)$ in $\overline{\mathbb{Z}}_p$. Then K is a finite field of order p^n for some positive integer n, and consists precisely of all zeros of $x^{p^n} - x$ in $\overline{\mathbb{Z}}_p$. Now $p(x)$ factors into linear factors in $K[x]$, and these linear factors are among the linear factors of $x^{p^n} - x$ in $K[x]$. Thus $p(x)$ is a divisor of $x^{p^n} - x$.

11. Because $\alpha \in F$, we have $\mathbb{Z}_p(\alpha) \subseteq F$. But because α is a generator of the multiplicative group F^*, we see that $\mathbb{Z}_p(\alpha) = F$. Because $|F| = p^n$, the degree of α over $\mathbb{Z}_p$ must be n.

12. Let F be a finite field of p^n elements containing (up to isomorphism) the prime field $\mathbb{Z}_p$. Let m be a divisor of n, so that $n = mq$. Let $\overline{F} = \overline{\mathbb{Z}}_p$ be an algebraic closure of F. If $\alpha \in \overline{\mathbb{Z}}_p$ and $\alpha^{p^m} = \alpha$, then $\alpha^{p^n} = \alpha^{p^{mq}} = (\alpha^{p^m})^{p^{m(q-1)}} = \alpha^{p^{m(q-1)}} = (\alpha^{p^m})^{p^{m(q-2)}} = \alpha^{p^{m(q-2)}} = \cdots = \alpha^{p^m} = \alpha$. By Theorem 33.3, the zeros of $x^{p^m} - x$ in $\overline{\mathbb{Z}}_p$ form the *unique* subfield of $\overline{\mathbb{Z}}_p$ of order p^m. Our computation shows that the elements in this subfield are also zeros of $x^{p^n} - x$, and consequently all lie in the field F, which by Theorem 33.3 consists of all zeros of $x^{p^n} - x$ in $\overline{\mathbb{Z}}_p$.

13. Let F be the extension of $\mathbb{Z}_p$ of degree n, consisting of all zeros of $x^{p^n} - x$ by Theorem 33.3. Each $\alpha \in F$ is algebraic over $\mathbb{Z}_p$ and has degree that divides n by Theorem 31.4. Thus each $\alpha \in F$ is a zero of a monic irreducible polynomial of a degree dividing n. Conversely, a zero β of an irreducible monic polynomial having degree m dividing n lies in a field $\mathbb{Z}_p(\beta)$ of p^m elements that is contained in F by Exercise 12. Thus the elements of F are precisely the zeros of all monic irreducible polynomials in $\mathbb{Z}_p[x]$ of degree dividing n, as well as precisely all zeros of $x^{p^n} - x$. Factoring into linear factors in $F[x]$, we see that both $x^{p^n} - x$ and the product g(x) of all monic polynomials in $\mathbb{Z}_p[x]$ of degree d dividing n have the factorization $\prod_{\alpha \in F}(x - \alpha)$, so $x^{p^n} - x = g(x)$.

14. **a.** Now $x^2 \equiv a \pmod{p}$ has a solution in $\mathbb{Z}$ if and only if $x^2 = b$ has a zero in $\mathbb{Z}_p$ where b is the remainder of a modulo p. Now $\mathbb{Z}_p^*$ is cyclic of order $p - 1$. The elements b of a cyclic group that are squares are those that are even powers of a generator, and these are precisely the elements b satisfying $b^{(p-1)/2} = 1$. Thus we see that $x^2 \equiv a \pmod{p}$, for a not congruent to zero modulo p, has a solution in $\mathbb{Z}$ if and only if $a^{(p-1)/2} \equiv 1 \pmod{p}$.

b. We know that $x^2 - 6$ is irreducible in $\mathbb{Z}_{17}[x]$ if and only if it has no zero in $\mathbb{Z}_{17}$, so that $6 \neq b^2$ for any $b \in \mathbb{Z}_{17}$. By Part(**a**), we test by computing whether $6^{(17-1)/2} = 6^8$ is congruent to 1 modulo 17. Computing in $\mathbb{Z}_{17}$, we have $6^2 = 2, 6^4 = 2^2 = 4$, and $6^8 = 4^4 = 16$, so 6 is not a square in $\mathbb{Z}_{17}$ and $x^2 - 6$ is irreducible.

34. Isomorphism Theorems

1. (See the answer in the text.)

2. **a.** $K = \{0, 6, 12\}$

$\quad 0 + K = \{0, 6, 12\}, \quad 1 + K = \{1, 7, 13\},$

b. $\quad 2 + K = \{2, 8, 14\}, \quad 3 + K = \{3, 9, 15\},$

$\quad 4 + K = \{4, 10, 16\}, \quad 5 + K = \{5, 11, 17\}$

c. $\phi[\mathbb{Z}_{18}]$ is the subgroup $\{0, 2, 4, 6, 8, 10\}$ of $\mathbb{Z}_{12}$.

d. $\mu(0 + K) = 0, \mu(1 + K) = 10, \mu(2 + K) = 8, \mu(3 + K) = 6, \mu(4 + K) = 4, \mu(5 + K) = 2$

3. (See the text answer.)

4. a. $HN = \{0, 3, 6, 9, 12, 15, 18, 21, 24, 27, 30, 33\}$

   b. $0 + N = \{0, 9, 18, 27\}, \quad 3 + N = \{3, 12, 21, 30\}, \quad 6 + N = \{6, 15, 24, 33\}$

   c. $0 + (H \cap N) = \{0, 18\}, 6 + (H \cap N) = \{6, 24\}, 12 + (H \cap N) = \{12, 30\}$

   d. $\phi(0 + N) = 0 + (H \cap N), \phi(3 + N) = 12 + (H \cap N), \phi(6 + N) = 6 + (H \cap N)$

5. (See the text answer.)

6. a.
$0 + H = \{0, 9, 18, 27\}, \quad 1 + H = \{1, 10, 19, 28\}$
$2 + H = \{2, 11, 20, 29\}, \quad 3 + H = \{3, 12, 21, 30\}$
$4 + H = \{4, 13, 22, 31\}, \quad 5 + H = \{5, 14, 23, 32\}$
$6 + H = \{6, 15, 24, 33\}, \quad 7 + H = \{7, 16, 25, 34\}$
$8 + H = \{8, 17, 26, 35\}$

   b.
$0 + K = \{0, 18\}, \quad 1 + K = \{1, 19\}, \quad 2 + K = \{2, 20\}$
$3 + K = \{3, 21\}, \quad 4 + K = \{4, 22\}, \quad 5 + K = \{5, 23\}$
$6 + K = \{6, 24\}, \quad 7 + K = \{7, 25\}, \quad 8 + K = \{8, 26\}$
$9 + K = \{9, 27\}, \quad 10 + K = \{10, 28\}, \quad 11 + K = \{11, 29\}$
$12 + K = \{12, 30\}, \quad 13 + K = \{13, 31\}, \quad 14 + K = \{14, 32\}$
$15 + K = \{15, 33\}, \quad 16 + K = \{16, 34\}, \quad 17 + K = \{17, 35\}$

   c. $0 + K = \{0, 18\}, \quad 9 + K = \{9, 27\}$

   d.
$(0 + K) + H/K = \{0 + K, 9 + K\} = \{\{0, 18\}, \{9, 27\}\}$
$(1 + K) + H/K = \{1 + K, 10 + K\} = \{\{1, 19\}, \{10, 28\}\}$
$(2 + K) + H/K = \{2 + K, 11 + K\} = \{\{2, 20\}, \{11, 29\}\}$
$(3 + K) + H/K = \{3 + K, 12 + K\} = \{\{3, 21\}, \{12, 30\}\}$
$(4 + K) + H/K = \{4 + K, 13 + K\} = \{\{4, 22\}, \{13, 31\}\}$
$(5 + K) + H/K = \{5 + K, 14 + K\} = \{\{5, 23\}, \{14, 32\}\}$
$(6 + K) + H/K = \{6 + K, 15 + K\} = \{\{6, 24\}, \{15, 33\}\}$
$(7 + K) + H/K = \{7 + K, 16 + K\} = \{\{7, 25\}, \{16, 34\}\}$
$(8 + K) + H/K = \{8 + K, 17 + K\} = \{\{8, 26\}, \{17, 35\}\}$

   e.
$\phi(0 + H) = (0 + K) + H/K, \quad \phi(1 + H) = (1 + K) + H/K$
$\phi(2 + H) = (2 + K) + H/K, \quad \phi(3 + H) = (3 + K) + H/K$
$\phi(4 + H) = (4 + K) + H/K, \quad \phi(5 + H) = (5 + K) + H/K$
$\phi(6 + H) = (6 + K) + H/K, \quad \phi(7 + H) = (7 + K) + H/K$
$\phi(8 + H) = (8 + K) + H/K$

7. Let $x \in H \cap N$ and let $h \in H$. Because $x \in H$ and H is a subgroup, we know that $hxh^{-1} \in H$. Because $x \in N$ and N is normal in G, we also know that $hxh^{-1} \in N$. Thus $hxh^{-1} \in H \cap N$, so $H \cap N$ is a normal subgroup of H.

8. a. Let $\gamma : G \to G/H$ be the natural homomorphism of a group onto its factor group. Then $\gamma[K] = K/H = B$ is a normal subgroup of $A = G/H$ by Theorem 15.16. Similarly $\gamma[L] = L/H = C$ is a normal subgroup of A. It is clear that $B = K/H$ is a subgroup of $C = L/H$ because K is a subgroup of L.

b. Theorem 34.7 shows that

$$(A/B)/(C/B) \simeq A/C = (G/H)/(L/H) \simeq G/L.$$

9. By Lemma 34.4, we know that $K \vee L = KL = LK$, so $G = KL = LK$. By Theorem 34.5, $G/L = KL/L \simeq K/(K \cap L) = K/\{e\} \simeq K$. Similarly, $G/K = LK/K \simeq L/(L \cap K) = L/\{e\} \simeq L$.

35. Series of Groups

1. (See the answer in the text.)

2. We have to insert subgroups to produce additional factor groups of orders 5 and 49 in the first series, and we create the refinement

$$\{0\} < 14700\mathbb{Z} < 300\mathbb{Z} < 60\mathbb{Z} < 20\mathbb{Z} < \mathbb{Z}$$

of the series $\{0\} < 60\mathbb{Z} < 20\mathbb{Z} < \mathbb{Z}$. We have to insert subgroups to produce additional factor groups of orders 3 and 20 in the second series, and we create the refinement

$$\{0\} < 14700\mathbb{Z} < 4900\mathbb{Z} < 245\mathbb{Z} < 49\mathbb{Z} < \mathbb{Z}$$

of the series $\{0\} < 245\mathbb{Z} < 49\mathbb{Z} < \mathbb{Z}$. These two refinements are isomorphic, producing cyclic factor groups of orders 3, 20, 5, 49, and an infinite cyclic factor group.

3. The given series are already isomorphic, with factor groups of orders 3 and 8.

4. The first series has cyclic factor groups of orders 4, 6, and 3 while the second has cyclic factor groups of orders 3, 2, and 12. Thus we break the 4 into two 2's and break the 12 into a 6 and a 2. We obtain the refinement

$$\{0\} < \langle 36 \rangle < \langle 18 \rangle < \langle 3 \rangle < \mathbb{Z}_{72}$$

of the series $\{0\} < \langle 18 \rangle < \langle 3 \rangle < \mathbb{Z}_{72}$, and the refinement

$$\{0\} < \langle 24 \rangle < \langle 12 \rangle < \langle 6 \rangle < \mathbb{Z}_{72}$$

of the series $\{0\} < \langle 24 \rangle < \langle 12 \rangle < \mathbb{Z}_{72}$.

5. (See the answer in the text.)

6. $\{0\} < \langle 30 \rangle < \langle 15 \rangle < \langle 5 \rangle < \mathbb{Z}_{60}$, $\quad \{0\} < \langle 30 \rangle < \langle 15 \rangle < \langle 3 \rangle < \mathbb{Z}_{60}$,
   $\{0\} < \langle 30 \rangle < \langle 10 \rangle < \langle 5 \rangle < \mathbb{Z}_{60}$, $\quad \{0\} < \langle 30 \rangle < \langle 10 \rangle < \langle 2 \rangle < \mathbb{Z}_{60}$,
   $\{0\} < \langle 30 \rangle < \langle 6 \rangle < \langle 3 \rangle < \mathbb{Z}_{60}$, $\quad \{0\} < \langle 30 \rangle < \langle 6 \rangle < \langle 2 \rangle < \mathbb{Z}_{60}$,
   $\{0\} < \langle 20 \rangle < \langle 10 \rangle < \langle 5 \rangle < \mathbb{Z}_{60}$, $\quad \{0\} < \langle 20 \rangle < \langle 10 \rangle < \langle 2 \rangle < \mathbb{Z}_{60}$,
   $\{0\} < \langle 20 \rangle < \langle 4 \rangle < \langle 2 \rangle < \mathbb{Z}_{60}$, $\quad \{0\} < \langle 12 \rangle < \langle 6 \rangle < \langle 3 \rangle < \mathbb{Z}_{60}$,
   $\{0\} < \langle 12 \rangle < \langle 6 \rangle < \langle 2 \rangle < \mathbb{Z}_{60}$, $\quad \{0\} < \langle 12 \rangle < \langle 4 \rangle < \langle 2 \rangle < \mathbb{Z}_{60}$.

   For each series, the factor groups are isomorphic to $\mathbb{Z}_2, \mathbb{Z}_2, \mathbb{Z}_3$, and $\mathbb{Z}_5$ in some order.

7. (See the text answer for the series.) For each series, the factor groups are isomorphic to $\mathbb{Z}_2, \mathbb{Z}_2, \mathbb{Z}_2, \mathbb{Z}_2$, and $\mathbb{Z}_3$ in some order.

8. There are six possible series $\{(0,0)\} < \langle(m,n)\rangle < \mathbb{Z}_5 \times \mathbb{Z}_5$ where (m,n) is either $(0, 1)$, $(1, 0)$, $(1, 1)$, $(1, 2)$, $(1, 3)$, or $(1, 4)$. There are two factor groups isomorphic to $\mathbb{Z}_5$.

9. (See the text answer for the series.) The factor groups are isomorphic to $\mathbb{Z}_3, \mathbb{Z}_2$, and $\mathbb{Z}_2$ in some order.

10. $\{(0,0,0)\} < \mathbb{Z}_2 \times \{0\} \times \{0\} < \mathbb{Z}_2 \times \mathbb{Z}_5 \times \{0\} < \mathbb{Z}_2 \times \mathbb{Z}_5 \times \mathbb{Z}_7,$
$\{(0,0,0)\} < \mathbb{Z}_2 \times \{0\} \times \{0\} < \mathbb{Z}_2 \times \{0\} \times \mathbb{Z}_7 < \mathbb{Z}_2 \times \mathbb{Z}_5 \times \mathbb{Z}_7,$
$\{(0,0,0)\} < \{0\} \times \mathbb{Z}_5 \times \{0\} < \{0\} \times \mathbb{Z}_5 \times \mathbb{Z}_7 < \mathbb{Z}_2 \times \mathbb{Z}_5 \times \mathbb{Z}_7,$
$\{(0,0,0)\} < \{0\} \times \mathbb{Z}_5 \times \{0\} < \mathbb{Z}_2 \times \mathbb{Z}_5 \times \{0\} < \mathbb{Z}_2 \times \mathbb{Z}_5 \times \mathbb{Z}_7,$
$\{(0,0,0)\} < \{0\} \times \{0\} \times \mathbb{Z}_7 < \{0\} \times \mathbb{Z}_5 \times \mathbb{Z}_7 < \mathbb{Z}_2 \times \mathbb{Z}_5 \times \mathbb{Z}_7,$
$\{(0,0,0)\} < \{0\} \times \{0\} \times \mathbb{Z}_7 < \mathbb{Z}_2 \times \{0\} \times \mathbb{Z}_7 < \mathbb{Z}_2 \times \mathbb{Z}_5 \times \mathbb{Z}_7.$

The factor groups are isomorphic to $\mathbb{Z}_2, \mathbb{Z}_5$, and $\mathbb{Z}_7$ in some order.

11. $\{\rho_0\} \times \mathbb{Z}_2$

12. $\{\rho_0\} \times \{\rho_0, \rho_2\}$

13. $\{\rho_0\} \times \mathbb{Z}_4 \leq \{\rho_0\} \times \mathbb{Z}_4 \leq \{\rho_0\} \times \mathbb{Z}_4 \leq \cdots$

14. $\{\rho_0\} \times \{\rho_0, \rho_2\} \leq \{\rho_0\} \times D_4 \leq \{\rho_0\} \times D_4 \leq \cdots$

15. The definition is correct.

16. The definition is incorrect. It is the factor groups of the series, not the series groups themselves, that need to be abelian.

   A **solvable group** G is a group that has a composition series $\{e\} = H_0 < H_1 < \cdots < H_n = G$ such that the quotient groups H_{i+1}/H_i are abelian for $i = 1, 2, \cdots, n-1$.

17. T F T F F T F F T T

18. $\{\rho_0\} \times \{\rho_0\} \leq A_3 \times \{\rho_0\} \leq S_3 \times \{\rho_0\} \leq S_3 \times A_3 \leq S_3 \times S_3$ is a composition series. Yes, $S_3 \times S_3$ is solvable because all the factor groups in this series are of order either 2 or 3 and hence are abelian.

19. Yes, D_4 is solvable, for $\{\rho_0\} \leq \{\rho_0, \rho_2\} \leq \{\rho_0, \rho_1, \rho_2, \rho_3\} \leq D_4$ is a composition series with all factor groups of order 2 and hence abelian.

20. *Chain* (3)   *Chain* (4)
   $\{0\} \leq \{0\} \leq \langle12\rangle$ $\{0\} \leq \{0\} \leq \langle18\rangle \leq \langle18\rangle$
   $\leq \langle6\rangle \leq \langle3\rangle$ $\leq \langle6\rangle \leq \langle3\rangle \leq \mathbb{Z}_{36}$
   $\leq \langle3\rangle \leq \mathbb{Z}_{36}$

Isomorphisms:
$\{0\}/\{0\} \simeq \{0\}/\{0\} \simeq \{0\}$, $\langle12\rangle/\{0\} \simeq \langle6\rangle/\langle18\rangle \simeq \mathbb{Z}_3$,
$\langle6\rangle/\langle12\rangle \simeq \langle18\rangle/\{0\} \simeq \mathbb{Z}_2$, $\langle3\rangle/\langle6\rangle \simeq \langle3\rangle/\langle6\rangle \simeq \mathbb{Z}_2$,
$\langle3\rangle/\langle3\rangle \simeq \langle18\rangle/\langle18\rangle \simeq \{0\}$, $\mathbb{Z}_{36}/\langle3\rangle \simeq \mathbb{Z}_{36}/\langle3\rangle \simeq \mathbb{Z}_3$

21. *Chain* (3)   *Chain* (4)
   $\{0\} \leq \langle12\rangle \leq \langle12\rangle \leq \langle12\rangle$ $\{0\} \leq \langle12\rangle \leq \langle12\rangle \leq \langle6\rangle$
   $\leq \langle12\rangle \leq \langle12\rangle \leq \langle4\rangle$ $\leq \langle6\rangle \leq \langle6\rangle \leq \langle3\rangle$
   $\leq \langle2\rangle \leq \mathbb{Z}_{24} \leq \mathbb{Z}_{24}$ $\leq \langle3\rangle \leq \mathbb{Z}_{24} \leq \mathbb{Z}_{24}$

Isomorphisms:
$\langle12\rangle/\{0\} \simeq \langle12\rangle/\{0\} \simeq \mathbb{Z}_2$, $\langle12\rangle/\langle12\rangle \simeq /\langle6\rangle/\langle6\rangle \simeq \{0\}$,

$\langle 12 \rangle / \langle 12 \rangle \simeq \langle 3 \rangle / \langle 3 \rangle \simeq \{0\}, \quad \langle 12 \rangle / \langle 12 \rangle \simeq \langle 12 \rangle / \langle 12 \rangle \simeq \{0\},$

$\langle 12 \rangle / \langle 12 \rangle \simeq \langle 6 \rangle / \langle 6 \rangle \simeq \{0\}, \quad \langle 4 \rangle / \langle 12 \rangle \simeq \mathbb{Z}_{24} / \langle 3 \rangle \simeq \mathbb{Z}_3,$

$\langle 2 \rangle / \langle 4 \rangle \simeq \langle 6 \rangle / \langle 12 \rangle \simeq \mathbb{Z}_2, \quad \mathbb{Z}_{24} / \langle 2 \rangle \simeq \langle 3 \rangle / \langle 6 \rangle \simeq \mathbb{Z}_2,$

$\mathbb{Z}_{24} / \mathbb{Z}_{24} \simeq \mathbb{Z}_{24} / \mathbb{Z}_{24} \simeq \{0\}$

22. Let $a \in H^* \cap K$ and let $b \in H \cap K$. Then $b \in H$ and $a \in H^*$ so $bab^{-1} \in H^*$ because H^* is a normal in H. Also $b \in K$ and $a \in K$ so $bab^{-1} \in K$. Thus $bab^{-1} \in H^* \cap K$, so $H^* \cap K$ is a normal subgroup of $H \cap K$.

23. We use induction to show that $|H_i| = s_1 s_2 \cdots s_i$ for $i = 1, 2, \cdots n$. For $n = 1, s_1 = |H_1 / H_0| = |H_1 / \{e\}| = |H_1|$ because each coset in $H_1 / \{e\}$ has only one element. Now suppose that $|H_k| = s_1 s_2 \cdots s_k$ for $k < i \leq n$. Now H_i / H_{i-1} consists of s_i cosets of H_{i-1}, each having $|H_{i-1}|$ elements. By our induction assumption, $|H_{i-1}| = s_1 s_2 \cdots s_{i-1}$. Thus $|H_i| = |H_{i-1}| s_i = s_1 s_2 \cdots s_{i-1} s_i$. Our induction is complete and the desired assertion follows by taking $i = n$.

24. By Definition 35.1, a composition series for G contains a *finite* number of subgroups of G. If G is infinite and abelian, and $\{e\} = H_0 < H_1 < H_2 < \cdots < H_n = G$ is a subnormal series, the factor groups H_i / H_{i-1} cannot all be of finite order for $i = 1, 2, \cdots, n$, or $|G|$ would be finite by Exercise 23. Suppose $|H_k / H_{k-1}|$ is infinite. Now every infinite abelian group has a proper nontrivial subgroup and hence a normal subgroup. To see this we need only consider the cyclic subgroup $\langle a \rangle$ for some $a \neq e$ in the group. If $\langle a \rangle$ is finite, we are done. If $\langle a \rangle$ is infinite cyclic, then $\langle a^2 \rangle$ is a proper subgroup. Thus we see that H_k / H_{k-1}, as an infinite abelian group, has a proper nontrivial subgroup, so it is not simple and our series is not a composition series.

25. Let $G = G_1 \times G_2 \times \cdots \times G_m$ and suppose that G_i is solvable for $i = 1, 2, \cdots, m$. We form a composition series for G as follows: Start with $H_0 = \{e_1\} \times \{e_2\} \times \cdots \times \{e_m\}$ where e_i is the identity of G_i. Let $H_1 = H_{11} \times \{e_2\} \times \cdots \times \{e_m\}$ where H_{11} is the smallest nontrivial subgroup of G_1 in a composition series $\{e_1\} < H_{11} < H_{12} < \cdots < H_{1n_1} = G_1$ for G_1. Continue to build the composition series for G by putting these subgroups H_{1i} in sequence in the first factor of the direct product series until you arrive at $G_1 \times \{e_2\} \times \cdots \times \{e_n\}$. Then start putting the sequence of subgroups $H_{21}, H_{22}, \cdots, H_{2n_2}$ in a composition series for G_2 into the second factor until you arrive at $G_1 \times G_2 \times \{e_3\} \times \cdots \times \{e_m\}$. Continue in this way across the factors in the direct product until you arrive at

$$G = H_{1n_1} \times H_{2n_2} \times \cdots \times H_{mn_m} = G_1 \times G_2 \times \cdots \times G_m.$$

A factor group formed from two consecutive terms of this series for G is naturally isomorphic to one of the factor groups in a composition series for one of the groups G_i by our construction. Thus these factor groups are all simple so we have indeed constructed a composition series for G. Because all the factor groups in the composition series for G_i are abelian for $i = 1, 2, \cdots, m$ we see that the factor groups of the composition series for G are abelian, so G is a solvable group.

26. Following the hint, Exercise 22 shows that $K \cap H_i$ is a normal subgroup of $K \cap H_{i+1}$ for $i = 0, 1, \cdots, n-1$, so the subgroups $K \cap H_{i+1}$ form a subnormal series for K. Taking $N = H_{i-1}$ and $H = K \cap H_i$ as subgroups of H_i, and applying Theorem 34.5, we see that $HN/N = [(K \cap H_i) H_{i-1}] / H_{i-1} \simeq H/(H \cap N) = (K \cap H_i) / (K \cap H_{i-1})$. Now $(K \cap H_i) H_{i-1} \leq (K \cap H_i) H_i = H_i$ so $[(K \cap H_i)] / H_{i-1}$ can be viewed as a subgroup of H_i / H_{i-1}. Because H_i / H_{i-1} is a simple abelian group, we see that $[(K \cap H_i) H_{i-1}] / H_{i-1}$ is either the trivial group or is isomorphic to H_i / H_{i-1}, and hence is simple and abelian because G is solvable. Thus the distinct groups among the $K \cap H_i$ form a composition series for K with abelian factor groups, and consequently K is solvable.

27. Following the hint, we show that $H_{i-1}N$ is normal in H_iN. Let $h_{i-1}n_1 \in H_{i-1}N$ and $h_in_2 \in H_iN$ where the elements belong to the obvious sets. Using the fact that H_{i-1} is normal in H_i and that N is normal in G, we obtain $(h_in_2)h_{i-1}n_1(h_in_2)^{-1} = h_in_2h_{i-1}n_1n_2^{-1}h_i^{-1} = h_ih_{i-1}n_3n_1n_2^{-1}h_i^{-1} = h'_{i-1}h_in_4h_i^{-1} = h'_{i-1}h_ih_i^{-1}n_5 = h'_{i-1}n_5 \in H_{i-1}N$. Thus $H_{i-1}N$ is a normal sugroup of H_iN.

The hint does the rest of the work for us, except to observe at the end that H_i/H_{i-1} being simple implies that $[H_i \cap (H_{i-1}N)]/H_{i-1}$ is either trivial or isomorphic to H_i/H_{i-1}. Thus $(H_iN)/(H_{i-1}N)$ is either isomorphic to H_i/H_{i-1} or is trivial. Because $N = H_0N$ is itself simple, it follows at once that the distinct groups among the H_iN for $i = 0, 1, 2, \cdots, n$ form a composition series for G.

28. First we show that ψ is well defined. Let h_in_1 and h'_in_2 be the same elements of H_iN. Then $h_in_1 = h'_in_2$ so $h_i = h'_in_2n_1^{-1} = h'_in_3$. Because γ has kernel N, we see that $\psi(h_in_1) = \gamma(h_in_1)\gamma[H_{i-1}] = (h_iN)\gamma[H_{i-1}] = (h'_in_3N)\gamma[H_{i-1}] = (h'_1N)\gamma[H_{i-1}] = \gamma(h'_1n_2)\gamma[H_{i-1}]$, so ψ is well defined.

We show that ψ is a homomorphism. This follows from $\gamma(h_in_1h'_in_2) = \gamma(h_in_1)\gamma(h'_in_2)$ because γ is a homomorphism.

The kernel of ψ consists of all $x \in H_iN$ such that $\gamma(x) \in \gamma[H_{i-1}] = H_{i-1}N$ and ψ is clearly an onto map. By Theorem 34.2, $\gamma[H_i]/\gamma[H_{i-1}]$ is isomorphic to $(H_iN)/(H_{i-1}N)$. Exercise 27 shows that these factor groups are simple, and the desired result follows immediately.

29. Let $H_0 = e < H_1 < H_2 < \cdots < H_n = G$ be a composition series for G, and let $\phi : G \to G'$ be a group homomorphism of G onto G' with kernel N. Then $G' \simeq G/N$. Exercise 28 shows that the distinct groups among the groups H_iN for $i = 0, 1, \cdots, n$ form a composition series for G/N, and Exercise 27 shows that a factor group of this composition series is isomorphic to one of the factor groups in the composition series of groups H_i for G. Because G is a solvable group, it follows at once that all the factor groups in this composition series for G/N, composed of some of the groups H_iN, are also abelian, so that G/N is solvable

36. Sylow Theorems

1. 3   **2.** 27

3. A Sylow 2-subgroup of a group of order $24 = 8 \cdot 3$ has order 8. By Theorem 36.11, the number of them must be conguent to 1 modulo 2, and hence is an odd number. It must also divide 24, and the only odd divisors of 24 are 1 and 3 so the group has either one or three Sylow 2-subgroups.

4. The only numbers congruent to 1 modulo 3 that divide $255 = 3 \cdot 5 \cdot 17$ are 1 and $5 \cdot 17 = 85$. The only numbers congruent to 1 modulo 5 that divide 255 are 1 and $3 \cdot 17 = 51$.

5. Because $|S_4| = 24$, the Sylow 3-subgroups have order 3 and are thus cyclic and generated by a single 3-cycle. The possibilities are
$\langle(1,2,3)\rangle$,
$\langle(1,2,4)\rangle = \langle(3,4)(1,2,3)(3,4)\rangle$,
$\langle(1,3,4)\rangle = \langle(2,4)(1,2,3)(2,4)\rangle$, and
$\langle(2,3,4)\rangle = \langle(1,4)(1,2,3)(1,4)\rangle$.

6. A Sylow 2-subgroup of S_4 has order 8 and by Theorem 36.11, there inust be either 1 or 3 of them. The group of symmetries of the square has order 8 and can be viewed as a subgroup of S_4 if we number the vertices 1, 2, 3, and 4.

If we number the four vertices in order 1, 2, 3, 4 counterclockwise, we obtain the group $H = \{(1), (1, 2, 3, 4), (1, 3)(2, 4), (1, 4, 3, 2), (1, 3), (2, 4), (1, 2)(3, 4), (1, 4)(2, 3)\}$.

If we number the four vertices in order 1, 3, 2, 4 counterclockwise, we obtain the group $K = \{(1), (1, 3, 2, 4), (1, 2)(3, 4), (1, 4, 2, 3), (1, 2), (3, 4), (1, 4)(2, 3), (1, 3)(2, 4)\}$.

If we number the four vertices in order 1, 3, 4, 2 counterclockwise, we obtain the group $L = \{(1), (1, 3, 4, 2), (1, 4)(2, 3), (1, 2, 4, 3), (1, 4), (2, 3), (1,3)(2, 4), (1, 2)(3, 4)\}$.

Because there can be at most three of them, we have found them all. We see that $K = (2, 3)H(2, 3)$ and $L = (3,4)H(3,4)$

7. The definition is incorrect. The order may be a power of p.

   Let p be a prime. A **p-group** is a group with the property that the order of each element is some power of p.

8. The definition is incorrect. Elements of the normalizer are elements of G, not maps of G onto G.

   The **normalizer** $N[H]$ of a subgroup H of a group G is the set of all $g \in G$ such that $\{ghg^{-1} \mid h \in H\} = H$.

9. The definition is misleading. A Sylow p-subgroup need not be unique, so replace the first two occurrences of "the" by "a". Also refer to the group as G after it is defined. The word "largest" also implies uniqueness, so replace it by "maximal".

   Let G be a group whose order is divisible by a prime p. A **Sylow p-subgroup** of G is a maximal subgroup P of G with the property that P has some power of p as its order.

10. T T T F T F T T F F

11. *Closure:* Let $a, b \in G_H$. Then $aHa^{-1} = H$ and $bHb^{-1} = H$. Thus $(ab)H(ab)^{-1} = a(bHb^{-1})a^{-1} = aHa^{-1} = H$, so $ab \in G_H$.

    Identity: For all $h \in H, ehe^{-1} = ehe = h$ so $eHe^{-1} = H$ and $e \in G_H$.

    Inverses: Let $a \in G_H$. Then $aHa^{-1} = H$. Therefore $H = eHe = (a^{-1}a)H(a^{-1}a) = a^{-1}(aHa^{-1})a = a^{-1}Ha$, so $a^{-1} \in G_H$. Thus G_H is a subgroup of G.

12. Let H be a Sylow p-subgroup of G. Because q divides $|G|$, we know that $H \neq G$. For each $g \in G$, the conjugate group gHg^{-1} is also a Sylow p-subgroup of G. Because G has only one Sylow p-subgroup, it must be that $gHg^{-1} = H$ for all $g \in G$, so that H is a proper normal subgroup of G, and G is thus not a simple group.

13. The divisors of 45 are 1, 3, 5, 9, 15, and 45. Of these, only 1 is congruent to 1 modulo 3, so by Theorem 36.11, there is only one Sylow 3-subgroup of a group of order 45. By the argument in Exercise 12, this subgroup must be a normal subgroup.

14. Let G be a p-group, so that by definition, every element of G has order a power of p. If a prime $q \neq p$ divides $|G|$ then G has an element of order q by Cauchy's theorem, contradicting that G is a p-group. Thus the order of G must be a power of p.

    Conversely, if the order of G is a power of p, then the order of each element of G is also a power of p by the Theorem of Lagrange. Thus G is a p-group.

15. Because $N[P]$ is a normal subgroup of $N[N[P]]$, conjugation of P by an element of $N[N[P]]$ yields a subgroup of $N[P]$ that is a Sylow p-subgroup of G, and also of $N[P]$. Such a Sylow p-subgroup must be conjugate to P under conjugation by an element of $N[P]$ by Theorem 36.10, and must therefore be P because P is a normal subgroup of $N[P]$. Thus P is invariant under conjugation by every element of $N[N[P]]$, so $N[N[P]]$ is contained in $N[P]$. Because $N[P]$ is contained in $N[N[P]]$ by definition, we see that $N[N[P]] = N[P]$.

16. By Theorem 36.8, H is contained in a Sylow p-subgroup K of G. By Theorem 36.10, there exists $g \in G$ such that $gKg^{-1} = P$. Consequently $gHg^{-1} \leq P$.

17. The divisors of $(35)^3$ that are not divisible by 5 are 1, 7, 49, and 343, which are congruent to 1, 2, 4, and 3 respectively modulo 5. By Theorem 36.11, there is only one Sylow 5-subgroup of a group of order $(35)^3$, and it must be a normal subgroup by the argument in Exercise 12.

18. The divisors of 255 that are not multiples of 17 are 1, 3, 5, and 15. Of these, only 1 is congruent to 1 modulo 17. By Theorem 36.11, there is only one Sylow 17-subgroup of a group of order 255, and it must be a normal subgroup by the argument in Exercise 12. Thus no group of order 255 can be simple.

19. The divisors of $p^r m$ that are not divisible by p are 1 and m. Because $m < p$, of these two divisors of $p^r m$, only 1 is congruent to 1 modulo p. By Theorem 36.11, there is a unique Sylow p-subgroup of a group of order $p^r m$ where $m < p$, and this must be a normal subgroup by the argument in Exercise 12. Thus such a group cannot be simple.

20. a. As a G-set under conjugation, we have $G_G = \{g \in G \mid gxg^{-1} = x \text{ for all } x \in G\} = \{g \in G \mid gx = xg \text{ for all } x \in G\} = Z(G)$ by definition of $Z(G)$.

b. Let G be a nontrivial p-group, so that $|G| = p^r$ for $r \geq 1$. By Theorem 36.1, we see that $|G| \equiv |G_G| \pmod{p}$. Thus p is a divisor of $|G_G|$, and hence is a divisor of $|Z(G)|$ by Part(**a**). Thus $Z(G)$ is nontrivial.

21. We proceed by induction on n. If $n = 1$, the statement is obviously true, for $H_0 = \{e\}$ and the entire group $G = H_1$ are the required subgroups. If $n = 2$, the subgroups are supplied by Theorem 36.8. Suppose the statement is true for $n = k$, and let G have order p^{k+1}. Let $Z(G)$ have order p^j where $j \geq 1$ by Exercise 20. If $j - k + 1$, then G is abelian and the subgroups provided by Theorem 36.8 are all normal subgroups of G and we are done. If $j < k + 1$, apply the induction hypothesis to $Z(G)$ to find its desired normal subgroups $H_0 < H_1 < \cdots < H_{j-1} < H_j = Z(G)$. Then form the factor group $G/Z(G)$ which has order p^{k+1-j} and find normal subgroups $K_1 < K_2 < \cdots < K_{k+1-j}$ of $G/Z(G)$ where the order of K_i is p^i. If $\gamma : G \to G/Z(G)$ is the canonical homomorphism, then $H_{j+i} = \gamma^{-1}[K_i]$ is a normal subgroup of G of order p^{i+j}, and all these subgroups H_i form the desired chain of normal subgroups of G.

22. Let P be a normal p-subgroup of G. By Theorem 36.8, there exists a Sylow p-subgroup H of G containing P. By Theorem 36.20, every Sylow p-subgroup of G is of the form gHg^{-1} for some $g \in G$. Because $gPg^{-1} = P$, we see that P is contained in every Sylow p-subgroup of G.

37. Applications of the Sylow Theory

1. a. The conjugate classes are $\{\rho_0\}$, $\{\rho_2\}$, $\{\rho_1, \rho_2\}$, $\{\mu_1, \mu_2\}$, and $\{\delta_1, \delta_2\}$.

b. The class equation is $8 = 2 + 2 + 2 + 2$.

2. As p-groups, groups of these orders are not simple by Theorem 36.8: 4, 8, 9, 16, 25, 27, 32, 49.

As groups of order pq, groups of these orders are not simple by Theorem 37.7: 6, 10, 14, 15, 21, 22, 26, 33, 34, 35, 38, 39, 46, 51, 55, 57, 58.

As groups of order $p^r m$ with $m < p$, groups of these orders are not simple by Exercise 19 of Section 36: 18, 20, 21, 28, 42, 44, 50, 52, 54.

The text showed that groups of these orders are not simple: 30, 36, 48.

Order 12: Such a group has either 1 or 3 Sylow 2-subgroups of order 4 and either 1 or 4 Sylow 3-subgroups of order 3. To have 3 subgroups of order 4 and 4 subgroups of order 3 would require at least 4 elements of order divisible by 2 and at least 8 elements of order 3, which would require 12 elements other than the identity. Thus there is either only one subgroup of order 4 or only one of order 3, which must be normal.

Order 24: Such a group has either 1 or 3 subgroups of order 8 and either 1 or 4 subgroups of order 3. If there is a unique subgroup of either order, we are done. Suppose that H and K are different subgroups of order 8. By Lemma 37.8, $H \cap K$ must have order 4, and is normal in both H and K, being of index 2. Thus $N[H \cap K]$ contains both H and K so it has order a multiple > 1 of 8 and a divisor of 24. Hence $N[H \cap K]$ is of order 24 and $H \cap K$ is a normal subgroup.

Order 40: Theorem 36.11 shows that there is a unique subgroup of order 5, which must be normal.

Order 45: Theorem 35.11 shows that there is a unique subgroup of order 9, which must be normal.

Order 56: Such a group has either 1 or 7 subgroups of order 8 and either 1 or 8 subgroups of order 7. If there is a unique subgroup of either order, we are done. Eight subgroups of order 7 require 48 elements of order 7, and 7 subgroups of order 8 require at least 8 elements of order divisible by 2, which is impossible in a group of 56 elements.

All orders from 2 to 59 that are not prime have been considered. We know that A_5, which has order 60, is simple.

3. T T F T T T T T F F

4. By Theorem 36.11, a group G of order $5 \cdot 7 \cdot 47$ contains a unique subgroup H of order 47, which must be normal in G. By the same arguments, there exist unique normal subgroups K and L of orders 7 and 5 respectively. By Lemma 37.8, LK has order 35 because $L \cap K = \{e\}$. By the proof of Lemma 37.8. $(LK)H$ has order $35 \cdot 47$, so $(LK)H = G$. Now LK must be the unique subgroup of G of order 35, because another subgroup would lead to subgroups of orders 7 and 5 other than L and K, which is impossible. By Lemma 37.5, G is isomorphic to $LK \times H$ and consequently to $L \times K \times H$ which is abelian and cyclic.

5. A group of order 96 has either 1 or 3 subgroups of order 32. If there is only one such subgroup, it is normal and we are done. If not, let H and K be distinct subgroups of order 32. By Lemma 37.8, $H \cap K$ must have order 16, and is normal in both H and K, being of index 2. Thus $N[H \cap K]$ has order a multiple > 1 of 32 and a divisor of 96, so the order must be 96. Thus $H \cap K$ is normal in the whole group.

6. A group G of order 160 has either 1 or 5 subgroups of order 32 and either 1 or 16 subgroups of order 5. If there is only one of order 32 or only one of order 5, it is a normal sbgroup and we are done. Let us suppose that this is not the case. Let H and K be distinct subgroups of order 32. By Lemma

37.8, $H \cap K$ must have order either 16 or 8. If $|H \cap K| = 16$, then it is normal in both H and K, so $N[H \cap K]$ has order a multiple > 1 of 32 and a divisor of 160, so $N[H \cap K] = G$ and $H \cap K$ is a normal subgroup of the group G. If $|H \cap K| = 8$, then HK has order $(32)(32)/8 = 128$ by Lemma 37.8, so G has at least 127 elements of order divisible by 2. Then 16 subgroups of order 5 would contribute 64 elements of order 5, and $127 + 64 > 160$, which is impossible. Thus G is not simple.

7. By Example 37.12, a group G of order 30 has a normal subgroup of order 5 or of order 3. Suppose that G has a normal subgroup H of order 5. Then G/H is a group of order 6, which has a normal subgroup K of order 3 by Sylow theory. If $\lambda : G \to G/H$ is the canonical homomorphism, then $\lambda^{-1}[K]$ is a normal subgroup of G of order $3 \cdot 5 = 15$. If G has no normal subgroup of order 5, then it has a normal subgroup N of order 3, so G/N has order 10 and has a normal subgroup L of order 5. Applying to L the inverse of the canonical homomorphism mapping G onto G/N gives a normal subgroup of G of order 15.

8. **a.** We have $\tau \sigma \tau^{-1}(\tau(a_i)) = \tau \sigma(a_i) = \begin{cases} \tau(a_{i+1}) & \text{if } i < m, \\ \tau(a_1) & \text{if } i = m. \end{cases}$
   For any element b not of the form $\tau(a_i)$, we have $\tau \sigma \tau^{-1}(b) = \tau \tau^{-1}(b) = b$. Thus $\tau \sigma \tau^{-1}$ has the desired action on each element of $\{1, 2, \cdots, n\}$.

   b. Let $\sigma = (a_1, a_1, \cdots, a_m)$ and $\mu = (b_1, b_2, \cdots, b_m)$ be cycles of length m in S_n. Let τ be any permutation in S_n such that $\tau(a_i) = b_i$. Part**a** then shows that $\tau \sigma \tau^{-1} = \mu$.

   c. Let σ and μ be products of s disjoint cycles with the ith cycle in each product of length r_i. Let τ be any permutation in S_n that carries the jth element of the ith cycle of σ into the jth element of the ith cycle of μ for $1 \le j \le r_i$ for each i where $1 \le i \le s$. (That is, if we write μ directly under σ and erase all the parentheses at the ends of the cycles, we get a 2-rowed notation for the action of τ on the elements moved by σ.) Fill in the action of τ on the other elements in any way that yields an element of S_n. Repetition of the computation in Part(a) shows that $\tau \sigma \tau^{-1} = \mu$.

   d. Let σ be any permutation in S_n. Express σ as a product of disjoint cycles, supplying cycles of length 1 for all elements not moved by σ. The sum of the lengths of these cycles is then n, and the sum yields a partition of n. By Part(c), σ is conjugate to any other permutation that can be expressed similarly as a product of disjoint cycles that yield the same partition of n. The preceding parts of this exercise show that this correspondence between partitions of n and conjugate classes is one to one.

   e. $1 = 1$, so $p(1) = 1$

   $2 = 2 = 1 + 1$, so $p(2) = 2$

   $3 = 3 = 1 + 2 = 1 + 1 + 1$, so $p(3) = 3$

   $4 = 4 = 1 + 3 = 2 + 2 = 1 + 1 + 2 = 1 + 1 + 1 + 1$, so $p(4) = 5$

   If $p_1(n)$ is the number of partitions of n having 1 as a summand, then $p_1(n) = p(n-1)$ for $n \ge 2$, for all such partitions of n can be obtained by putting $+1$ after each partition of $n - 1$. Thus for $n \ge 2, p(n) =$ (number of partitions without 1 as a summand) $+ p(n-1)$.

   $5 = 5 = 2 + 3$, so $p(5) = 2 + p(4) = 2 + 5 = 7$

   $6 = 6 = 2 + 4 = 3 + 3 = 2 + 2 + 2$, so $p(6) = 4 + p(5) = 4 + 7 = 11$

   $7 = 7 = 2 + 5 = 3 + 4 = 2 + 2 + 3$, so $p(7) = 4 + p(6) = 4 + 11 = 15$

9. Using Exercise 8, we see the conjugate classes are
   $\{\iota\}$,
   $\{(1, 2), (1, 3), (1, 4), (2, 3), (2, 4), (3, 4)\}$,

$\{(1,2)(3,4),\ (1,3)(2,4),\ (1,4)(2,3)\},$
$\{(1,2,3),(1,2,4),(1,3,4),(1,3,2),(1,4,2),(1,4,3),(2,3,4),(2,4,3)\},$
$\{(1,2,3,4),(1,2,4,3),(1,3,2,4),(1,3,4,2),(1,4,2,3),(1,4,3,2)\}$
The class equation is $24 = 1 + 6 + 3 + 8 + 6$.

10. *Class Equation for S_5* : Rather than list as we did for Exercise 9, we use combinatorics for S_5. All we want is the class equation.

There are $\binom{5}{2} = 10$ transpositions.

There are $\binom{5}{2}\binom{3}{2}/2 = (10)(3)/2 = 15$ products of two disjoint transpositions.

There are $\binom{5}{3} \cdot 2 = 10 \cdot 2 = 20$ cycles of length 3.

By the previous computation, there are $20 \cdot 1 = 20$ products of disjoint cycles of lengths 3 and 2.

There are $5 \cdot 6 = 30$ cycles of length 4. (Five choices for the number not moved by the cycle, and then 6 cycles using the remaining four numbers as shown in Exercise 9.)

There are $4 \cdot 3 \cdot 2 \cdot 1 = 24$ cycles of length 5. (Put 1 in left position, and fill the remaining four positions in 4! ways.)

The class equation is $120 = 1 + 10 + 15 + 20 + 20 + 30 + 24$.

Class Equation for S_6 : We continue to use combinatorics.

There are $\binom{6}{2} = 15$ transpositions.

There are $\binom{6}{2}\binom{4}{2}/2 = (15)(6)/2 = 45$ products of two disjoint 2-cycles.

There are $\binom{6}{2}\binom{4}{2}\binom{2}{2}/6 = (15)(6)(1)/6 = 15$ products of three disjoint 2 cycles.

There are $\binom{6}{3} \cdot 2 = 20 \cdot 2 = 40$ cycles of length 3

There are $40 \cdot \binom{3}{2} = 120$ products of a 3-cycle and a disjoint 2-cycle. (40 choices for the 3 cycle by the last computation, times $\binom{3}{2}$ choices for a tranposition from the remaining three elements.)

There are $(40 \cdot 2)/2 = 40$ products of two disjoint 3-cycles. (Choose one of the 40 3-cycles, there are only two choices for the other 3 cycle, and divide by 2 because the order of their choice doesn't matter.)

There are $\binom{6}{4} \cdot 6 = 15 \cdot 6 = 90$ 4-cycles. (Choose 4 of the 6 numbers as those to be moved, and there are 6 different 4-cycles moving on them as shown in the solution of Exercise 9.)

There are $90 \cdot 1 = 90$ products of a 4-cycle and a disjoint 2-cycle. (Choose 1 of the 90 4-cycles, and there is only one choice for the disjoint 2-cycle.)

There are $\binom{6}{5} \cdot 4 \cdot 3 \cdot 2 \cdot 1 = 6 \cdot 24 = 144$ 5-cycles. (Choose the 5 elements to be moved, list the smallest of them at the left, and then fill in the remaining 4 position in 4! ways.)

There are $5! = 120$ 6-cycles. (Put 1 in left position, and fill the remaining 5 positions in 5! ways.)

The class equation is $720 = 1 + 15 + 45 + 15 + 40 + 120 + 40 + 90 + 90 + 144 + 120$.

11. Exercise 8 shows that the number of conjugate classes in S_n is the number $p(n)$ of partitions of n. By Theorem 11.12, the number of abelian groups of order p^n is also the number of partitions of n; it is the number of ways that p^n can be split up into a product of powers of p, where the order of the product doesn't matter. It is determined by how split the exponent n into a sum of exponents for the factors; that is, by how to partition n.

12. Each element of the center of a group G gives rise to a 1-element conjugate class of G. It is clear from Exercise 8a that for $n > 2$, every permutation in S_n having an orbit with more than one element is conjugate to some other permutation in S_n. Thus if $n > 2$, the identity in S_n is the only element that is conjugate only to itself.

38. Free Abelian Groups

1. $\{1, 1, 1), (1, 2, 1), (1, 1, 2)\}$ is a basis. (Note that the 2nd - 1st gives $(0, 1, 0)$ and the 3rd - 1st gives $(0, 0, 1)$, so it is clear that this set generates, and it has the right number of elements for a basis by Theorem 38.6.

2. Yes, $\{(2, 1), (3, 1)\}$ is a basis. Now $(1, 0) = (3, 1) + (-1)(2, 1)$ and $(0, 1) = 3(2, 1) + (-2)(3, 1)$ so $\{(2, 1), (3, 1)\}$ generates $\mathbb{Z} \times \mathbb{Z}$. If $m(2,1) + n(3,1) + = (0,0)$, then $2m + 3n = 0$ and $m + n = 0$. Then $m = -n$ so $2(-n) + 3n = 0, n = 0$, and $m = 0$. Thus the conditions for a basis in Theorem 38.1 are satisfied.

3. (See the answer in the text.)

4. By Cramer's rule, the equations

$$
\begin{aligned}
ax + cy &= e \\
bx + dy &= f
\end{aligned}
$$

have a unique solution in $\mathbb{R}$ if and only if $ad - bc \neq 0$. The solution is then

$$
x = \frac{ed - fc}{ad - bc} \quad \text{and} \quad y = \frac{af - be}{ad - bc}.
$$

These values x and y are integers for *all* choices of e and f if and only if $D = ad - bc$ divides each of $a, b, c,$ and d. Let $a = r_a D, b = r_b D, c = r_c D$, and $d = r_d D$. Then $D = ad - bc = (r_a r_d - r_b r_c)D^2$ so that D is an integer which is an integer multiple of its square. The only possible values for D are ± 1, so we obtain the condition $|ad - bc| = 1$.

5. The definition is incorrect. Change "generating set" to "basis".

The **rank** of a free abelian group G is the number of elements in a basis for G.

6. The definition is correct.

7. $2\mathbb{Z}$ is a proper subgroup of rank $r = 1$ of the free abelian group $\mathbb{Z}$ of rank $r = 1$.

8. T T T T T F F T T F

9. Let $\phi : G \to \underbrace{\mathbb{Z} \times \mathbb{Z} \times \cdots \times \mathbb{Z}}_{r \text{ factors}}$ be the map described before the statement of Theorem 38.5. Note that ϕ is well defined because each $a \in G$ has a *unique* expression in the form $n_1 x_1 + n_2 x_2 + \cdots + n_r x_r$ where each $n_i \in \mathbb{Z}$. Suppose $b \in G$, and $b = m_1 x_1 + m_2 x_2 + \cdots + m_r x_r$. Then

$$
\begin{aligned}
\phi(a + b) &= \phi[(n_1 + m_1)x_1 + (n_2 + m_2)x_2 + \cdots + (n_r + m_r)x_r] \\
&= (n_1 + m_1, n_2 + m_2, \cdots, n_r + m_r) \\
&= (n_1, n_2, \cdots, n_r) + (m_1, m_2, \cdots, m_r) \\
&= \phi(a) + \phi(b)
\end{aligned}
$$

so ϕ is a homomorphism. If $\phi(a) = \phi(b)$, then $n_i = m_i$ for $i = 1, 2, \cdots, r$ so $a = b$; this shows that ϕ is one to one. Clearly ϕ is an onto map because $n_1 x_1 + n_2 x_+ \cdots + n_r x_r$ is in G for all integer choices of the coefficients n_i, for $i = 1, 2, \cdots, r$. Thus ϕ is an isomorphism.

10. Let G be free abelian with a basis X. Let $a \neq 0$ in G be given by $a = n_1 x_1 + n_2 x_2 + \cdots + n_r x_r$ where $x_i \in X$ and $n_i \in \mathbb{Z}$ for $i = 1, 2, \cdots, r$. If a has finite order $m > 0$, then $ma = mn_1 x_1 + mn_2 x_2 + \cdots + mn_r x_r = 0$. Because G is free abelian and X is a basis, we deduce that $mn_1 = mn_2 = \cdots = mn_r = 0$, so $n_1 = n_2 = \cdots = n_r = 0$ and $a = 0$, contradicting our choice of a. Thus no element of G has finite order > 0.

11. Suppose that G and G' are free abelian with bases X and X' respectively. Let $\overline{X} = \{(x, 0) \mid x \in X\}$ and $\overline{X}' = \{(0, x') \mid x' \in X'\}$. We claim that $Y = \overline{X} \cup \overline{X}'$ is a basis for $G \times G'$. Let $(g, g') \in G \times G'$. Then

$$g = n_1 x_1 + \cdots + n_r x_r \quad \text{and} \quad g' = m_1 x_1' + \cdots + m_s x_s'$$

for *unique* choices of n_i and m_j, except for possible zero coefficients. Thus

$$(g, g') = n_1(x_1, 0) + \cdots + n_r(x_r, 0) + m_1(0, x_1') + \cdots + m_s(0, x_s')$$

for *unique* choices of the n_i and m_j, except for possible zero coefficients. This shows that Y is a basis for $G \times G'$, which is thus free abelian.

12. If G is free abelian of finite rank, then G is of course finitely generated, and by Exercise 10, G has no elements of finite order. Conversely, if G is a finitely generated torsion-free abelian group, then Theorem 11.12 shows that G is isomorphic to a direct product of the group $\mathbb{Z}$ with itself a finite number of times, so G is free abelian of finite rank.

13. Because $\mathbb{Q}$ is not cyclic, any basis for $\mathbb{Q}$ must contain at least two elements. Suppose n/m and r/s are in a basis for $\mathbb{Q}$ where n, m, r, and s are nonzero integers. Then

$$mr\frac{n}{m} + (-ns)\frac{r}{s} = rn - nr = 0,$$

which is an impossible relation in a basis. Thus $\mathbb{Q}$ has no basis, so it is not a free abelian group.

14. Suppose $p^r a = 0$ and $p^s b = 0$. Then $p^{r+s}(a + b) = p^s(p^r a) + p^r(p^s b) = p^s 0 + p^r 0 = 0 + 0 = 0$, so $a + b$ is also of p-power order. Also $0 = p^r 0 = p^r[a + (-a)] = p^r a + p^r(-a) = 0 + p^r(-a)$, so $-a$ also has p-power order. Thus all elements of T of p-power order, together with zero, form a subgroup T_p of T.

15. Given the decomposition in Theorem 11.12, it is clear that the elements of T of p-power order are precisely those having 0 in all components except those of the form $\mathbb{Z}_{p^r}$. (Recall that the order of an element in a direct product is the least common multiple of the orders of its components in the individual groups.) Thus T_p is isomorphic to the direct product of those factors having p-power order.

16. Suppose that $na = nb = 0$ for $a, b \in G$. Then $n(a + b) = na + nb = 0 + 0 = 0$. This shows that $G[n]$ is closed under the group addition. If $na = 0$, then $0 = na = n[a + (-a)] = na + n(-a) = 0 + n(-a)$, so $n(-a) = 0$ also. Of course $n0 = 0$. Thus $G[n]$ is a subgroup of G.

17. Let $x \in \mathbb{Z}_{p^r}$. If $px = 0$, then px, computed in $\mathbb{Z}$, is a multiple of p^r. The possibilities for x are

$$0, 1p^{r-1}, 2p^{r-1}, 3p^{r-1}, \cdots, (p-1)p^{r-1}.$$

Clearly these elements form a subgroup of $\mathbb{Z}_{p^r}$ that is isomorphic to $\mathbb{Z}_p$.

18. This follows at one from Exercise 17 and the fact that for abelian groups G_i, we have

$$(G_1 \times G_2 \times \cdots \times G_m)[p] = G_1[p] \times G_2[p] \times \cdots \times G_m[p].$$

This relation follows at once from the fact that computation in a direct prodct is performed in the component groups.

19. a. By Exercise 18, both m and n are $\log_p |T_p[p]|$.

b. Suppose that $r_1 < s_1$. Then the prime-power decomposition of the subgroup $p^{r_1}T_p$ computed using the first decomposition of T_p would have less than m factors, while the decomposition of the same subgroup computed using the second decomposition of T_p would still have $m = n$ factors. But applying Part(**a**) to this subgroup, we see that this is an impossible situation; the number of factors in the prime-power decomposition of an abelian p-power group H is well defined as $\log_p |H[p]|$. Thus $r_1 = s_1$.

Proceeding by induction, suppose that $r_i = s_i$ for all $i < j$, and suppose $r_j < s_j$. Multiplication of elements of T_p by p^{r_j} annihilates all components in the first decomposition given of T_p through at least component j, while the component $\mathbb{Z}_{p^{s_j}}$ of the second decomposition given is not annihilated. This would contradict the fact that by Part(**a**), the number of factors in the prime-power decomposition of any abelian p-power group, in particular of $p^{r_j}T_p$, is well defined. Thus $r_j = s_j$ and our induction proof is complete.

20. If $m = p_1^{r_1} p_2^{r_2} \cdots p_k^{r_k}$ for distinct primes p_i, then we know that $\mathbb{Z}_m \simeq \mathbb{Z}_{p_1^{r_1}} \times \mathbb{Z}_{p_2^{r_2}} \times \cdots \times \mathbb{Z}_{p_k^{r_k}}$ from Section 11. If we form this decomposition of each factor in a torsion-coefficient decomposition, we obtain the unique (up to order of factors) prime-power decomposition .

21. Following the notation defined in the exercise, we know that if the unique prime-power decomposition is formed from a torsion-coefficient decomposition, as described in Exercise 20, then cyclic factors of order $p_i^{h_i}$ must appear for each $i = 1, \cdots, t$. Because each torsion coefficient except the final one must divide the following one, the final one must contain as factors all these prime powers $p_i^{h_i}$ for $i = 1, \cdots, t$. Because the p_i for $i = 1, \cdots, t$ are the only primes that divide $|T|$, we see that m_r and n_r must both be equal to $p_1^{h_1} p_2^{h_2} \cdots p_t^{h_t}$.

22. If we cross off the last factors $\mathbb{Z}_{m_r}$ and $\mathbb{Z}_{n_r}$ in the two given torsion-coefficient decompositions of T, we get torsion-coefficient decompositions of isomorphic groups, because both decompositions must, by Exercises 19 and 20, be isomorphic to the group obtained by crossing off from the prime-power decomposition of T, one factor of order $p_i^{h_i}$ for $i = 1, \cdots, t$. (We are using the notation of Exercise 21 here.) We now apply the argument of Exercise 21 to these torsion-coefficient decompositions of this group, and deduce that $m_{r-1} = n_{r-1}$. Continuing to cross off identical final factors, we see that we must have the same number of factors, that is, $r = s$, and $m_{r-i} = n_{r-i}$ for $i = 0, \cdots, r - 1$.

39. Free Groups

1. a. We obtain $a^2b^2a^3c^3b^{-2}$ whose inverse is $b^2c^{-3}a^{-3}b^{-2}a^{-2}$.

b. We obtain $a^{-1}b^3a^4c^6a^{-1}$ whose inverse is $ac^{-6}a^{-4}b^{-3}a$.

2. For the product in Part(**a**) of Exercise 1, it reduces to a^5c^3 and its inverse reduces to $a^{-5}c^{-3}$ in the abelian case. For Part(**b**), the product reduces to $a^2b^3c^6$ and its inverse reduces to $a^{-2}b^{-3}c^{-6}$.

3. **a.** There are $4 \cdot 4 = 16$ homomorphisms, because each of the two generator can be mapped into any one of four elements of $\mathbb{Z}_4$ by Theorem 39.12.

   b. There are $6 \cdot 6 = 36$ homomorphisms by reasoning analagous to that in Part(**a**).

   c. There are $6 \cdot 6 = 36$ homomorphisms by reasoning analagous to that in Part(**a**).

4. **a.** Let the free group have generators x and y. By Theorem 39.12, x and y can be mapped into any elements to give a homomorphism. The homomorphism will be onto $\mathbb{Z}_4$ if and only if not both x and y are mapped into the subgroup $\{0, 2\}$. Because 4 of the 16 possible homomorphisms map x and y into $\{0, 2\}$, there are 16 - 4 = 12 homomorphisms onto $\mathbb{Z}_4$.

   b. Arguing as in Part(**a**), we eliminate the 4 homomorphisms that map x and y into $\{0, 3\}$ and the 9 that map x and y into $\{0, 2, 4\}$. The homomorphism mapping both x and y into $\{0\}$ is counted in both cases, so there are a total of 12 of the possible homomorphisms to eliminate, so 36 - 12 = 24 are onto $\mathbb{Z}_6$.

   c. Arguing as in Part(**a**), we eliminate the 4 homomorphisms that map x and y into $\{\rho_0, \mu_1\}$, the 4 that map x and y into $\{\rho_0, \mu_2\}$, the 4 that map x and y into $\{\rho_0, \mu_3\}$, and the 9 that map x and y into $\{\rho_0, \rho_1, \rho_2\}$. The homomorphism that maps x and y into $\{0\}$ is counted four times, so we have found a total of 4 + 4 + 4 + 9 - 3 = 18 homomorphisms to eliminate, leaving 36 - 18 = 18 that are onto S_3.

5. **a.** There are 16 homomorphisms by the count in Exercise 3a.

   b. There are 36 homomorphisms by count in Exercise 3b.

   c. Because a homomorphic image of an abelian group is abelian, the image must be $\{0\}, \{\rho_0, \mu_1\}$, $\{\rho_0, \mu_2\}, \{\rho_0, \mu_3\}$, or $\{\rho_0, \rho_1, \rho_2\}$. The count made in Exercise 4c shows that there are 18 such homomorphisms.

6. **a.** There are 12 homomorphisms onto $\mathbb{Z}_4$ as in Exercise 4a.

   b. There are 24 homomorphisms onto $\mathbb{Z}_6$ as in Exercise 3b.

   c. There are no homomorphisms onto S_3, because the homomorphic image of an abelian group must be abelian, and S_3 is not abelian.

7. The definition is correct.

8. The definition is incorrect. The group must be free on the set of generators.

   The **rank** of a free group G is the number of elements in a generating set A such that G is free on A.

9. Our reaction to these instances was given in the text. *You* have to give your reaction.

10. T F F T F F F T F T

11. **a.** We have $3(2) + 2(3) = 0$ but $3(2) \neq 0$ and $2(3) \neq 0$. A basis for $\mathbb{Z}_4$ is $\{1\}$.

   b. We see that $\{1\}$ is a basis for $\mathbb{Z}_6$ because the group is cyclic with generator 1, and because $m1 = 0$ if and only if $m1 = 0$.

   If $m_1 2 + m_2 3 = 0$ in $\mathbb{Z}_6$, then in $\mathbb{Z}$, we know that 6 divides $m_1 2 + m_2 3$. Thus 3 divides $m_1 2 + m_2 3$, and hence 3 divides $m_1 2$. Because 3 is prime and does not divide 2, it must be that 3 divides m_1. Thus 6 divides $m_1 2$ in $\mathbb{Z}$, so $m_1 2 = 0$ in $\mathbb{Z}_6$. A similar argument starting with the fact that 2 divides $m_1 2 + m_2 3$ shows that $m_2 3 = 0$ in $\mathbb{Z}_6$. Thus $\{2, 3\}$ is a basis for $\mathbb{Z}_6$.

c. Yes it is, for if x_i is an element of a basis of a free abelian group, then $n_i x_i = 0$ if and only if $n_i = 0$, so we stated the "independence condition" in that form there.

d. By Theorem 38.12, a finite abelian group G is isomorphic to a direct product $\mathbb{Z}_{m_1} \times \mathbb{Z}_{m_2} \times \cdots \times \mathbb{Z}_{m_r}$ where m_i divides m_{i+1} for $i = 1, 2, \cdots, r-1$. Let b_i be the element of this direct product having 1 in the ith component and 0 in the other components. The computation by components in a direct product shows at once that $\{b_1, b_2, \cdots, b_r\}$ is a basis, and because the order of each b_i is m_i, we see that the orders have the desired divisibility property.

12. a. We proceed as suggested by the hint, and use the notation given there. Let $x \in G_1^*$. Because ϕ_1 is onto, we have $x = \phi_1(y)$ for some $y \in G$. Then $\theta_2(x) = \theta_2\phi_1(y) = \phi_2(y)$ and $\theta_1\theta_2(x) = \theta_1\phi_2(y) = \phi_1(y) = x$. In a similar fashion, starting with $z \in G_2^*$, we can show that $\theta_2\theta_1(z) = z$. Thus both $\theta_1\theta_2$ and $\theta_2\theta_1$ are identity maps. Because $\theta_2\theta_1$ is the identity, θ_2 must be an onto map and θ_1 must be one to one. Because $\theta_1\theta_2$ is the identity, θ_1 is an onto map and θ_2 is one to one. Thus both θ_1 and θ_2 are one to one and onto, and hence are isomorphisms. Thus G_1^* and G_2^* are isomorphic groups.

b. Let C be the commutator subgroup of G and let $\phi : G \to G/C$ be the canonical homomorphism (which we have usually called γ). Let $\psi : G \to G'$ be a homomorphism of G into an abelian group G'. Then the kernel K of ψ contains C by Theorem 15.20. Let $\theta : G/C \to G'$ be defined by $\theta(aC) = \psi(a)$ for $aC \in G/C$. Now θ is well defined, for if $bC = aC$, then $b = ac$ for some $c \in C$ and $\theta(bC) = \psi(b) = \psi(ac) = \psi(a)\psi(c) = \psi(a)e' = \psi(a)$ because C is contained in the kernel K of ψ. Now $\theta((aC)(bC)) = \theta((ab)C) = \psi(ab) = \psi(a)\psi(b) = \theta(aC)\theta(bC)$, so θ is a homomorphism. Finally, for $a \in G$, we have $\theta(\phi(a)) = \theta(aC) = \psi(a)$, so $\theta\phi = \psi$, which is the desired factorization. Thus $G^* = G/C$ is a blip group of G.

c. A blip group of G is isomorphic to the *abelianized version* of G, that is, to G modulo its commutator subgroup.

13. a. Consider the blop group G_1 on S and let G' be the free group $F[S]$, with $f : S \to F[S]$ given by $f(s) = s$ for $s \in S$. Because f is one to one and $f = \phi_f g_1$, we see that g_1 must be one to one. By a similar argument, g_2 must be one to one.

To see that $g_1[S]$ generates G_1, we take $G' = G_1$ and let $f(s) = g_1(s)$ for all $s \in S$. Clearly the identity map $\iota : G_1 \to G_1$ is a homomorphism and $f(s) = g_1(s) = \iota(g_1(s)) = (\iota g_1)(s)$, so $f = \iota g_1$, and the *unique* homomorphism ϕ_f mapping G_1 into G_1 is ι. Let H_1 be the subgroup of G_1 generated by $g_1[S]$. Thinking of H_1 as G' for a moment, we see that by hypothesis, there exists a homomorphism, $\phi_{H_1} : G_1 \to H_1$ and satisfying $f = \phi_{H_1} g_1$. Now ϕ_{H_1} also maps G_1 into G_1, and can also serve as the required homomorphism ϕ_f for the case where $G' = G_1$. By the uniqueness of ϕ_f, we see that $\phi_{H_1} = \iota$. Because $\phi_{H_1}[G_1] = H_1$ and $\iota[G_1] = G_1$, this can only be the case if $H_1 = G_1$. Therefore $g_1[S]$ generates G_1, and of course changing subscripts from 1 to 2 shows that $g_2[S]$ generates G_2.

Taking $G' = G_2$ and $f = g_2$, we obtain a homomorphism $\phi_{g_2} : G_1 \to G_2$ such that $\phi_{g_2} g_1 = g_2$. Taking $G' = G_1$ and $f = g_1$, we obtain a homomorphism $\phi_{g_1} : G_2 \to G_1$ such that $\phi_{g_1} g_2 = g_1$. Then for $s \in S$, we have $\phi_{g_1}\phi_{g_2} g_1(s) = \phi_{g_1} g_2(s) = g_1(s)$. Thus $\phi_{g_1}\phi_{g_2}$ is a homomorphism mapping G_1 into itself and acts as the identity on a generating set $g_1[S]$ of G_1, so it is the identity map of G_1 onto G_1. By a symmetric argument, $\phi_{g_2}\phi_{g_1}$ is the identity map of G_2 onto G_2. As in Exercise 12, we conclude that ϕ_{g_1} and ϕ_{g_2} are isomorphisms, so that $G_1 \simeq G_2$.

b. Let $G = F[S]$, the free group on S, and let $g : S \to G$ be defined by $g(s) = s$ for all $s \in S$. Let a group G' and a function $f : S \to G'$ be given. Let $\phi_f : G \to G'$ be the unique homomorphism given by Theorem 39.12 such that $\phi_f(s) = f(s)$. Then $\phi_f g(s) = \phi_f(s) = f(s)$ for all $s \in S$, so $\phi_f g = f$.

c. A blop group on S is isomorphic to the *free group* $F[S]$ on S.

14. The characterization is just like that in Exercise 13 with the requirement that both G and G' be abelian groups.

40. Group Presentations

1. Three presentations of $\mathbb{Z}_4$ are $(a : a^4 = 1)$, $(a, b : a^4 = 1, b = a^2)$, and $(a, b, c : a = 1, b^4 = 1, c = a)$.

2. Thinking of $a = \rho_1, b = \mu_1$, and $c = \rho_2$, we obtain the presentation $(a, b, c : a^3 = 1, b^2 = 1, c = a^2, ba = cb)$. Starting with this presentation, the relations can be used to express every word in one of the forms $1, a, b, a^2, ab$, or a^2b, so a group with this presentation has at most 6 elements. Because the relations are satisfied by S_3, we know it must be a presentation of a group isomorphic to S_3. (Many other answers are possible.)

3. (See the answer in the text.)

4. Let G be a nonabelian group of order 14. By Sylow theory, there exists a normal subgroup H of order 7. Let b be an element of G that is not in H. Because $G/H \simeq \mathbb{Z}_2$, we see that $b^2 \in H$. If $b^2 \neq 1$, then b has order 14 and G is cyclic and abelian. Thus $b^2 = 1$. Let a be a generator for the cyclic group H. Now $bHb^{-1} = H$ so $bab^{-1} \in H$, so $ba = a^r b$ for some value of r where $1 \leq r \leq 6$. If $r = 1$, then $ba = ab$ and G is abelian. By Exercise 13, Part(**b**), the presentation

$$(a, b : a^7 = 1, b^2 = 1, ba = a^r b)$$

gives a group of order $2 \cdot 7 = 14$ if and only if $r^2 \equiv 1 \pmod 7$. Of the possible values $r = 2, 3, 4,$ 5, 6, only $r = 6$ satisfies this condition. Thus every nonabelian group of order 14 is isomorphic to the group with presentation $(a, b : a^7 = 1, b^2 = 1, ba = a^6 b)$. We know then that this group must be isomorphic to the dihedral group D_7. Of course, $\mathbb{Z}_{14}$ is the only abelian group of order 14.

5. Let G be nonabelian of order 21. By Sylow theory, there exists a normal subgroup H of order 7. Let b be an element of G that is not in H. Because $G/H \simeq \mathbb{Z}_3$, we see that $b^3 \in H$. If $b^3 \neq 1$, then b has order 21 and G is cyclic and abelian. Thus $b^3 = 1$. Let a be a generator for the cyclic group H. Now $bHb^{-1} = H$ so $bab^{-1} \in H$, so $ba = a^r b$ for some value of r where $1 \leq r \leq 6$. If $r = 1$, then $ba = ab$ and G is abelian. By Exercise 13, Part(**b**), the presentation

$$(a, b : a^7 = 1, b^3 = 1, ba = a^r b)$$

gives a group of order $3 \cdot 7 = 21$ if and only if $r^3 \equiv 1 \pmod 7$. Of the possible values $r = 2, 3,$ 4, 5, 6, both $r = 2$ and $r = 4$ satisfy this condition. To see that the presentations with $r = 2$ and $r = 4$ yield isomorphic groups, consider the group having this presentation with $r = 2$, and let us form a new presentation of it, taking the same a but replacing b by $c = b^2$. We then have $a^7 = 1$ and $c^3 = 1$, but now $ca = b^2 a = a^4 b^2 = a^4 c$. Thus in terms of the elements a and c, this group has presentation $(a, c : a^7 = 1, c^3 = 1, ca = a^4 c)$. This shows the two values $r = 2$ and $r = 4$ lead to isomorphic presentations. Thus every group of order 21 is isomorphic to either $\mathbb{Z}_{21}$ or to the group with presentation

$$(a, b : a^7 = 1, b^3 = 1, ba = a^2 b).$$

6. The definition is incorrect. A consequence may be any element of the normalizer of the group generated by the relators in the free group on the generators. Also, one should say, "the relators in a presentation of a group".

A **consequence** of the set of relators in a group presentation is any element of the least normal subgroup, containing the relators, of the free group on the generators of the presentation.

7. The definition is incorrect. See Example 40.3 and the definition (in bold type within the text) that follows.

Two group presentations are **isomorphic** if and only if the groups G and G' presented by them are isomorphic.

8. T T F F F T T F T F (Concerning the answer to Part(**a**), for any group $\bar{G}$, the presentation $F[G]$ with relators the elements of the kernel of the homomorphism $\phi : F[G] \rightarrow G$ where $\phi(g) = g$ for $g \in G$, as described in Theorem 39.12, is a presentation of G. Concerning the answer to Part(**j**), the presentation $(a, b, c : c = b)$ is a free group on two generators.)

9. Let G be nonabelian of order 15. By Sylow theory, there exists a normal subgroup H of order 5. Let b be an element of G that is not in H. Because $G/H \simeq \mathbb{Z}_3$, we see that $b^3 \in H$. If $b^3 \neq 1$, then b has order 15 and G is cyclic and abelian. Thus $b^3 = 1$. Let a be a generator for the cyclic group H. Now $bHb^{-1} = H$ so $bab^{-1} \in H$, so $ba = a^r b$ for some value of r where $1 \leq r \leq 4$. If $r = 1$, then $ba = ab$ and G is abelian. By Exercise 13, Part(**b**), the presentation

$$(a, b : a^5 = 1, b^3 = 1, ba = a^r b)$$

gives a group of order $3 \cdot 5 = 15$ if and only if $r^3 \equiv 1 \pmod 5$. But none of $2^3, 3^3$, or 4^3 is congruent to 1 modulo 5, so there are no nonabelian groups of order 15.

10. Exercise 13, Part(**b**), shows that the given presentation is a group of order $2 \cdot 3 = 6$ if and only if $2^2 \equiv 1 \pmod 3$, which is the case. Thus we do have a group of order 6. Because the elements $1, a, b, a^2, ab, a^2 b$ are all distinct and because $ba = a^2 b$, the group is not abelian, for if $ba = ab$, then $a^2 b = ab$ from which we deduce that $a = 1$.

11. Let G be nonabelian of order 6. By Sylow theory, there exists a normal subgroup H of order 3. Let b be an element of G that is not in H. Because $G/H \simeq \mathbb{Z}_2$, we see that $b^2 \in H$. If $b^2 \neq 1$, then b has order 6 and G is cyclic and abelian. Thus $b^2 = 1$. Let a be a generator for the cyclic group H. Now $bHb^{-1} = H$ so $bab^{-1} \in H$, so $ba = a^r b$ for some value of r where $1 \leq r \leq 2$. If $r = 1$, then $ba = ab$ and G is abelian. Thus $ba = a^2$. The preceding exercise shows that the presentation $(a, b : a^3 = 1, b^2 = 1, ba = a^2 b)$ gives a nonabelian group of order 6, and this exercise shows that a nonabelian group of order 6 is isomorphic to one with this presentation. Thus every nonabelian group of order 6 is isomorphic to S_3.

12. Every element of A_4 can be written as a product of disjoint cycles involving some of the numbers 1, 2, 3, 4 and each element is also an even permutation. Because no product of such disjoint cycles can give an element of order 6, we see that A_4 has no elements of order 6, and hence no subgroup isomorphic to $\mathbb{Z}_6$, the only possibility for an abelian subgroup of order 6. Therefore, any subgroup of order 6 of A_4 must be nonabelian, and hence isomorphic to S_3 by the preceding exercise. Now S_3 has two elements of order 3 and three elements of order 2. The only *even* permutations in S_4 of order 2 are products of two disjoint transpositions, and the only such permutations are (1,2)(3,4) and (1,3)(2,4) and (1,4)(2,3). Thus a subgroup of A_4 isomorphic to S_3 must contain all three of these elements. It must also contain an element of order 3: we might as well assume that it is the 3-cycle $(1, 2, 3)$. Then it must contain $(1, 2, 3)^2 = (1, 3, 2)$, and the identity would be the sixth element. But this set is not closed under multiplication, for $(1, 2)(3, 4)(1, 2, 3) = (2, 4, 3)$. Thus A_4 has no nonabelian subgroup of order 6 either.

13. **a.** We know that when computing integer sums modulo n, we may either reduce modulo n after each addition, or add in $\mathbb{Z}$ and reduce modulo n at the end. The same is true for products, as we now show. Suppose $c = nq_1 + r_1$ and $d = nq_2 + r_2$, both in accord with the division algorithm. Then

$$cd = n(nq_1q_2) + n(q_1r_2 + r_1q_2) + r_1r_2,$$

showing that the remainder of cd modulo n is the remainder of r_1r_2 modulo n. That is, it does not matter whether we first reduce modulo n and then multiply and reduce, or whether we multiply in $\mathbb{Z}$ and then reduce.

Turning to our problem and delaying reduction modulo m and n of sums and products *in exponents* to the end, we have

$$a^s b^t[(a^u b^v)(a^w b^z)] = a^s b^t[a^{u+wr^v} b^{v+z}] = a^{s+(u \mid wr^v)r^t} b^{t+v+z} \tag{1}$$

and

$$[(a^s b^t)(a^u b^v)]a^w b^z = [a^{s+ur^t} b^{t+v}]a^w b^z. \tag{2}$$

Before we can continue this last computation, we must reduce the exponent $t + v$ modulo n, for in the next step $t + v$ will appear as an exponent of an exponent, rather than as a sum or product of first exponents where we are allowed to delay our reduction modulo n to the end. Let $t + v = nq_1 + r_1$ by the division algorithm. Note that because both t and v lie in the range from 0 to n - 1, either $q_1 = 1$ or $q_1 = 0$. Continuing, we see the expression (2) is equal to

$$a^{s+ur^t+wr^{r_1}} b^{t+v+z}. \tag{3}$$

Comparing (1) and (3), we see the associative law holds if and only if

$$s + (u + wr^v)r^t \equiv s + ur^t + wr^{r_1} \pmod{m}$$

and

$$t + v + z \equiv t + v + z \pmod{m}.$$

Of course this second condition is true, and the first one reduces to

$$wr^{v+t} \equiv wr^{r_1} \pmod{m}.$$

Now this relation must hold for all w where $0 \leq w < m$ and for all v and t from 0 to n - 1. Taking $w = 1$ and $v + t = n$ so that $r_1 = 0$, we see that we must have $r^n \equiv 1 \pmod{m}$. On the other hand, if this is true, then

$$wr^{v+t} = wr^{nq_1+r_1} = w(r^n)^{q_1}r^{r_1} \equiv wr^{r_1} \pmod{m}.$$

This completes the proof.

b. Part(a) proved the associative law, and $a^0 b^0$ is the identity for multiplication. Given $a^u b^v$, we can find $a^s b^t$ such that $(a^s b^t)(a^u b^v) = a^0 b^0$ by determining t and s is succession so that

$$t \equiv -v \pmod{n} \quad \text{and} \quad s \equiv -u(r^t) \pmod{m}.$$

The "left group axioms" hold, so we have a group of order mn.

14. Let G be a group of order pq for p and q primes, $q > p$, and $q \equiv 1 \pmod{p}$. By Sylow theory, G contains a normal subgroup H of order q which is cyclic, being of prime order. Let a be a generator of H and let $b \in G, b \notin H$. Now G/H has order p, so $b^p \in H$. If $b^p \neq 1$, then b is of order pq and G is cyclic and thus abelian, so for nonabelian G, we must have $b^p = 1$. Now $bab^{-1} \in H$. If $bab^{-1} = a$, then $ba = ab$ and G is abelian. Thus bab^{-1} must be one of $a^2, a^3, \cdots, a^{q-1}$. By Exercise 13, Part(**b**), the exponents x from 2 to $q-1$ such that the presentation $(a, b : a^q = 1, b^p = 1, ba = a^x b)$ gives a group of order pq are those such that $x^p \equiv 1 \pmod{q}$. By Corollary 23.6, the integers $1, 2, 3, \cdots, q-1$ form a cyclic group $\langle \mathbb{Z}_q^*, \cdot \rangle$ of order $q-1$ under multiplication modulo q. Because $q \equiv 1 \pmod{p}$ and $p < q$, we see that p divides $q-1$, so there is a cyclic subgroup $\langle r \rangle$ of $\langle \mathbb{Z}_q^*, \cdot \rangle$ having order p, so that $(r^j)^p \equiv 1 \pmod{q}$ for $j = 0, 1, \cdots, p-1$. Thus by Exercise 13, the presentations

$$(a, b : a^q = 1, b^p = 1, ba = a^{(r^j)}b)$$

give groups of order pq for $j = 1, 2, \cdots, p-1$. The hint in the exercise concludes with the demonstration that these $p-1$ presentations are isomorphic.

41. Simplicial Complexes and Homology Groups.

1. a. We have $\partial_2(c) = 2\partial_2(P_1 P_3 P_4) - 4\partial_2(P_3 P_4 P_6) + 3\partial_2(P_3 P_2 P_4) + \partial_2(P_1 P_6 P_4)$
$= 2(P_3 P_4 - P_1 P_4 + P_1 P_3) - 4(P_4 P_6 - P_3 P_6 + P_3 P_4) +$
$$3(P_2 P_4 - P_3 P_4 + P_3 P_2) + (P_6 P_4 - P_1 P_4 + P_1 P_6)$$
$= 2P_1 P_3 - 3P_1 P_4 + P_1 P_6 - 3P_2 P_3 + 3P_2 P_4 - 5P_3 P_4 + 4P_3 P_6 - 5P_4 P_6$.

b. No, $\partial_2(c)$ was just computed, and is nonzero.

c. Yes, it is a 1-cycle because $\partial^2 = 0$, that is, $\partial_1(\partial_2(c)) = 0$.

2. We have $\partial_2(\partial_3(P_1 P_2 P_3 P_4)) = \partial_2(P_2 P_3 P_4 - P_1 P_3 P_4 + P_1 P_2 P_4 - P_1 P_2 P_3)$
$= P_3 P_4 - P_2 P_4 + P_2 P_3 - P_3 P_4 + P_1 P_4 - P_1 P_3 + P_2 P_4 - P_1 P_4 \qquad + P_1 P_2 - P_2 P_3 + P_1 P_3 - P_1 P_2$
$= 0$

3. For $i > 0$, all four of these groups are zero. We have $B_0(P) = 0$ while $C_0(P) = Z_0(P) = \{nP \mid n \in \mathbb{Z}\} \simeq \mathbb{Z}$, and $H_0(P) = \{\{nP\} \mid n \in \mathbb{Z}\} \simeq \mathbb{Z}$.

4. For $i > 0$, all four of these groups are zero. We have $B_0(X) = 0$ while $C_0(X) = Z_0(X) = \{mP + nP' \mid m, n \in \mathbb{Z}\} \simeq \mathbb{Z} \times \mathbb{Z}$, and $H_0(X) = \{\{mP + nP'\} \mid m, n \in \mathbb{Z}\} \simeq \mathbb{Z} \times \mathbb{Z}$.

5. For $i > 1$, all four of these groups are zero. We have $C_1(X) = \{nP_1 P_2 \mid n \in \mathbb{Z}\} \simeq \mathbb{Z}$. Because $\partial_1(P_1 P_2) = -P_1 + P_2 \neq 0$, we have $Z_1(X) = B_1(X) = H_1(X) = 0$. For dimension 0, we have $C_0(X) = Z_0(X) = \{mP_1 + nP_2 \mid m, n \in \mathbb{Z}\} \simeq \mathbb{Z} \times \mathbb{Z}$ and $B_0(X) = \{n(P_2 - P_1) \mid n \in \mathbb{Z}\} \simeq \mathbb{Z}$. Because

$$rP_1 + sP_2 + (-s)(P_2 - P_1) = (r + s)P_1,$$

we see that every coset of $B_0(X)$ in $Z_0(X)$ contains a unique element of the form mP_1 for $m \in \mathbb{Z}$, so $H_0(Z) = Z_0 X / B_0(X) = \{mP_1 + B_0(X) \mid m \in \mathbb{Z}\} \simeq \mathbb{Z}$.

6. T F T T T T F T T

7. a. An **oriented n-simplex** $\sigma = P_1 P_2 \cdots P_{n+1}$ is an ordered sequence of $n+1$ vertices in $\mathbb{R}^m$ where $m \geq n$. If the n-simplex $\mu = P_{k_1} P_{k_2} \cdots P_{k_{n+1}}$ contains the same vertices as σ in a different order, then $\mu = \sigma$ if the permutation

$$\begin{pmatrix} 1 & 2 & \cdots & n+1 \\ k_1 & k_2 & \cdots & k_{n+1} \end{pmatrix}$$

is even and $\mu = -\sigma$ if the permutation is odd.

b. We have

$$\partial_n(P_1 P_2 \cdots P_{n+1}) = \sum_{i=1}^{n+1} (-1)^{i+1} P_1 P_2 \cdots P_{i-1} P_{i+1} \cdots P_{n+1}.$$

c. Each summand of the boundary of an n-simplex is a **face** of the simplex.

8. They are already defined in the text in terms of a general integer $n \geq 0$. Take them just as they stand.

9. Let $\sigma = P_1 P_2 \cdots P_{n+1}$. The ith face of $\partial_n(\sigma)$ is

$$(-1)^{i+1} P_1 \cdots P_{i-1} P_{i+1} \cdots P_{n+1}$$

and the jth face is $(-1)^{j+1} P_1 \cdots P_{j-1} P_{j+1} \cdots P_{n+1}$. Let us suppose that $i < j$. Applying ∂_{n-1} to the ith face produces an $(n\text{-}2)$-chain containing the term

$$m_i P_1 \cdots P_{i-1} P_{i+1} \cdots P_{j-1} P_{j+1} \cdots P_{n+1}$$

where $m_i = (-1)^{i+1}(-1)^j$, for the vertex P_j became the $(j-1)$st vertex in the ith face after its predecessor P_i was removed. On the other hand, applying ∂_{n-1} to the jth face produces this same $(n-2)$-simplex with coefficient $m_j = (-1)^{j+1}(-1)^{i+1}$, for P_i was still the ith vertex in the jth face since $i < j$. Thus these two terms cancel each other in the computation of $\partial_{n-1}(\partial_n(\sigma))$.

10. a. Because P_1 is a *summand* of ∂_1 of $P_2 P_1, P_3 P_1$, and $P_4 P_1$, we have $\delta^{(0)} P_1 = P_2 P_1 + P_3 P_1 + P_4 P_1$. Similarly, $\delta^{(0)}(P_4) = P_1 P_4 + P_2 P_4 + P_3 P_4$.

b. Note that $P_3 P_2$ is a *summand* of both ∂_2 of $P_1 P_3 P_2$ and ∂_2 of $P_3 P_2 P_1$. However, as stated in the text, these 2-simplexes are the same because we consider $P_1 P_3 P_2 = P_3 P_2 P_1 = P_2 P_1 P_3$. Thus $\delta^{(1)}(P_3 P_2) = P_1 P_3 P_2 + P_4 P_3 P_2$.

c. We have $\delta^{(2)}(P_3 P_2 P_4) = P_1 P_3 P_2 P_4$.

11. a. We define $\delta^{(n)} : C^{(n)} \to C^{(n+1)}$ by

$$\delta^{(n)} \left(\sum_i m_i \sigma_i \right) - \sum_i m_i \, \delta^{(n)}(\sigma_i).$$

b. It suffices to show that $\delta^{(n+1)}(\delta^{(n)}(\sigma)) = 0$ for every n-simplex σ. In the case where σ is not a face of a face of an $(n+2)$-simplex, the conclusion is obvious. Otherwise, let σ be a face of a face of an $(n+2)$-simplex that contains two additional vertices, P_i and P_j. Now $\delta^{(n)}(\sigma)$ contains both $P_i\sigma$ and $P_j\sigma$ as *summands*. Then $\delta^{(n+1)}(\delta^{(n)}(\sigma))$ contains both $P_j P_i\sigma$ and $P_i P_j\sigma$ as *summands*. Because the reordering of the vertices of $P_j P_i\sigma$ to produce $P_i P_j\sigma$ is accomplished by an odd permutation, namely the single transposition (i,j), we see that the second simplex is the negative the the first, so these two terms cancel each other in $C^{(n+2)}(X)$. Thus $\delta^2 = 0$.

12. We define the **group** $Z^{(n)}(X)$ **of n-cocycles** of X to be the kernel of the coboundary homomorphism $\delta^{(n)}$. We define the **group** $B^{(n)}$ **of n-coboundaries** of $C^{(n)}(X)$ to be the image of $\delta^{(n-1)}$, that is, it is $\delta^{(n-1)}[C^{(n-1)}(X)]$. Because $\delta^{(n)}[B^{(n)}] = \delta^{(n)}(\delta^{(n-1)}[C^{(n-1)}(X)]) = 0$, we see that $B^{(n)}(X) \leq Z^{(n)}(X)$.

13. The **n-dimensional cohomology group** $H^{(n)}(X)$ of X is $Z^{(n)}(X)/B^{(n)}(X)$.

Computing $H^{(0)}(S)$: We have $B^{(0)}(S) = 0$ because $C^{(-1)}(S) = 0$. Now $\delta^{(0)}(P_1) = P_2P_1 + P_3P_1 + P_4P_1$, and all these summands would have to be "cancelled" for a 0-cocycle. The term P_2P_1 can only be eliminated by another summand P_1P_2, which appears in $\delta^{(0)}(P_2) = P_1P_2 + P_3P_2 + P_4P_2$, and similar observations hold for the summands P_3P_1 and P_4P_1. Thus our only hope for a 0-cocycle is a multiple of $P_1 + P_2 + P_3 + P_4$. Computing, we have

$$\delta^{(0)}(P_1 + P_2 + P_3 + P_4) = P_2P_1 + P_3P_1 + P_4P_1 + P_1P_2 + P_3P_2 + P_4P_2 + P_1P_3 + P_2P_3 + P_4P_3 + P_1P_4 + P_2P_4 + P_3P_4 = 0.$$

Thus $H^{(0)}(S) \simeq \mathbb{Z}$ and is generated by $(P_1 + P_2 + P_3 + P_4) + \{0\}$.

Computing $H^{(1)}(S)$: If a coset of $B^{(1)}(S)$ in $Z^{(1)}(S)$ contains a 1-cochain c having mP_4P_1 as a summand, then it also contains $c - m\delta^{(0)}(P_1)$, which is a 1-cochain c that does not contain a multiple of P_4P_1. By a similar argument, we can adjust terms of c involving P_2P_4 and P_3P_4 by coboundaries of P_2 and P_3, and we see that the coset contains a 1-cochain of the form $c' = rP_1P_2 + sP_2P_3 + tP_3P_1$. Computing, $\delta^{(1)}(c') = r(P_3P_1P_2 + P_4P_1P_2) + s(P_1P_2P_3 + P_4P_2P_3) + t(P_2P_3P_1 + P_4P_3P_1)$. Now the three 2-simplexes containing the vertex P_4 are all different, so we see that $\delta^{(1)}(c') \neq 0$ unless $r = s = t = 0$, which means $c' = 0$. Thus there are no 1-cocycles that are not 1-coboundaries, so $Z^{(1)}(S) = B^{(1)}(S)$ and $H^{(1)}(S) = 0$.

Computing $H^{(2)}(S)$: Because $C^{(3)}(S) = 0$, every 2-cochain is a cocycle. Examining the coboundaries in $B^{(2)}(S)$, we take the 1-simplex P_1P_2 and find that $\delta^{(1)}(P_1P_2) = P_1P_2P_3 + P_1P_2P_4$. Thus if z is a 2-cycle having $mP_1P_2P_4$ as a summand, we can subtract $m\delta^{(1)}(P_1P_2)$ from Z and obtain another representative of the coset $z + B^{(2)}(S)$ in which $P_1P_2P_4$ does not appear. Now $\delta^{(1)}(P_1P_3) = P_1P_3P_2 + P_1P_3P_4$ and $\delta^{(1)}(P_2P_3) = P_1P_2P_3 + P_2P_3P_4$. Thus by subtracting suitable multiples of these coboundaries, we can find an element of the coset $z + B^{(2)}(S)$ in which no simplex involving P_4 appears. Thus we see that every coset of $B^{(2)}(S)$ contains a unique element of the form $mP_1P_2P_3$ for $m \in \mathbb{Z}$. Clearly $H^{(2)}(S)$ is then infinite cyclic, generated by $P_1P_2P_3 + B^{(2)}(S)$, and isomorphic to $\mathbb{Z}$.

42. Computations of Homology Groups

1. The space X is connected, so $H_0(X) \simeq \mathbb{Z}$. Each of the two tangent 1-spheres (circles) is a 1-cycle that is not a 1-boundary, because there is nothing of dimension 2. Thus $H_1(X) \simeq \mathbb{Z} \times \mathbb{Z}$. There is nothing of higher dimension, so $H_i(X) = 0$ for $i > 1$.

2. The space X is connected, so $H_0(X) \simeq \mathbb{Z}$. Every 1-cycle is also a 1-boundary. That is, if you cut along the 1-cycle, the surface falls into two disconnected pieces, and the 1-cycle is the 1-boundary of each piece. Thus $H_1(X) = 0$. Each of the two 2-spheres is a 2-cycle, which is not a 2-boundary because there is nothing of dimension 3, so $H_2X \simeq \mathbb{Z} \times \mathbb{Z}$. There is nothing of higher dimension, so $H_i(X) = 0$ for $i > 2$.

3. The space X consists of two disconnected surfaces, the 2-sphere and the annular ring. If P_1 is a vertex in the 2-sphere and P_2 is a vertex in the annular ring, there is no 1-chain having $P_2 - P_1$ as its boundary. Thus $H_0(X)$ is generated by two elements, $P_1 + B_0(X)$ and $P_2 + B_0(X)$; we see that $H_0(X) \simeq \mathbb{Z} \times \mathbb{Z}$. There are no 1-cycles on the 2-sphere that are not 1-boundaries. As we showed in Example 42.10, any 1-cycle of the annular ring that is not a 1-boundary can be "pushed to the outer rim 1-cycle z" in the homology group, so $H_1(X)$ is generated by $z + B_1(X)$. Thus $H_1(X) \simeq \mathbb{Z}$. Finally, there are no 2-boundaries, and the only 2-cycles are multiples of the 2-sphere, so $H_2(X) \simeq \mathbb{Z}$. There is nothing of higher dimension, so $H_i(X) = 0$ for $i > 2$.

4. The space X is connected, so $H_0(X) \simeq \mathbb{Z}$. This time the 1-cycle z which is the outer rim of the annular ring is the 1-boundary of X. There are no 1-cycles that are not 1-boundaries, so $H_1(X) = 0$. The only 2-cycles are multiples of the 2-sphere, so $H_2(X) \simeq \mathbb{Z}$. There is nothing of higher dimension, so $H_i(X) = 0$ for $i > 2$.

5. The space X is connected, so $H_0(X) \simeq \mathbb{Z}$. The only 1-cycles that are not 1-boundaries are multiples of the circle (1-sphere), so $H_1(X) \simeq \mathbb{Z}$. The only 2-cycles are multiples of the 2-sphere, so $H_2(X) \simeq \mathbb{Z}$. There is nothing of higher dimension, so $H_i(X) = 0$ for $i > 2$.

6. This space is homeomorphic to the torus in Fig. 42.13. (Just let air out of the sphere until it collapses down to be the other half of the "doughnut" surface.) Thus the homology groups are the same as the ones we computed in Example 42.12, namely, $H_0(X) = 0, H_1(X) \simeq \mathbb{Z} \times \mathbb{Z}$, and $H_2(X) \simeq \mathbb{Z}$. There is nothing of higher dimension, so $H_i(X) = 0$ for $i > 2$.

7. T F F F T F T F F T

8. If P_1 is a vertex on one torus and P_2 is a vertex on the other, then there is no 1-cycle with boundary $P_2 - P_1$, so the 0-cycles are the elements of the cosets $(mP_1 + nP_2) + B_0(X)$ for $m, n \in \mathbb{Z}$. Thus $H_0(X) \simeq \mathbb{Z} \times \mathbb{Z}$. Let a and b be the 1-cycles on the first torus as shown in Fig. 42.13, and let a' and b' be the corresponding 1-cycles on the second torus. The 1-cycles of X that are not 1-boundaries are the elements of cosets

$$(ma + nb + m'a' + m'b') + B_1(X)$$

for $m, n, m', n' \in \mathbb{Z}$ not all zero. Thus we see that $H_1(X) \simeq \mathbb{Z} \times \mathbb{Z} \times \mathbb{Z} \times \mathbb{Z}$. Each torus is a 2-cycle, and there are no 2-boundaries, so clearly $H_2(X) \simeq \mathbb{Z} \times \mathbb{Z}$. There is nothing of higher dimension, so $H_i(X) = 0$ for $i > 2$.

9. The space X is connected, so $H_0(X) \simeq \mathbb{Z}$. Let b be the 1-cycle which is the circle of intersection of torus 1 with torus 2. Every 1-cycle going that same long way around a torus, like the 1-cycle b in Fig. 42.13, is homologous to our 1-cycle b. (If you cut the surface apart on two circles of this type, the surface will fall into pieces, one of which will have boundary consisting of both circles, so their difference lies in $B_1(X)$. Equivalently, you can "push" one circle, keeping it on the surface, into the other.) We also have two 1-cycles a and a', one on each torus, going the short way around like the circle a in Fig. 42.13. These are not homologous to each other nor to b. Thus the elements of $Z_1(X)$ are the elements of the cosets $(ma + m'a' + nb) + B_1(X)$ for $m, m', n \in \mathbb{Z}$, so $H_1(X) \simeq \mathbb{Z} \times \mathbb{Z} \times \mathbb{Z}$. Each torus is a 2-cycle and there are no 2-boundaries. We see that $H_2(X) \simeq \mathbb{Z} \times \mathbb{Z}$. There is nothing of higher dimension, so $H_i(X) = 0$ for $i > 2$.

10. The space X is connected, so $H_0(X) \simeq \mathbb{Z}$. Let b be the 1-cycle that is the intersection of the torus with the 2-sphere. Now a every 1-cycle that is a circle going the long way around the torus, like the circle b in Fig. 42.13, is homologous to this 1-cycle that is the circle of intersection. (It can be "pushed" into this circle of intersection.) However, if you cut along this circle of intersection, you find that it is the 1-boundary of a hemisphere of S_2, so every 1-cycle of type b is homologous to 0. Thus the only one cycles are those in the cosets $ma + B_1(X)$ for $m \in \mathbb{Z}$, where a is a 1-cycle going the short way around the torus, like the 1-cycle a in Fig. 42.13. We see that $H_1(X) \simeq \mathbb{Z}$. Turning to $H_2(X)$, each

of the 2-sphere and the torus is a 2-cycle, and there are no 2-boundaries, so $H_2(X) \simeq \mathbb{Z} \times \mathbb{Z}$. There is nothing of higher dimension, so $H_i(X) = 0$ for $i > 2$.

11. The space X is connected, so $H_0(X) \simeq \mathbb{Z}$. Each of the two handles has a 1-cycle that is a circle around the handle, like the circle a in Fig. 42.13. We let a be such a 1-cycle on the left handle and a' an analogous one on the right handle. Then we have 1-cycles b on the left handle and b' on the right handle, which go along the handle, onto the sphere, and then back on the handle at its other end. These are analogous to the circle b on the annulus in Fig. 42.13. We should also consider the 1-cycle z that is the "equator" of the sphere in Fig. 42.18, going through the holes made by the handles. If we cut the space along this equatorial circle z, it does not fall into two pieces; the handles hold it together, and the boundary of the resulting 2-figure consists of two copies of the circle with opposite orientation, whose algebraic sum in $C_n(X)$ is then zero, so the equatorial circle z is not a 1-boundary. However, if we slice the entire figure into two pieces with a horizontal slash which cuts the sphere on the equatorial circle z and cuts the handles at their extreme left and right points, the space does fall into two pieces, and the boundary of each consists the equatorial circle z and two 1-cycles of type a and a'. Thus, assuming proper orientation, we have $(z + a + a') \in B_1(X)$, so z is in the coset $(-a - a') + B_1(X)$. Thus the cosets $(ma + m'a' + nb + n'b') + B_1(X)$ include the 1-cycles homologous to z, and we see that $H_1(X) \simeq \mathbb{Z} \times \mathbb{Z} \times \mathbb{Z} \times \mathbb{Z}$. The 2-cycles consist of sums of multiples of the entire space, so $H_2(X) \simeq \mathbb{Z}$. There is nothing of higher dimension, so $H_i(X) = 0$ for $i > 2$.

12. The space X is connected, so $H_0(X) \simeq \mathbb{Z}$. An analysis similar to the one we made in the solution of Exercise 42.11 indicates that $H_1(X) \simeq \underbrace{\mathbb{Z} \times \mathbb{Z} \times \cdots \times \mathbb{Z}}_{2n \text{ factors}}$. Multiples of the entire space are the only 2-cycles so $H_2((X) \simeq \mathbb{Z}$. There is nothing of higher dimension, so $H_i(X) = 0$ for $i > 2$.

43. More Homology Computations and Applications

1. *2-triangle triangulation:* $n_0 = 4, n_1 = 5, n_2 = 2, n_3 = 0$.
   $\chi(X) = 4 - 5 + 2 - 0 = 1$.

   6-triangle triangulation: $n_0 = 7, n_1 = 12, n_2 = 6, n_3 = 0$.
   $\chi(X) = 7 - 12 + 6 - 0 = 1$.

2. **a.** From Fig. 42.11 we have $n_0 = 10, n_1 = 20, n_2 = 10$, and $n_3 = 0$ so $\chi(X) = 10 - 20 + 10 - 0 = 0$. From the homology groups computed in Example 42.10, we have $\beta_0 = 1, \beta_1 = 1, \beta_2 = 0$, and $\beta_3 = 0$ so $\beta_0 - \beta_1 + \beta_2 - \beta_3 = 1 - 1 + 0 - 0 = 0$.

   b. From Fig. 42.14 we have $n_0 = 9, n_1 = 27, n_2 = 18$, and $n_3 = 0$ so $\chi(X) = 9 - 27 + 18 - 0 = 0$. From the homology groups computed in Example 42.13, we have $\beta_0 = 1, \beta_1 = 2, \beta_2 = 1$, and $\beta_3 = 0$ so $\beta_0 - \beta_1 + \beta_2 - \beta_3 = 1 - 2 + 1 - 0 = 0$.

   c. Taking the triangulation in Fig. 42.14 with the arrow on the top edge reversed for the Klein bottle, we see by Part(**b**) that $\chi(X) = 0$. From the homology groups computed in Example 43.1, we have $\beta_0 = 1, \beta_1 = 1, \beta_2 = 0$, and $\beta_3 = 0$, so $\beta_0 - \beta_1 + \beta_2 - \beta_3 = 1 - 1 + 0 - 0 = 0$.

3. The theorem will hold for a square region, for such a region is homeomorphic to E^2. It obviously does not hold for two disjoint 2-cells, for each can be mapped continuously into the other, and such a map has no fixed points.

4. The space X is connected so $H_0(X) \simeq \mathbb{Z}$. The 2-sphere contains no 1-cycles that are not 1-boundaries, while the Klein bottle has 1-cycles mb for all $m \in \mathbb{Z}$, none of which are 1-boundaries, and a 1-cycle a

which is not a 1-boundary although $2a$ is a 1-boundary, as shown in Example 43.1 Thus we see that $H_1(X) \simeq \mathbb{Z}_2 \times \mathbb{Z}$. The 2-sphere is a 2-cycle, while the Klein bottle is not, as shown in Example 43.1. Thus $H_2(X) \simeq \mathbb{Z}$. There is nothing of higher dimension, so $H_i(X) = 0$ for $i > 2$.

5. If P_1 is a vertex on one Klein bottle and P_2 is a vertex on the other, then there is no 1-cycle with boundary $P_2 - P_1$, so the 0-cycles are the elements of the cosets $(mP_1 + nP_2) + B_0(X)$ for $m, n \in \mathbb{Z}$. Thus $H_0(X) \simeq \mathbb{Z} \times \mathbb{Z}$. Each Klein bottle contributes a homology class $b + B_1(X)$ of infinite order in $H_1(X)$ and a homology class $a + B_1(X)$ of order 2, as explained in Example 43.1. Thus $H_1(X) \simeq \mathbb{Z}_2 \times \mathbb{Z}_2 \times \mathbb{Z} \times \mathbb{Z}$. Neither Klein bottle is a 2-cycle, as explained in Example 43.1, so $H_2(X) = 0$. There is nothing of higher dimension, so $H_i(X) = 0$ for $i > 2$.

6. F T T F T T T T T F

7. By Exercise 12 of Section 42, the nonzero homology groups of a 2-sphere with n handles are $H_0(X) \simeq \mathbb{Z}, H_1(X) \simeq \underbrace{\mathbb{Z} \times \mathbb{Z} \times \cdots \times \mathbb{Z}}_{2n \text{ factors}}$, and $H_2(X) \simeq \mathbb{Z}$. Using Theorem 43.7, we obtain $\beta_0 = 1, \beta_1 = 2n$, and $\beta_2 = 1$, so $\chi(X) = \beta_0 - \beta_1 + \beta_2 = 1 - 2n + 1 = -2n + 2$.

8. We describe a triangulation. Viewing the circle in Fig. 43.14 as the usual face of a 12 hour clock, mark points Q on the circle at 10:30 and 4:30, and mark points R at 1:30 and 7:30. Draw lines joining the two points P, joining the two points Q, and joining the two points R. They meet at a vertex C in the center of the circle. Then draw the lines from Q at 10:30 to P at 3:00, from P at 3:00 to R at 7:30, and from R at 7:30 to Q at 10:30; they contribute 3 more vertices where lines intersect. This gives us a triangulation of the projective plane.

Because the projective plane X is connected, $H_0(X) \simeq \mathbb{Z}$. Every 1-cycle can be "pushed" (by subtracting 1-boundaries) to the arcs on the circle. More specifically, starting with the triangle having as bottom edge the line from C to 3:00 P and going around counterclockwise, we can eliminate that bottom edge, the right edge of the next triangle having C as vertex, etc., until we have eliminated the edge from C to 7:30 R. This leaves the line from C to 4:30 Q. Then we start with some triangle having an arc of the circle as edge and go around counterclockwise in a similar way, eliminating the righthand edges. We are left with our 1-cycle having edges only on the arcs except for one or two single edges sticking in from the circle or out from the center, which also must have coefficient 0 in our 1-cycle, so really all edges are on the circle; note there are only three of them, PR, RQ and QP, not six of them. Because it is a 1-cycle, they must all occur with the same coefficients. Let $a = PR + RQ + QP$, as indicated on your figure. Now the boundary of the projective plane X is clearly $2a$, due to the same counterclockwise orientation of the two arcs labeled a. Thus $2a$ is a 1-boundary, and we see that $H_1(X) \simeq \mathbb{Z}_2$. Let c be the 2-chain consisting of the sum of all 2-simplexes oriented the same way. Because the boundary of $\partial_2(c) = 2a$, we see that $Z_2(X) = 0$ so $H_2(X) = 0$. There is nothing of higher dimension, so $H_i(X) = 0$ for $i > 2$.

9. The space X is connected so $H_0(X) \simeq \mathbb{Z}$. Computation of $H_1(X)$ is our toughest job so far. As suggested in the exercise, view the space as the disk in Fig. 21.8 with $q - 1$ circular holes in it to be sewn up by sewing diametrically opposite points. Let us number these holes from 1 to $q - 1$, and let a_i be the top semicircle of the ith hole with counterclockwise orientation. Note that in view of the identification to take place in the sewing, we can also consider the bottom semicircle of the ith hole to be a_i with counterclockwise orientation. We consider the top and bottom semicircles of the rim of the disk in Fig. 21.8 to be a_q rather than a. They will form the qth crosscap in the final sewing. Now consider a circle on the 2-sphere that encircles only the ith crosscap of X for $i < q$. This corresponds to a circle going around the ith hole but not containing any other hole in our disk model of X. Without

making a triangulation, we can consider this circle, oriented counterclockwise, to be a 1-cycle z_i. Now z_i is not a 1-boundary, for if we cut our sphere along this circle, it falls into two pieces, neither of which has boundary z_i on account of the crosscaps contained in each piece. The piece that contains the ith crosscap, corresponding to the piece containing the ith hole in our disk model, has boundary $z_i - 2a_i$. Doing this for each i where $1 \le i \le q - 1$, we obtain cycles z_i such that $(z_i - 2a_i) \in B_1(X)$ for $i = 1, 2, \cdots, q - 1$. (Remember that z_i could be *any* 1-cycle encircling only crosscap number i on the 2-sphere X.) Let c be the 2-chain consisting of the sum of all 2-simplexes oriented the same way. Another element of $B_1(X)$ is $\partial_2(c)$, and Exercise 8 indicates that $\partial_2(c) = 2a_1 + 2a_2 + \cdots + 2a_q$. As indicated in Exercise 8, each a_i is a cycle for $i = 1, 2, \cdots, q$, but no a_i is a 1-boundary, and indeed no ma_i is a 1-boundary. It is only the *sum* of all the $m(2a_i)$ that is a 1-boundary for all $m \in \mathbb{Z}$.

We take as generators for the 1-cycles in X the 1-cycles $a_1, a_2, \cdots, a_{q-1}$, and $a_1 + a_2 + \cdots + a_q$. Note that the coset $2ma_i + B_1(X)$ contains the 1-cycle mz_i for all $m \in \mathbb{Z}$, because $mz_i - 2ma_i = m(z_i - 2a_i) \in B_1(X)$. Since the generators a_i have infinite order for $1 \le i \le q - 1$ and the generator $a_1 + a_2 + \cdots + a_q$ has order 2, we see that $H_1(X) \simeq \underbrace{\mathbb{Z} \times \mathbb{Z} \times \cdots \times \mathbb{Z}}_{q\text{-}1 \text{ factors}} \times \mathbb{Z}_2$.

Let c be the 2-chain consisting of the sum of all 2-simplexes oriented the same way. Because $\partial_2(c) \ne 0$, there are no 2-cycles so $H_2(X) = 0$. There is nothing of higher dimension, so $H_i(X) = 0$ for $i > 2$.

You may be bothered by the apparent lack of symmetry in our computation of $H_1(X)$. What was special about the qth crosscap, that didn't have a 1-cycle z_q around it? The answer is, it did. Remember that z_i could be any 1-cycle containing the ith crosscap. We can take for z_1 and z_2 two 1-cycles that have an edge in common. In their sum, $z_1 + z_2$, that edge cancels and we get a 1-cycle containing both the first and second crosscaps. Continuing in the obvious way, we can get a cycle $z_1 + z_2 + \cdots + z_{q-1}$ containing all the crosscaps but the qth. Viewed on the 2-sphere, we realize that we can view this as a 1-cycle $-z_q$ that contains only the qth crosscap. Note that from the 1-boundaries $z_i - 2a_i$ for $i = 1, 2, \cdots, q - 1$ and $2a_1 + 2a_2 + \cdots + 2a_q$ that we found in $B_1(X)$, we can deduce that $B_1(X)$ contains $(z_1 - 2a_1) + (z_2 - 2a_2) + \cdots + (z_{q-1} - 2a_{q-1}) + (2a_1 + 2a_2 + \cdots + 2a_q) = -z_q + 2a_q$, so $z_q - 2a_q$ is in $B_1(X)$ and everything really is symmetric.

10. Let Q be a vertex of X on a, and let c be the 2-chain consisting of all 2-simplexes of X, all oriented the same way, so that $c \in Z_2(X)$.

a. $f_{*0} : H_0(X) \to H_0(X)$ is given by

$$f_{*0}(Q + B_0(X)) = Q + B_0(X),$$

that is, f_{*0} is the identity map.

$f_{*1} : H_1(X) \to H_1(X)$ is given by

$$f_{*1}((ma + nb) + B_1(X)) = (ma + 2nb) + B_1(X),$$

reflecting the fact that f maps X twice around itself in the θ direction.

$f_{*2} : H_2(X) \to H_2(X)$ is given by

$$f_{*2}(c + B_2(X)) = 2c + B_2(X),$$

reflecting the fact that each point of X is the image of two points of X under f.

b. f_{*0} is as in Part(**a**).

$f_{*1} : H_1(X) \to H_1(X)$ is given by

$$f_{*1}((ma + nb) + B_1(X)) = (2ma + nb) + B_1(X),$$

reflecting the fact that f maps X twice around itself in the ϕ direction.

f_{*2} is as in Part(**a**).

c. f_{*0} is as in Part(**a**).

$f_{*1} : H_1(X) \to H_1(X)$ is given by

$$f_{*1}((ma + nb) + B_1(X)) = (2ma + 2nb) + B_1(X),$$

reflecting the fact that f maps X twice around itself in both the θ and the ϕ directions.

$f_{*2} : H_2(X) \to H_2(X)$ is given by

$$f_{*2}(c + B_2(X)) = 4c + B_2(X),$$

reflecting the fact that each point of X is the image of four points of X under f.

11. Let Q be a vertex of b, and let c be the 2-chain consisting of all 2-simplexes of X, all oriented the same way, so that $c \in Z_2(X)$.

    a. $f_{*0} : H_0(X) \to H_0(b)$ is given by $f_{*0}(Q + B_0(X)) = Q + B_0(b)$.

    $f_{*1} : H_1(X) \to H_1(b)$ is given by $f_{*1}((ma + nb) + B_1(X)) = nb + B_1(b)$.

    $f_{*2} : H_2(X) \to H_2(X)$ is given by $f_{*2}(c + B_2(X)) = 0$.

    b. f_{*0} is as in Part(**a**).

    $f_{*1} : H_1(X) \to H_1(b)$ is given by $f_{*1}((ma + nb) + B_1(X)) = 2nb + B_1(b)$.

    f_{*2} is as in Part(**a**).

12. The answers are the same as for Exercise 11 with $B_i(b)$ replaced by $B_i(X)$.

13. Let Q be a vertex on b.

    $f_{*0} : H_0(X) \to H_0(b)$ is given by $f_{*0}(Q + B_0(X)) = Q + B_0(b)$.

    $f_{*1} : H_1(X) \to H_1(b)$ is given by $f_{*1}((ma + nb) + B_1(X)) = nb + B_1(b)$, where $m = 0, 1$.

    f_{*2} is trivial, because both $H_2(X)$ and $H_2(b)$ are 0.

44. Homological Algebra

1. Because the sequence is exact, the image of the map $0 \to A$, which must be 0, is the kernel of $f : A \to B$. Because all these maps are homomorphism, this means that f is one to one. The kernel of $B \to 0$ is certainly B, which, by exactness, must be the image $f[A]$ of A under f. Thus f is a homomorphism mapping A one to one onto B, and is thus an isomorphism.

2. **a.** The map $C \to 0$ had kernel C, which must, by exactness at C, be the image $j[B]$, that is, j maps B onto C.

   b. The map $0 \to A$ has image 0, which must be the kernel of i by exactness at A. Thus the map $i : A \to B$ has kernel 0, so i maps A one to one into B.

   c. Exactness at B means that the image $i[A]$ is the kernel of j. Because j is a homomorphism, we then know that $j[B] \simeq B/i[A]$. But $j[B] = C$ by Part(**a**), so we have $C \simeq B/i[A]$.

3. By exactness at B, the map i is onto B if and only if the kernel of j is B, which is true if and only if j maps B onto 0. We have shown (1) if and only if (2). By exactness at C, we see that j maps B onto 0 if and only if the kernel of the map k is 0, which holds if and only if k is a one to one map. This shows (2) if and only if (3).

4. Now exactness at C and means that h maps everything onto 0 if and only if kernel of i is 0 , which is true if and only if i is one to one. Now j maps everything onto zero if and only if the the kernel of j is D which is true, by exactness at D, if and only if the image of i is D. Thus h and j both map everything onto 0 if and only if i is one to one and maps C onto D, in other words, if and only if i is an isomorphism. We have shown (1) if and only if (2).

   Now h maps everthing onto 0 if and only the kernel of h is B, which, by exactness at B, holds if and only if $g[A] = B$ that is if and only if g maps A onto B. Also exactness at E means that j maps everything onto 0 if and only if the kernel of k is 0, which is true if and only if k is one to one. Thus h and j both map everything onto 0 if and only if g is onto B and k is one to one. We have shown (1) if and only if (3).

5. (See the answer in the text.)

6. Let X be the torus complex and let Y be the subcomplex consisting of the 1-cycle a. Lct P be a vertex on a and let Q be a vertex on the torus such that Q is not a vertex on a but PQ is a 1-simplex. Then $Q + C_0(Y)$ generates $Z_0(X, Y)$ and

$$\overline{\partial}_1(PQ + C_1(Y)) = \partial_1(PQ) + C_0(Y)$$
$$= (Q - P) + C_0(Y) = Q + C_0(Y)$$

because $P \in C_0(Y)$. Thus our generator of $Z_0(X, Y)$ is a relative 0-boundary, so $H_0(X, Y) = 0$.

   The generators of $H_1(X)$ are the cosets $a + B_1(X)$ and $b + B_1(X)$. Because $a \in Y$, we see that $b + B_1(X, Y)$ generates $H_1(X, Y)$, which is thus isomorphic to $\mathbb{Z}$.

   Let c be the 2-chain consisting of the sum of all 2-simplexes oriented the same way. Because $\partial_2(c) = 0$, we see that

$$\overline{\partial}_2(c + C_2(Y)) = \partial_2(c) + C1(Y) = 0 + C_1(Y),$$

and $C_1(Y) = 0$ in $C_1(X, Y)$. Thus $H_2(X, Y) \simeq \mathbb{Z}$. Of course $H_n(X, Y) = 0$ for $n > 2$.

7. (See the answer in the text.)

8. Let X be the Klein bottle complex and let Y be the subcomplex consisting of the 1-cycle a. Let P be a vertex on a and let Q be a vertex on the Klein bottle such that Q is not a vertex on a but PQ is a 1-simplex. Then $Q + C_0(Y)$ generates $Z_0(X, Y)$ and

$$\overline{\partial}_1(PQ + C_1(Y)) = \partial_1(PQ) + C_0(Y)$$
$$= (Q - P) + C_0(Y) = Q + C_0(Y)$$

because $P \in C_0(Y)$. Thus our generator of $Z_0(X, Y)$ is a relative 0-boundary, so $H_0(X, Y) = 0$.

The generators of $H_1(X)$ are the cosets $a + B_1(X)$ and $b + B_1(X)$. Because $a \in Y$, we see that $b + B_1(X, Y)$ generates $H_1(X, Y)$, which is thus isomorphic to $\mathbb{Z}$.

Let c be the 2-chain consisting of the sum of all 2-simplexes oriented the same way. Because $\partial_2(c) = 2a \in C_1(Y)$, we see that

$$\overline{\partial}_2(c + C_2(Y)) = \partial_2(c) + C_1(Y) = 2a + C_1(Y) = C_1(Y),$$

and $C_1(Y) = 0$ in $C_1(X, Y)$. Thus $H_2(X, Y) \simeq \mathbb{Z}$. Of course $H_n(X, Y) = 0$ for $n > 2$.

9. (See the answer in the text.)

10. Let P be a vertex in Y and let R be a vertex on the annular ring such that R is not a vertex in Y but PR is a 1-simplex. Then $R + C_0(Y)$ generates $Z_0(X, Y)$ and

$$\overline{\partial}_1(PR + C_1(Y)) = \partial_1(PR) + C_0(Y)$$
$$= (R - P) + C_0(Y) = R + C_0(Y)$$

because $P \in C_0(Y)$. Thus our generator of $Z_0(X, Y)$ is a relative 0-boundary, so $H_0(X, Y) = 0$.

We saw in Example 42.10 than any 1-cycle in $Z_1(X)$ could be pushed to the outer rim of the annulus. This outer rim is now part of Y, so this 1-cycle in $Z_1(X)$ becomes homologous to 0 in $Z_1(X, Y)$. However, the 1-simplex P_1Q_1 from the inner rim to the outer rim in Fig. 42.11, which was not a 1-cycle in $Z_1(X)$, now becomes a 1-cycle in $Z_1(X, Y)$, because its boundary $Q_1 - P_1$ lies in Y. In fact, for any triangulation, every sequence $R_1R_2 + R_2R_3 + \cdots + R_{m-1}R_m$ where the R_i are vertices with $R_1, R_m \in Y$ is in $Z_1(X, Y)$. However, if R_1 and R_m are both in the outer rim or both in the inner rim of the annulus, then the 1-cycle is in $B_1(X, Y)$. In terms of the triangulation in Fig. 42.11, We see that $H_1(X, Y)$ is generated by $P_1Q_1 + B_1(X, Y)$, so $H_1(X, Y) \simeq \mathbb{Z}$.

Finally, let c be the 2-chain consisting of the sum of all 2-simplexes oriented the same way. Then $\partial_2(c) \in C_1(Y)$, and $C_1(Y) = 0$ in $C_1(X, Y)$. Thus $c + C_2(Y)$ becomes a 2-cycle in $Z_2(X, Y)$, and $c + B_2(X, Y)$ is a generator of $H_2(X, Y)$ which is isomorphic to $\mathbb{Z}$. Of course $H_n(X, Y) = 0$ for $n > 2$.

11. Because $H_2(X) = 0$, the image of i_{*2} and kernel of j_{*2} are both zero, so we have exactness there.

Let a be the inner rim of the annular ring, oriented clockwise and let b be its outer rim oriented counterclockwise in Fig. 42.11. Let c be the 2-chain consisting of the sum of all 2-simplexes oriented counterclockwise. Then ∂_{*2} maps the generator $c + B_2(X, Y)$ of $H_2(X, Y)$ onto $(a + b) + B_1(Y)$ which is a nonzero element of $H_1(Y)$, so the kernel of ∂_{*2} and image of j_{*2} are both 0, and we have exactness at $H_2(X, Y)$.

We can take $-a + B_1(Y)$ and $b + B_1(Y)$ as generators of $H_1(Y)$, and i_{*1} maps these generators into $-a + B_1(X)$ and $b + B_1(X)$ respectively. Because $b + a = b - (-a)$ is in $B_1(X)$, we see that $-a + B_1(X)$ and $b + B_1(X)$ are the same homology class, which generates $H_1(X)$. Thus the image of ∂_{*2} consists of all $m(a + b) + B_1(Y)$ for $m \in \mathbb{Z}$, which are precisely the elements mapped into 0 in $H_1(X)$, because

$a + b$ generates $B_1(X)$. Thus we have exactness at $H_1(Y)$. (Note that in the identification of $H_1(Y)$ with $\mathbb{Z} \times \mathbb{Z}$ and the identification of $H_1(X)$ with $\mathbb{Z}$, the kernel of the homomorphism corresponding to i_{*1} is $\langle (1,1) \rangle$, so that $(\mathbb{Z} \times \mathbb{Z})/\langle (1,1) \rangle \simeq \mathbb{Z}$.)

Now j_{*1} maps the generator $b + B_1(X)$ of $H_1(X)$ into $0 + B_1(X,Y)$ because b is in Y. We saw that $\langle b + B_1(X) \rangle = H_1(X)$ is the image under i_{*1}, so we have exactness at $H_1(X)$.

Referring to Fig. 42.11, we see that $H_1(X,Y)$ is generated by $P_1 Q_1 + B_1(X,Y)$, and $\partial(P_1 Q_1) = Q_1 - P_1$, so ∂_{*1} maps $P_1 Q_1 + B_1(X,Y)$ into $(Q_1 - P_1) + B_0(Y)$, and $Q_1 - P_1 \notin B_0(Y)$ because $P_1 Q_1 \notin Y$. Thus the kernel of ∂_{*1} and the image of j_{*1} are both zero, so we have exactness at $H_1(X,Y)$.

Now $H_0(Y)$ has as generators $P_1 + B_0(Y)$ and $Q_1 + B_0(Y)$. These are mapped by i_{*0} into $P_1 + B_0(X)$ and $Q_1 + B_0(X)$ respectively. However, these are the same homology class in $H_0(X)$ because $(Q_1 - P_1) \in B_0(X)$. Thus the image $m(Q_1 - P_1) + B_0(Y)$ under ∂_{*1} is the kernel of i_{*0}, so we have exactness at $H_0(Y)$.

Finally, j_{*0} maps the generator $P_1 + B_0(X)$ of $H_0(X)$ onto $0 + B_0(X,Y)$ because $P_1 \in Y$. Thus the image $m P_1 + B_0(X)$ under i_{*0} is the kernel under j_{*0}, so we have exactness at $H_0(X)$.

12. Let $c \in A_k$. Then $(\bar{\partial} j_k)(c) = \bar{\partial}(j_k(c)) = \bar{\partial}(c + A'_k) = \partial(c) + A'_{k-1} = j_{k-1}(\partial(c)) = (j_{k-1}\partial)(c)$, so $\bar{\partial} j_k = j_{k-1}\partial$.

13. Let $h \in H_k(A/A')$. Let $z_1, z_2 \in Z_k(A/A')$ be such that $h = z_1 + B_k(A/A') = z_2 + B_k(A/A')$, so that $(z_2 - z_1) \in B_k(A/A')$. Let $z_1 = c_1 + A'_k$ and $z_2 = c_2 + A'_k$ To show that ∂_{*k} is well defined, we must show that $\partial_k(c_1) + B_{k-1}(A') = \partial_k(c_2) + B_{k-1}(A')$. Now $z_2 - z_1 = c_2 - c_1 + A'_k$ is in $B_k(A/A')$. Consequently there is some $r \in A_{k+1}$ such that $\partial_{k+1}r = c_2 - c_1 + a'$ for some $a' \in A'_k$. Then $0 = \partial_k \partial_{k+1}(r) = \partial_k(c_2 - c_1 + a') = \partial_k(c_2) - \partial_k(c_1) + \partial_k(a')$. Now $\partial_k(a') \in B_{k-1}(A')$, so we see that $\partial_k(c_2) \in \partial_k(c_1) + B_{k-1}(A')$. This shows that ∂_{*k} is well defined.

Let $h_1, h_2 \in H_k(A/A')$, and now let $z_1, z_2 \in Z_k(A/A')$ be such that $h_1 = z_1 + B_k(A/A')$ and $h_2 = z_2 + B_k(A/A')$, so that $h_1 + h_2 = (z_1 + z_2) + B_k(A/A')$. Let $z_1 = c_1 + A'_k$ and $z_2 = c_2 + A'_k$, so that $z_1 + z_2 = c_1 + c_2 + A'_k$. Then

$$
\begin{aligned}
\partial_{*k}(h_1 + h_2) &= \partial_k(c_1 + c_2) + B_{k-1}(A') \\
&= (\partial_k(c_1) + \partial_k(c_2)) + B_{k-1}(A') \\
&= (\partial_k(c_1) + B_{k-1})(A')) + (\partial_k(c_2) + B_{k-1}(A')) \\
&= \partial_{*k}(h_1) + \partial_{*k}(h_2).
\end{aligned}
$$

Thus ∂_{*k} is a homomorphism.

14. Let $z' + B_k(A')$ be an element of $H_k(A')$. Then $i_{*k}(z' + B_k(A')) = z' + B_k(A)$. Now $j_{*k}(z' + B_k(A)) = (z' + A') + B_k(A/A') = 0$ because $z' \in A'$. Thus $j_{*k}i_{*k} = 0$.

Let $z + B_k(A)$ be an element of $H_k(A)$. Then $j_{*k}(z + B_k(A)) = (z + A') + B_k(A/A')$. Now $\partial_{*k}((z + A') + B_k(A/A')) = \partial_k(z) + B_{k-1}(A') = B_{k-1}(A')$ because z is a k-cycle in A_k, and $B_{k-1}(A')$ is the zero element of $H_{k-1}(A')$. Thus $\partial_{*k}j_{*k} = 0$.

Let $h = (c + A') + B_k(A/A')$ be an element of $H_k(A/A')$. We have $\partial_{*k}(h) = \partial_k(c) + B_{k-1}(A')$ where $\partial_k(c) \in A'$. Then we have $i_{*k-1}(\partial_k(c) + B_{k-1}(A')) = \partial_k(c) + B_{k-1}(A) = B_{k-1}(A)$ because $c \in A$ implies $\partial_k(c)$ is in $B_{k-1}(A)$. Now $B_{k-1}(A)$ is the zero element of $H_k(A)$, so $i_{*k-1}\partial_{*k} = 0$.

15. a. We must show that $j_{*k}i_{*k} = 0$. Let $h' \in H_k(A')$. Then $h' = z' + B_k(A')$ for some $z' \in A'_k$, and $i_{*k}(h') = z' + B_k(A)$, and $j_{*k}(z' + B_k(A)) = (z' + A'_k) + B_k(A/A')$. But $z' \in A'_k$ so $(z' + A'_k) \in B_k(A/A')$, because

$$
B_k(A/A') = \{\partial_{k+1}(a_{k+1}) + A'_k \mid a_{k+1} \in A_{k+1}\},
$$

and taking $a_{k+1} = 0$, we get $A'_k \subseteq B_k(A/A')$. Thus $j_{*k}(i_{*k}(h')) = 0$ in $H_k(A/A')$.

b. Let $h \in H_k(A)$ and let $j_{*k}(h) = 0$ in $H_k(A/A')$. Now if $h = z + B_k(A)$, we have $j_{*k}(h) = (z + A'_k) + B_k(A/A')$, and $j_{*k}(h) = 0$ implies that $(z + A'_k) \in B_k(A/A')$. Now

$$B_k(A/A') = \{\partial_{k+1}(a_{k+1}) + A'_k \mid a_{k+1} \in A_{k+1}\}.$$

Thus $z = \partial_{k+1}(a_{k+1}) + a'_k$ for some $a_{k+1} \in A_{k+1}$ and $a'_k \in A'_k$. Because z is a k-cycle, we have

$$0 = \partial_k(z) = \partial_k(\partial_{k+1}(a_{k+1}) + \partial_k(a'_k)) = 0 + \partial_k(a'_k),$$

so a'_k is a k-cycle in A'. Therefore we see that $i_{*k}(a'_k + B_k(A')) = a'_k + B_k(A) = (z - \partial_{k+1}(a_{k+1})) + B_k(A) = z + B_k(A) = h$, because $\partial_{k+1}(a_{k+1}) \in B_k(A)$. We have shown that $h \in (\text{kernel } j_{*k})$ is also in (image i_{*k}).

c. We must show that $\partial_{*k}j_{*k} = 0$. Let $(z + b_k(A)) \in H_k(A)$. Then $\partial_{*k}(j_{*k}(z + B_k(A))) = \partial_{*k}((z + A') + B_k(A/A')) = \partial_k(z) + B_{k-1}(A') = B_{k-1}(A')$ because z is a k-cycle so $\partial_k(z) = 0$. Because $B_{k-1}(A')$ is the 0-element of $H_{k-1}(A')$, we are done.

d. Let $h \in H_k(A/A')$ be such that $\partial_{*k}(h) = 0$, and let $h = (z + A') + B_k(A/A')$. Then $\partial_{*k}(h) = \partial_k(z) + B_{k-1}(A')$, which must be the zero element of $H_{k-1}(A')$, so $\partial_k(z) \in B_{k-1}(A')$. Let $a'_k \in A'_k$ be such that $\partial_k(a'_k) = \partial_k(z)$. Then $\partial_k(z - a'_k) = \partial_k(z) - \partial_k(a'_k) = \partial_k(z) - \partial_k(z) = 0$, so $z - a'_k$ is a k-cycle in A_k. Then we have $j_{*k}((z - a'_k) + B_k(A)) = (z - a'_k) + B_k(A/A')$. We saw in Part(**a**) that $A'_k \subseteq B_k(A/A')$ so $(z - a'_k) + B_k(A/A') = z + B_k(A/A') = h$. Thus $h \in (\text{image } j_{*k})$.

e. We must show that $i_{*k-1}\partial_{*k} = 0$. Let $h \in H_k(A/A')$ and let $h = (z + A') + B_k(A/A')$. Now $\partial_{*k}(h) = \partial_k(z) + B_{k-1}(A')$, and $i_{*k-1}(\partial_k(z) + B_{k-1}(A')) = \partial_k(z) + B_{k-1}(A)$. But $\partial_k(z) \in B_{k-1}(A)$ and $B_{k-1}(A)$ is the zero element of $H_{k-1}(A)$. Thus $i_{*k-1}(\partial_{*k}(h)) = 0$.

f. Let $h' \in H_{k-1}(A')$ and suppose that $i_{*k-1}(h') = 0$. Let $h' = z' + B_{k-1}(A')$. Then $i_{*k-1}(h') = z' + B_{k-1}(A)$, and this is zero in $H_{k-1}(A)$ if and only if $z' \in B_{k-1}(A)$. Let $c \in A_k$ be such that $\partial_k(c) = z'$. Then $c + A'$ is a k-cycle in $H_k(A/A')$ because $\partial_k(c + A') = z' + A' = A'$. Then $\partial_{*k}((c + A') + B_k(A/A')) = z' + B_{k-1}(A') = h'$, which is what we wished to show.

16. Let z_k be a k-cycle in A_k. Using the relation in the text, we have

$$f_k(z_k) - g_k(z_k) = \partial'_{k+1}(D_k(z_k)) + D_{k-1}(\partial_k(z_k)).$$

Because z_k is a k-cycle, we have $\partial_k(z_k) = 0$, and because D_{k-1} is a homomorphism, we then know that $D_{k-1}(0) = 0'$ in A'_k. Also, $\partial'_{k+1}(D_k(z_k))$ is some element b'_k of B'_k. Thus we have $f_k(z_k) - g_k(z_k) - b'_k$, so $f_k(z_k) = g_k(z_k) + b'_k$ for some $b'_k \in B'_k$. Consequently

$$\begin{aligned} f_{*k}(z_k + B_k) &= f_k(z_k) + B'_k = (g_k(z_k) + b'_k) + B'_k \\ &= g_k(z_k) + (b'_k + B'_k) = g_k(z_k) + B'_k \\ &= g_{*k}(z_k + B_k), \end{aligned}$$

showing that f_{*k} and g_{*k} are the same homomorphism of $H_k(A)$ into $H_k(A')$.

45. Unique Factorization Domains

1. Yes, 5 is an irreducible in $\mathbb{Z}$.

2. Yes, -17 is an irreducible in $\mathbb{Z}$.

3. No, $14 = 2 \cdot 7$ is not an irreducible in $\mathbb{Z}$.

4. Yes, $2x - 3$ is an irreducible in $\mathbb{Z}[x]$.

5. No, $2x - 10 = 2(x - 5)$ is not an irreducible in $\mathbb{Z}[x]$.

6. Yes, $2x - 3$ is an irreducible in $\mathbb{Q}[x]$.

7. Yes, $2x - 10$ is an irreducible in $\mathbb{Q}[x]$, for 2 is a unit there.

8. Yes, $2x - 10$ is an irreducible in $\mathbb{Z}_{11}[x]$, for 2 is a unit there.

9. (See the answer in the text.)

10. In $\mathbb{Z}[x], 4x^2 - 4x + 8 = (2)(2)(x^2 - x + 2)$. The quadratic polynomial is irreducible because its zeros are complex numbers.

In $\mathbb{Q}[x], 4x^2 - 4x + 8$ is already irreducible because 4 is a unit and the zeros of the polynomial are complex numbers.

In $\mathbb{Z}_{11}[x], 4x^2 - 4x + 8 = (4x + 2)(x + 4)$. We found the factorization by discovering that -4 and 5 are zeros of the polynomial. Note that 2 is a unit.

11. We proceed by factoring the smallest number into irreducibles, and using a calculator, discover which irreducibles divide the larger numbers. We find that $234 = 2 \cdot 117 = 2 \cdot 9 \cdot 13$. Our calculator shows that 9 does not divide 3250, but 2 and 13 do, and both 2 and 13 divide 1690. Thus the gcd's are 26 and -26.

12. We proceed by factoring the smallest number into irreducibles, and using a calculator, discover which irreducibles divide the larger numbers. We find that $448 = 4 \cdot 112 = 4 \cdot 4 \cdot 28 = 2^6 \cdot 7$. Our calculator shows that 7 divides both 784 and 1960, and that the highest power of 2 dividing 784 is 16 while the highest power dividing 1960 is 8. Thus the gcd's are $8 \cdot 7 = 56$ and -56.

13. We proceed by factoring the smallest number into irreducibles, and using a calculator, discover which irreducibles divide the larger numbers. We find that $396 = 6 \cdot 66 = 6 \cdot 6 \cdot 11 = 2^2 \cdot 3^2 \cdot 11$. Our calculator shows that both 11 and 9 divide divides the other 3 numbers, but 2178 and 594 are not divisible by 4, but are divisible by 2. Thus the gcd's are $11 \cdot 9 \cdot 2 = 198$ and -198.

14. $18x^2 - 12x + 48 = 6(3x^2 - 2x + 8)$.

15. Because every nonzero $q \in \mathbb{Q}$ is a unit in $\mathbb{Q}[x]$, we can "factor out" any nonzero rational constant as the (unit) content of this polynomial. For example,

$$(1)(18x^2 - 12x + 48) \quad \text{and} \quad \frac{1}{2}(36x^2 - 24x + 96)$$

are two of an infinite number of possible answers.

16. The factorization is $(1)(2x^2 - 3x + 6)$ because the polynomial is primitive.

17. Because every nonzero $a \in \mathbb{Z}_7$ is a unit in $\mathbb{Z}_7[x]$, we can "factor out" any nonzero constant as the (unit) content of this polynomial. For example,

$$(1)(2x^2 - 3x + 6) \quad \text{and} \quad (5)(6x^2 + 5x + 4)$$

are two of an infinite number of possible answers.

18. The definition is incorrect. Quotients may not exist in D.

Two elements a and b in an integral domain D are **associates** in D if and only if there exists a unit $u \in D$ such that $au = b$.

19. The definition is incorrect. Neither factor can be a unit.

A nonzero element of an integral domain D is an **irreducible** of D if and only if it cannot be factored into a product of two elements of D, neither of which is a unit.

20. The definition is incorrect; there may be no notion of *size* for elements of D.

A nonzero element p of an integral domain D is a **prime** of D if and only if p is not a unit, and p does not divide a product of two elements in D unless p divides one of those two elements.

21. T T T F T F F T F T

22. The irreducibles of $D[x]$ are the irreducibles of D, together with the irreducibles of $F[x]$ which are in $D[x]$ and are furthermore *primitive* polynomials in $D[x]$. (See the paragraph following Lemma 45.26.)

23. The polynomial $2x + 4$ is irreducible in $\mathbb{Q}[x]$ but not in $\mathbb{Z}[x]$.

24. Not every nonzero nonunit of $\mathbb{Z} \times \mathbb{Z}$ has a factorization into irreducibles. For example $(1, 0)$ is not a unit, and every factorization of $(1,0)$ has a factor of the form $(\pm 1, 0)$, which is not irreducible because $(\pm 1, 0) = (\pm 1, 0)(1, 9)$. The only irreducibles of $\mathbb{Z} \times \mathbb{Z}$ are $(\pm 1, p)$ and $(q, \pm 1)$, where p and q are irreducibles in $\mathbb{Z}$.

25. Let p be a prime of D, and suppose that $p = ab$ for some $a, b \in D$. Then $ab = (1)p$, so p divides ab and thus divides either a or b, because p is a prime. Suppose that $a = pc$. Then $p = (1)p = pcb$ and cancellation in the integral domain yields $1 = cb$, so b is a unit of D. Similarly, if p divides b, we conclude that a is a unit in D. Thus either a or b is a unit, so p is an irreducible.

26. Let p be an irreducible in a UFD, and suppose that p divides ab. We must show that either p divides a or p divides b. Let $ab = pc$, and factor ab into irreducibles by first factoring a into irreducibles, then factoring b into irreducibles, and finally taking the product of these two factorizations. Now ab could also be factored into irreducibles by taking p times a factorization of c into irreducibles. Because factorization into irreducibles in a UFD is unique up to order and associates, it must be that an associate of p appears in the first factorization, formed by taking factors of a times factors of b. Thus an associate of p, say up, appears in the factorization of a or in the factorization of b. It follows at once that p divides either a or b.

27. *Reflexive:* $a = a \cdot 1$, so $a \sim a$.

Symmetric: Suppose $a \sim b$, so that $a = bu$ for a unit u. Then u^{-1} is a unit and $b = au^{-1}$, so $b \sim a$.

Transitive: Suppose that $a \sim b$ and $b \sim c$. Then there are units u_1 and u_2 such that $a = bu_1$ and $b = cu_2$. Substituting, we have $a = cu_2u_1 = c(u_2u_1)$. Because the product u_2u_1 of two units is again a unit, we find the $a \sim c$.

28. Let a and b be nonunits in $D^* - U$. Suppose that ab is a unit, so that $(ab)c = 1$ for some $c \in D$. Then $a(bc) = 1$ and a is a unit, contrary to our choice for a. Thus ab is again a nonunit, and $ab \ne 0$ because D has no divisors of zero. Hence $ab \in (D^* - U)$ also.

We see that $D^* - U$ is not a group, for the multiplicative identity is a unit, and hence is not in $D^* - U$.

29. Let $g(x)$ be a nonconstant divisor of the primitive polynomial $f(x)$ in $D[x]$. Suppose that $f(x) = g(x)q(x)$. Because D is a UFD, we know that $D[x]$ is a UFD also. Factor $f(x)$ into irreducibles by factoring each of $g(x)$ and $q(x)$ into irreducibles, and then taking the product of these factorizations. Each nonconstant factor appearing is an irreducible in $D[x]$, and hence is a primitive polynomial. Because the product of primitive polynomials is primitive by Corollary 45.26, we see that the content of $g(x)q(x)$ is the product of the content of $g(x)$ and the content of $q(x)$, and must be the same (up to a unit factor) as the content of $f(x)$. But $f(x)$ has content 1 because it is primitive. Thus $g(x)$ and $q(x)$ both have content 1. Hence $g(x)$ is a product of primitive polynomials, so it is primitive by Corollary 45.26.

30. Let N be an ideal in a PID D. If N is not maximal, then there is a proper ideal N_1 of D such that $N \subset N_1$. If N_1 is not maximal, we find a proper ideal N_2 such that $N_1 \subset N_2$. Continuing this process, we construct a chain $N \subset N_1 \subset N_2 \subset \cdots \subset N_i$ of proper ideals, each properly contained in the next except for the last ideal. Because a PID satisfies the ascending chain condition, we cannot extend this to an infinite such chain, so after some finite number of steps we must encounter a proper ideal N_r that contains N and that is not properly contained in any proper ideal of D. That is, we attain a maximal ideal N_r of D that contains N.

31. We have $x^3 - y^3 = (x - y)(x^2 + xy + y^2)$. Of course $x - y$ is irreducible. We claim that $x^2 + xy + y^2$ is irreducible in $\mathbb{Q}[x, y]$. Suppose that $x^2 + xy + y^2$ factors into a product of two polynomials that are not units in $\mathbb{Q}[x, y]$. Such a factorization would have to be of the form $x^2 + xy + y^2 = (ax + by)(cx + dy)$ with $a, b, c,$ and d all nonzero elements of $\mathbb{Q}$. Consider the evaluation homomorphism $\phi_1 : (\mathbb{Q}[x])[y] \to \mathbb{Q}[x]$ such that $\phi(y) = 1$. Applying ϕ_1 to both sides of such a factorization would yield $x^2 + x + 1 = (ax + b)(cx + d)$. But $x^2 + x + 1$ is irreducible in $\mathbb{Q}[x]$ because its zeros are complex, so no such factorization exists. This shows that $x^2 + xy + y^2$ is irreducible in $(\mathbb{Q}[x])[y]$ which isomorphic to $\mathbb{Q}[x, y]$ under an isomorphism that identifies $y^2 + yx + x^2$ and $x^2 + xy + y^2$.

32. We show that ACC implies MC, that MC implies FBC, and that FBC implies ACC.

ACC *implies* MC: Suppose that MC does not hold for some set S of ideals of R; that is, suppose it is not true that S contains an ideal not properly contained in any other ideal of S. Then every ideal of S is properly contained in another ideal of S. We can then start with any ideal N_1 of S and find an ideal N_2 of S properly containing it, then find $N_3 \in S$ properly containing N_2, etc. Thus we could construct an infinite chain of ideals $N_1 \subset N_2 \subset N_3 \subset \cdots$ which contradicts the ACC. Hence the ACC implies the MC.

MC *implies* FBC: Suppose the FBC does not hold, and let N be an ideal of R having no finite generating set. Let $b_1 \in N$ and let $N_1 = \langle b_1 \rangle$ be the smallest ideal of N containing b_1. Now $N_1 \neq N$ or $\{b_1\}$ would be a generating set for N, so find $b_2 \in N$ such that $b_2 \notin N_1$. Let N_2 be the intersection of all ideals containing b_1 and b_2. Because N contains b_1 and b_2, we see that $N_2 \subseteq N$, but $N_2 \neq N$ because $\{b_1, b_2\}$ cannot be a generating set for N. We then choose $n_3 \in N$ but not in N_2, and let N_3 be the intersection of all ideals containing $b_1, b_2,$ and b_3. Continuing this process, using the fact that N has no finite generating set, we can construct an infinite chain of ideals $N_1 \subset N_2 \subset N_3 \subset \cdots$ of R. But then the set $S = \{N_i \mid i \in \mathbb{Z}^+\}$ is a set of ideals, each of which is properly contained in another ideal of the set, for $N_i \subset N_{i+1}$. This would contradict the MC, so the FBC is true.

FBC *implies* ACC: Let $N_1 \subseteq N_2 \subseteq N_3 \subseteq \cdots$ be a chain of ideals in R, and let $N = \bigcup_{i=1}^{\infty} N_i$. It is easy to see that N is an ideal of R. Let $B_N = \{b_1, b_2, \cdots, b_n\}$ be a finite basis for N. Let $b_j \in N_{i_j}$. If r is the maximm of the subscripts i_j, then $B_N \subseteq N_r$. Because $N_r \subseteq N$ and N is the intersection of all ideals containing B_N, we must have $N_r = N$. Hence $N_r = N_{r+1} = N_{r+2} = \cdots$ so the ACC is satisfied.

33. *DCC implies mC:* Suppose that mC does not hold in R, and let S be a set of ideals in R where the mC fails, so that every ideal in S does properly contain another ideal of S. Then starting with any ideal $N_1 \in S$, we can find an ideal $N_2 \in S$ properly contained in N_1, and then an ideal N_3 of S properly contained in N_2, etc. This leads to an unending chain of ideals, each properly containing the next, which would contradict the DCC. Thus the DCC implies the mC.

mC *implies DCC:* Let $N_1 \supseteq N_2 \supseteq N_3 \supseteq \cdots$ be a descending chain of ideal in R. Let $S = \{N_i \mid i \in \mathbb{Z}^+\}$. By the mC, there is some ideal N_r of S that does not properly contain any other ideal in the set. Thus $N_r = N_{r+1} = N_{r+2} = \cdots$ so the DCC holds.

34. Now $\mathbb{Z}$ is a ring in which the ACC holds, because $\mathbb{Z}$ is PID. However,

$$\mathbb{Z} \supset 2\mathbb{Z} \supset 4\mathbb{Z} \supset \cdots \supset 2^{i-1}\mathbb{Z} \supset \cdots$$

is an infinite descending chain of ideals, so the DCC does not hold in $\mathbb{Z}$.

46. Euclidean Domains

1. Yes, it is a Euclidean norm. To see this, remember that we know $|\ |$ is a Euclidean norm on $\mathbb{Z}$. For Condition 1, find q and r such that $a = bq + r$ where either $r = 0$ or $|r| < |b|$. Then surely we have either $\nu(r) = 0$ or $\nu(r) = r^2 < b^2 = \nu(b)$, because r and b are integers. For Condition 2, note that $\nu(a) = a^2 \leq a^2b^2 = \nu(ab)$ for nonzero a and b, because a and b are integers.

2. No. ν is not a Euclidean norm. Let $a = x$ and $b = 2x$ in $\mathbb{Z}[x]$. There are no $q(x), r(x) \in \mathbb{Z}[x]$ satisfying $x = (2x)q(x) + r(x)$ where the degree of $r(x)$ is less than 1.

3. No. ν is not a Euclidean norm. Let $a = x$ and $b = x + 2$ in $\mathbb{Z}[x]$. There are no $q(x), r(x) \in \mathbb{Z}[x]$ satisfying $x = (x+2)q(x) + r(x)$ where the absolute value of the coefficient of the highest degree term in $r(x)$ is less than 1.

4. No. it is not a Euclidean norm. Let $a = 1/2$ and $b = 1/3$. Then $\nu(a) = (1/2)^2 = 1/4 > 1/36 = \nu(1/6) = \nu(ab)$, so Condition 2 is violated.

5. Yes. it is a Euclidean norm, but not a useful one. Let $a, b, \in \mathbb{Q}$. If $b \neq 0$, let $q = a/b$. Then $a = bq + 0$, which satisfies Condition 1. For Condition 2, if both a and b are nonzero, then $\nu(a) = 50 \leq 50 = \nu(ab)$.

6. We have $23 = 3(138) - 1(391)$, but $138 = 3{,}266 - 8(391)$, so

$$23 = 3[3{,}266 - 8(391)] - 1(391) = 3(3{,}266) - 25(391).$$

Now $391 = 7(3{,}266) - 22{,}471$, so

$$\begin{aligned} 23 &= 3(3{,}266) - 25[7(3{,}266) - 22{,}471] \\ &= 25(22{,}471) - 172(3{,}266). \end{aligned}$$

7. Performing the division algorithm, we obtain

$$\begin{aligned} 49{,}349 &= (15{,}555)3 + 2{,}684 \\ 15{,}555 &= (2{,}684)6 - 549 \\ 2{,}684 &= (549)5 - 61 \\ 549 &= (61)9 + 0 \end{aligned}$$

so the gcd is 61.

8. We have $61 = 5(549) - 2,684$, but $549 = 6(2,684) - 15,555$, so

$$61 = 5[6(2,684) - 15,555] - 2,684 = 29(2,684) - 5(15,555).$$

Now $2,684 = 49,349 - 3(15,555)$, so

$$
\begin{aligned}
61 &= 29[49,349 - 3(15,555)] - 5(15,555) \\
&= 29(49,349) - 92(15,555).
\end{aligned}
$$

9. We use the division algorithm.

$$
\begin{array}{r}
x^4 - 2x \\ \hline
x^6 - 3x^5 + 3x^4 - 9x^3 + 5x^2 - 5x + 2 \,\big)\, x^{10} - 3x^9 + 3x^8 - 11x^7 + 11x^6 - 11x^5 + 19x^4 - 13x^3 + 8x^2 - 9x + 3 \\
\underline{x^{10} - 3x^9 + 3x^8 - 9x^7 + 5x^6 - 5x^5 + 2x^4} \\
-2x^7 + 6x^6 - 6x^5 + 17x^4 - 13x^3 + 8x^2 - 9x + 3 \\
\underline{-2x^7 + 6x^6 - 6x^5 + 18x^4 - 10x^3 + 10x^2 - 4x} \\
-x^4 - 3x^3 - 2x^2 - 5x + 3
\end{array}
$$

$$
\begin{array}{r}
-x^2 + 6x - 19 \\ \hline
-x^4 - 3x^3 - 2x^2 - 5x + 3 \,\big)\, x^6 - 3x^5 + 3x^4 - 9x^3 + 5x^2 - 5x + 2 \\
\underline{x^6 + 3x^5 + 2x^4 + 5x^3 - 3x^2} \\
-6x^5 + x^4 - 14x^3 + 8x^2 - 5x \\
\underline{-6x^5 - 18x^4 - 12x^3 - 30x^2 + 18x} \\
19x^4 - 2x^3 + 38x^2 - 23x + 2 \\
\underline{19x^4 + 57x^3 + 38x^2 + 95x - 57} \\
-59x^3 - 118x + 59
\end{array}
$$

Multiply by the unit $-1/59$.

$$
\begin{array}{r}
-x - 3 \\ \hline
x^3 + 2x - 1 \,\big)\, -x^4 - 3x^3 - 2x^2 - 5x + 3 \\
\underline{-x^4 - 2x^2 + x} \\
-3x^3 - 6x + 3 \\
\underline{-3x^3 - 6x + 3} \\
0
\end{array}
$$

A gcd is $x^3 + 2x - 1$.

10. Use the Euclidean algorithm to find the gcd d_2 of a_2 and a_1. Then use it to find the gcd d_3 of a_3 and d_2. Then use it again to find the gcd d_4 of a_4 and d_3. Continue this process until you find the gcd d_n of a_n and d_{n-1}. The gcd of the n members $a_1, a_2, \cdots, a_n$ is d_n.

11. We use the notation of the solution of the preceding exercise with $a_1 = 2178, a_2 = 396, a_3 = 792$, and $a_4 = 726$. We have $2178 = 5(396) + 198$ and $396 = 2(198) + 0$, so $d_2 = 198$. We have $792 = 4(198) + 0$ so $d_3 = 198$. We have $726 = 3(198) + 132$, $198 = 1(132) + 66$, and $132 = 2(66) + 0$. Thus the gcd of 2178, 396, 792, and 726 is $d_4 = 66$.

12. a. Yes, $\mathbb{Z}[x]$ is a UFD because $\mathbb{Z}$ is a UFD and Theorem 45.29 tells us that if D is a UFD, then $D[x]$ is a UFD.

b. The set described consists of all polynomials in $\mathbb{Z}[x]$ that have a constant term in $2\mathbb{Z}$, that is, an even number as constant term. This property of a polynomial is obviously satisfied by $0 \in 2\mathbb{Z}$ and is preserved under addition, subtraction, and multiplication by every element of $\mathbb{Z}[x]$, so the set is an ideal of $\mathbb{Z}[x]$.

c. $\mathbb{Z}[x]$ is not a PID because the ideal described in Part(**b**) is not a principal ideal; a generating polynomial would to have constant term 2, but could not yield, under multiplication by elements of $\mathbb{Z}[x]$, all the polynomials of the form $2 + nx$ in the ideal because n can be odd as well as even.

d. $\mathbb{Z}[x]$ is not an Euclidean domain, because every Euclidean domain is a PID by Theorem 46.4, but Part(**c**) shows that $\mathbb{Z}[x]$ is not a PID.

13. T F T F T T T F T T

14. No, it does not. For any integral domain, its arithmetic structure is completely determined by the binary operations of addition and of multiplication. A Euclidean norm, if one exists, can be used to *study* the arithmetic structure, but it in no way changes it

15. Let a and b be associates of a Euclidean domain D with Euclidean norm ν. Because a and b are associates, there exists a unit u in D such that $a = bu$. Then u^{-1} is also a unit of D and $b = au^{-1}$. Condition 2 of a Euclidean norm yields $\nu(b) \leq \nu(bu) = \nu(a) \leq \nu(au^{-1}) = \nu(b)$. Thus $\nu(a) = \nu(b)$.

16. Suppose that $\nu(a) < \nu(ab)$. If b were a unit, then a and ab would be associates, and by Exercise 15, we would have $\nu(a) = \nu(ab)$. Thus b is not a unit.

For the converse, suppose that $\nu(a) = \nu(ab)$. We claim that then $\langle a \rangle = \langle ab \rangle$, for the proof of Theorem 46.4 shows that a nonzero ideal in a Euclidean domain is generated by any element having minimum norm in the ideal. Thus ab also generates $\langle a \rangle$, so $a = (ab)c$ for some $c \in D$. Then, cancelling the a in the integral domain, we find that $1 = bc$ so b is a unit.

17. The statement is false. For example, let $D = \mathbb{Z}$ and let the norm be $|\ |$. Then $|2| > 1$ and $|-3| > 1$, but $|2 + (-3)| = |-1| = 1$, so the given set is not closed under addition.

18. Let F be a field and let $\nu(a) = 1$ for all $a \in F, a \neq 0$. Because every nonzero element of F is a unit, we see that for nonzero $b \in F$ we have $a = b(a/b) + 0$, which satisfies Condition 1 for a Euclidean norm. Also, $\nu(a) = \nu(ab) = 1$, so Condition 2 is satisfied.

19. a. For Condition 1, let $a, b \in D^*$ where $b \neq 0$. There exist $q, r \in D$ such that $a = bq + r$ where either $r = 0$ or $\nu(r) < \nu(b)$. But then either $r = 0$ or $\eta(r) = \nu(r) + s < \nu(b) + s = \eta(b)$, so Condition 1 holds. For $a, b \neq 0$, we have $\eta(a) = \nu(a) + s \leq \nu(ab) + s = \eta(ab)$, so Condition 2 holds. The hypothesis that $\nu(1) + s > 0$ guarantees that $\eta(a) > 0$ for all $a \in D^*$, because $\nu(1)$ is minimal among all $\nu(a)$ for $a \in D^*$, by Theorem 46.6.

b. For Condition 1, let $a, b \in D^*$ where $b \neq 0$. There exist $q, r \in D$ such that $a = bq + r$ where either $r = 0$ or $\nu(r) < \nu(b)$. But then either $r = 0$ or $\lambda(r) = t \cdot \nu(r) < t \cdot \nu(b) = \lambda(b)$, so Condition 1 holds. For $a, b \neq 0$, we have $\lambda(a) = t \cdot \nu(a) \leq t \cdot \nu(ab) = \lambda(ab)$, so Condition 2 holds. The hypothesis that $t \in \mathbb{Z}^+$ guarantees that $\lambda(a) > 0$ and $\lambda(a)$ is an integer for all $a \in D^*$.

c. Let ν be a Euclidean norm on D. Then $\lambda(a) = 100 \cdot \nu(a)$ for $a \in D^*$ is a Euclidean norm on D by Part(**b**). Let $s = 1 - \lambda(1)$. Then $\mu(a) = \lambda(a) + s$ for $a \in D^*$ is a Euclidean norm on D by Part(**a**), and $\mu(1) = \lambda(1) + s = \lambda(1) + [1 - \lambda(1)] = 1$. If $a \neq 0$ is a nonunit in D, then $\nu(a) \geq \nu(1) + 1$ so $\lambda(a) = 100 \cdot \nu(a) \geq 100 \cdot [\nu(1) + 1]$ and $\mu(a) = \lambda(a) + s \geq 100 \cdot [\nu(1) + 1] + 1 - \lambda(1) = 100 \cdot [\nu(1) + 1] + 1 - 100 \cdot \nu(1) = 100 + 1 = 101$.

20. We know that all multiples of $a \in D$ form the principal ideal $\langle a \rangle$ and all multiples of b form the principal ideal $\langle b \rangle$. By Exercise 27 of Section 26, the intersection of ideals in a ring is an ideal, so $\langle a \rangle \cap \langle b \rangle$ is an ideal, and consists of all common multiples of a and b. Because a Euclidean domain is a PID, this ideal has a generator c. Now $a \mid c$ and $b \mid c$ because c is a common multiple of a and b.

Because every common multiple of a and b is in $\langle a \rangle \cap \langle b \rangle = \langle c \rangle$, we see that every common multiple is of the form dc, that is, every common multiple is a multiple of c. Thus c is an lcm of a and b.

21. The subgroup of $\langle \mathbb{Z}, + \rangle$ generated by two integers $r, s \in \mathbb{Z}$ is $H = \{mr + ns \mid n, m \in \mathbb{Z}\}$. Now $H = \mathbb{Z}$ if and only if $1 \in H$, so $H = \mathbb{Z}$ if and only if $1 = mr + ns$ for some $m, n \in \mathbb{Z}$. If r and s are relatively prime, then 1 is a gcd of r and s, and the Euclidean algorithm together with the last statement in Theorem 46.9, show that 1 can be expressed in the form $1 = mr + ns$ for some $m, n \in \mathbb{Z}$. Conversely, if $1 = mr + ns$, then every integer dividing both m and n divides the righthand side of this equation and thus divides 1, so 1 is a gcd of r and s.

22. If a and n are relatively prime, then a gcd of a and n is 1. By Theorem 46.9, we can express 1 in the form $1 = m_1 a + m_2 n$ for some $m_1, m_2 \in \mathbb{Z}$. Multiplying by b, we get $b = a(m_1 b) + (bm_2)n$. Thus $x = m_1 b$ is a solution of $ax \equiv b \pmod{n}$.

23. Suppose that the positive gcd d of a and n in $\mathbb{Z}$ divides b. By Theorem 46.9, we can express d in the form $d = m_1 a + m_2 n$ for some $m_1, m_2 \in \mathbb{Z}$. Multiplying by b/d, we obtain $b = a(m_1 b/d) + (bm_2/d)n$. Thus $x = m_1 b/d$ is a solution of $ax \equiv b \pmod{n}$. Conversely, suppose that $ac \equiv b \pmod{n}$ so that n divides $ac - b$, say $ac - b = nq$. Then $b = ac - nq$. Because the positive gcd d of a and n divides the righthand side of $b = ac - nq$, it must be that d divides b also.

In $\mathbb{Z}_n$, this result has the following interpretation: $ax = b$ has a solution in $\mathbb{Z}_n$ for nonzero $a, b \in \mathbb{Z}_n$ if and only if the positive gcd of a and n in $\mathbb{Z}$ divides b.

24. *Step 1.* Use the Euclidean algorithm to find the positive gcd d of a and n.

Step 2. Use the technique of Exercise 6 to express d in the form $d = m_1 a + m_2 n$.

Step 3. A solution of $ax \equiv b \pmod{n}$ is $x = m_1 b/d$.

We now illustrate with $22x \equiv 18 \pmod{42}$.

Step 1. Find the gcd of 22 and 42:

$$
\begin{aligned}
42 &= 1(22) + 20, \\
22 &= 1(20) + 2, \\
20 &= 10(2),
\end{aligned}
$$

so 2 is a gcd of 22 and 42.

Step 2. Express 2 in the form $m_1(22) + m_2(42)$:

$$
\begin{aligned}
2 &= 22 - 1(20) \quad \text{but} \quad 20 = 42 - 1(22), \\
2 &= 22 - 1[42 - 1(22)] = 2(22) + (-1)(42),
\end{aligned}
$$

so $m_1 = 2$.

Step 3. A solution of $22x \equiv 18 \pmod{42}$ is $x = m_1 b/d = (2)(18)/2 = 18$.

47. Gaussian Integers and Multiplicative Norms

1. Example 47.8 showed that $5 = (1 + 2i)(1 - 2i)$ is a factorization of 5 into irreducibles. Because $\mathbb{Z}[i]$ is a UFD, this factorization of 5 is unique up to unit factors. For example, $[i(1 + 2i)][-i(1 - 2i)] = (-2 + i)(-2 - i)$ is another factorization of 5 into irreducibles.

2. If α is a factor of 7 in $\mathbb{Z}[i]$, then $N(\alpha)$ must divide $N(7) = 49$, so $N(\alpha)$ must be 1, 7, or 49. If $N(\alpha) = 1$, then α is a unit and if $N(\alpha) = 49$, then the other factor must be a unit, and we are not interested in these cases. Thus we must have $N(\alpha) = 7$ if α is an irreducible dividing 7 and 7 is not irreducible. Because the equation $a^2 + b^2 = 7$ has no solutions in integers, it must be that 7 is already irreducible in $\mathbb{Z}[i]$.

3. Proceeding as in the answer to Exercise 2, if α is an irreducible factor of $4 + 3i$, then $N(\alpha)$ must be a divisor of $4^2 + 3^2 = 25$, and 5 is the only possibility if $4 + 3i$ is not irreducible. Thus we must have $\alpha = a + bi$ where a is ± 1 and b is ± 2 or where a is ± 2 and b is ± 1. A bit of trial and errror shows that $4 + 3i = (1 + 2i)(2 - i)$, and $1 + 2i$ and $2 - i$ are irreducible because they have the prime 5 as norm.

4. Proceeding as in the answer to Exercise 2, if α is an irreducible factor of $6 - 7i$, then $N(\alpha)$ must be a divisor of $6^2 + 7^2 = 85$, so $N(\alpha)$ must be either 5 or 17 if $6 - 7i$ is not irreducible. If $N(\alpha) = 5$, then $\alpha = a + bi$ where $a^2 + b^2 = 5$, so $a = \pm 1$ and $b = \pm 2$, or $a = \pm 2$ and $b = \pm 1$. We compute

$$\frac{6 - 7i}{1 + 2i} = \frac{6 - 7i}{1 + 2i} \cdot \frac{1 - 2i}{1 - 2i} = \frac{-8 - 19i}{5}$$

and find that the answer is not in $\mathbb{Z}[i]$. There is no use trying $\pm 1(1 + 2i)$ or $\pm i(1 + 2i)$, so we try $1 - 2i$. We obtain

$$\frac{6 - 7i}{1 - 2i} = \frac{6 - 7i}{1 - 2i} \cdot \frac{1 + 2i}{1 + 2i} = \frac{20 + 5i}{5} = 4 + i.$$

Thus $6 - 7i = (1 - 2i)(4 + i)$, and $1 - 2i$ and $4 + i$ are irreducibles because their norms are the primes 5 and 17 respectively.

5. We have $6 = 2 \cdot 3 = (-1 + \sqrt{-5})(-1 - \sqrt{-5})$. The numbers 2 and 3 are both irreducible in $\mathbb{Z}[\sqrt{-5}]$ because the equations $a^2 + 5b^2 = 2$ and $a^2 + 5b^2 = 3$ have no solutions in integers. (See Example 47.9 in the text.) If $-1 + \sqrt{-5}$ were not irreducible, then it would be a product $\alpha\beta$ where neither α nor β is a unit and $N(\alpha\beta) = N(\alpha)N(\beta) = 6$. This means that we would have to have $N(\alpha) = 2$ or $N(\alpha) = 3$, which is impossible as we have just seen. Because the only units in $\mathbb{Z}[\sqrt{-5}]$ are ± 1, we have two essentially different factorizations of 6.

6. We compute $\alpha\beta$:

$$\frac{7 + 2i}{3 - 4i} = \frac{7 + 2i}{3 - 4i} \cdot \frac{3 + 4i}{3 + 4i} = \frac{13 + 34i}{25}.$$

Now we take the integer 1 closest to $13/25$ and the integer 1 closest to $34/25$, and let $\sigma = 1 + 1i = 1 + i$. Then we compute

$$\begin{aligned}
\rho = \alpha - \beta\sigma &= (7 + 2i) - (3 - 4i)(1 + i) \\
&= (7 + 2i) - (7 - i) = 3i.
\end{aligned}$$

Then $\alpha = \sigma\beta + \rho$ where $N(\rho) = N(3i) = 9 < 25 = N(3 - 4i) = N(\beta)$.

7. We let $\alpha = 5 - 15i$ and $\beta = 8 + 6i$, and compute σ and ρ as in the answer to Exercise 6:

$$\frac{5 - 15i}{8 + 6i} = \frac{5 - 15i}{8 + 6i} \cdot \frac{8 - 6i}{8 - 6i} = \frac{-50 - 150i}{100}.$$

We take $\sigma = -i$, and $\rho = \alpha - \sigma\beta = (5 - 15i) - (-i)(8 + 6i) = -1 - 7i$, so

$$5 - 15i = (8 + 6i)(-i) + (-1 - 7i).$$

Continuing the Euclidean algorithm, we now take $\alpha = 8 + 6i$ and $\beta = -1 - 7i$, and obtain

$$\frac{8+6i}{-1-7i} = \frac{8+6i}{-1-7i} \cdot \frac{-1+7i}{-1+7i} = \frac{-50+50i}{50} = -1 + i.$$

Because $-1 + i \in \mathbb{Z}[i]$, we are done, and a gcd of $5 - 15i$ and $8 + 6i$ is $-1 - 7i$. Of course, the other gcd's are obtained by multiplying by the units -1, $\pm i$, so $1 + 7i, -7 + i$, and $7 - i$ are also acceptable answers.

8. T T T F T T T F T T

9. Suppose that $\pi = \alpha\beta$. Then $N(\pi) = N(\alpha)N(\beta)$. Because $|N(\pi)|$ is the minimal norm > 1, one of $|N(\alpha)|$ and $|N(\beta)|$ must be $|N(\pi)|$ and the other must be ± 1. Thus either α or β has norm ± 1, and is thus a unit by hypothesis. Therefore π is an irreducible in D.

10. a. We know that in $\mathbb{Z}[i]$, the units are precisely the elements $\pm 1, \pm i$ of norm 1. By Theorem 47.7, every element of $\mathbb{Z}[i]$ having as norm a prime in $\mathbb{Z}$ is an irreducible. Because $N(1 + i) = 1^2 + 1^2 = 2$, we see that $1 + i$ is an irreducible. The equation $2 = -i(1+i)^2$ thus gives the desired factorization of 2.

b. Every odd prime in $\mathbb{Z}$ is congruent to either 1 or 3 modulo 4. If $p \equiv 1 \pmod 4$, then Theorem 47.10 shows that $p = a^2 + b^2 = (a+ib)(a-ib)$ where neither $a + ib$ nor $a - ib$ is a unit because they each have norm $a^2 + b^2 = p > 1$, so p is not an irreducible.

Conversely, if p is not an irreducible, then $p = (a + ib)(c + di)$ in $\mathbb{Z}[i]$ where neither factor is a unit, so that both $a + bi$ and $c + di$ have norm greater than 1. Taking the norm of both sides of the equation, we obtain $p^2 = (a^2 + b^2)(c^2 + d^2)$, so we must have $p = a^2 + b^2 = c^2 + d^2$. Theorem 47.10 then shows that we must have $p \equiv 1 \pmod 4$.

We have shown that an odd prime p is not irreducible if and only if $p \equiv 1 \pmod 4$ so an odd prime p is irreducible if and only if $p \equiv 3 \pmod 4$.

11. *Property 1:* Let $\alpha = a + bi$. Then $N(\alpha) = a^2 + b^2$. As a sum of squares, $a^2 + b^2 \geq 0$.

Property 2: Continuing the argument for Property 1, we see that $a^2 + b^2 = 0$ if and only if $a = b = 0$, so $N(\alpha) = 0$ if and only if $\alpha = 0$.

Property 3: Let $\beta = c + di$. Then $\alpha\beta = (a + bi)(c + di) = (ac - bd) + (ad + bc)i$, so

$$\begin{aligned} N(\alpha\beta) &= (ac - bd)^2 + (ad + bc)^2 \\ &= a^2c^2 - 2abcd + b^2d^2 + a^2d^2 + 2abcd + b^2c^2 \\ &= a^2c^2 + b^2d^2 + a^2d^2 + b^2c^2 \\ &= (a^2 + b^2)(c^2 + d^2) = N(\alpha)N(\beta). \end{aligned}$$

[Of course it also follows from the fact that $|\alpha\beta|^2 = |\alpha|^2|\beta|^2$ for all $\alpha, \beta \in \mathbb{C}$.]

12. Let $\alpha = a + b\sqrt{-5}$ and $\beta = c + d\sqrt{-5}$. Then

$$\alpha\beta = (a + b\sqrt{-5})(c + d\sqrt{-5}) = (ac - 5bd) + (ad + bc)\sqrt{-5},$$

so

$$\begin{aligned} N(\alpha\beta) &= (ac - 5bd)^2 + 5(ad + bc)^2 \\ &= a^2c^2 - 10abcd + 25b^2d^2 + 5a^2d^2 + 10abcd + 5b^2c^2 \\ &= a^2c^2 + 25b^2d^2 + 5a^2d^2 + 5b^2c^2 \\ &= (a^2 + 5b^2)(c^2 + 5d^2) = N(\alpha)N(\beta). \end{aligned}$$

[Of course it also follows from the fact that $|\alpha\beta|^2 = |\alpha|^2|\beta|^2$ for all $\alpha, \beta \in \mathbb{C}$.]

13. Let $\alpha \in D$. We give a proof by induction on $|N(\alpha)|$, starting with $|N(\alpha)| = 2$, that α has a factorization into irreducibles. Let $|N(\alpha)| = 2$. Then α itself is an irreducible by Theorem 47.7, and we are done.

Suppose that every element of absolute norm > 1 but $< k$ has a factorization into irreducibles, and let $|N(\alpha)| = k$. If α is an irreducible, then we are done. Otherwise, $\alpha = \beta\gamma$ where neither β nor γ is a unit, so $|N(\beta)| > 1$ and $|N(\gamma)| > 1$. From $|N(\beta\gamma)| = |N(\beta)N(\gamma)| = |N(\alpha)| = k$, we then see that $1 < |N(\beta)| < k$ and $1 < |N(\gamma)| < k$, so by the induction assumption, both β and γ have factorizations into a product of irreducibles. The product of these two factorizations then provides a factorization of α into irreducibles.

14. Now

$$\frac{16+7i}{10-5i} = \frac{16+7i}{10-5i} \cdot \frac{10+5i}{10+5i} = \frac{125+150i}{125} = 1 + \frac{6}{5}i$$

so we let $\sigma = 1 + i$. Then $16 + 7i = (10-5i)(1+i) + (1+2i)$. We have

$$\frac{10-5i}{1+2i} = \frac{10-5i}{1+2i} \cdot \frac{1-2i}{1-2i} = \frac{0-25i}{5} = -5i,$$

so $10 - 5i = (1+2i)(-5i)$. Thus $1 + 2i$ is a gcd of $16 + 7i$ and $10 + 5i$. Other possible answers are $-1 - 2i, -2 + i$, and $2 - i$.

15. a. Let $\gamma + \langle\alpha\rangle$ be a coset of $\mathbb{Z}[i]/\langle\alpha\rangle$. By the division algorithm, $\gamma = \alpha\sigma + \rho$ where either $\rho = 0$ or $N(\rho) < N(\alpha)$. Then $\gamma + \langle\alpha\rangle = (\rho + \sigma\alpha) + \langle\alpha\rangle$. Now $\sigma\alpha \in \langle\alpha\rangle$, so $\gamma + \langle\alpha\rangle = \rho + \langle\alpha\rangle$. Thus every coset of $\langle\alpha\rangle$ contains a representative of norm less than $N(\alpha)$. Because there are only a finite number of elements of $\mathbb{Z}[i]$ having norm less than $N(\alpha)$, we see that $\mathbb{Z}[i]/\langle\alpha\rangle$ is a finite ring.

b. Let π be an irreducible in $\mathbb{Z}[i]$, and let $\langle\mu\rangle$ be an ideal in $\mathbb{Z}[i]$ such that $\langle\pi\rangle \subseteq \langle\mu\rangle$. (Remember that $\mathbb{Z}[i]$ is a PID so every ideal is principal.) Then $\pi \in \langle\mu\rangle$ so $\pi = \mu\beta$. Because π is an irreducible, either μ is a unit, in which case $\langle\mu\rangle = \mathbb{Z}[i]$, or β is a unit, in which case $\mu = \pi\beta^{-1}$ so $\mu \in \langle\pi\rangle$ and $\langle\mu\rangle = \langle\pi\rangle$. We have shown that $\langle\pi\rangle$ is a maximal ideal of $\mathbb{Z}[i]$, so $\mathbb{Z}[i]/\langle\pi\rangle$ is a field.

c. i. Because $\langle 3\rangle$ contains both 3 and $3i$, we see that each coset contains a unique representative of the form $a + bi$ where a and b are both in the set $\{0, 1, 2\}$. Thus there are 9 elements in all, and the ring has characteristic 3 because $1 + 1 + 1 = 0$.

ii. By Part(**a**), each coset contains a representative of norm less that $N(1 + i) = 2$. The only nonzero elements of $\mathbb{Z}[i]$ of norm less than 2 are ± 1 and $\pm i$. Because $i = -1 + (1 + i)$ and $-i = 1 - (1 + i)$, we can reduce our list of possible nonzero cosets to $1 + \langle 1+i\rangle$ and $-1 + \langle 1+i\rangle$. But $[1 + \langle 1+i\rangle] - [-1 + \langle 1+i\rangle] = 2 + \langle 1+i\rangle$, and $2 = (1+i)(1-i)$ is in $\langle 1+i\rangle$. Thus the only cosets are $\langle 1+i\rangle$ and $1 + \langle 1+i\rangle$, so the order of the ring is 2, and the characteristic is 2.

iii. By Part(**a**), each coset contains a representative of norm less than $N(1 + 2i) = 5$. The only nonzero elements of $\mathbb{Z}[i]$ of norm less than 5 are of the form $a + bi$ where a and b are in the set $\{0, 1, -1\}$ or where one of a and b is ± 2 and the other is zero. These elements are $1, 1, i, i, 1+i, 1-i, -1+i, -1-i, 2, -2, 2i$, and $-2i$. Because

$$
\begin{aligned}
i &= 2 + (1+2i)i, \\
-i &= -2 + (1+2i)(-i), \\
2i &= -1 + (1+2i), \\
-2i &= 1 + (1+2i)(-1), \\
1+i &= -2 + (1+2i)(1-i),
\end{aligned}
$$

$$1 - i = -1 + (1 + 2i)(-i),$$
$$-1 + i = 1 + (1 + 2i)(i),$$
$$-1 - i = 2 + (1 + 2i)(-1 + i),$$

we see that every coset contains either 0, 1, -1, 2, or -2 as a representative. The ring has 5 elements and characteristic 5.

16. a. *Property 1:* Because $n > 0$, we see that $a^2 + nb^2 = 0$ if and only if $a = b = 0$.

Property 2: Let $\alpha = a + b\sqrt{-n}$ and $\beta = c + d\sqrt{-n}$. Then

$$
\begin{aligned}
N(\alpha\beta) &= N((ac - bdn) + (ad + bc)\sqrt{-n}) \\
&= (ac - bdn)^2 + n(ad + bc)^2 \\
&= a^2c^2 - 2abcdn + b^2d^2n^2 + a^2d^2n + 2abcdn + b^2c^2n \\
&= a^2c^2 + b^2d^2n^2 + a^2d^2n + b^2c^2n \\
&= (a^2 + nb^2)(c^2 + nd^2) = N(\alpha)N(\beta).
\end{aligned}
$$

[Of course it also follows from the fact that $|\alpha\beta|^2 = |\alpha|^2|\beta|^2$ for all $\alpha, \beta \in \mathbb{C}$.]

b. By Theorem 47.7, if $\alpha \in \mathbb{Z}[\sqrt{-n}]$ is a unit, then $N(\alpha) = 1$. Conversely, suppose that $N(\alpha) = 1$. Now $a^2 + nb^2 = 1$ where $n \in \mathbb{Z}^+$ if and only if either $a = \pm 1$ and $b = 0$, or $a = 0$ and $n = 1$ and $b = \pm 1$. In the former case, $\alpha = \pm 1$, and of course 1 and -1 are units. In the latter case with $n = 1$, we are in the Gaussian integers which are a Euclidean domain, and Theorem 46.6 tells us that the elements of norm 1 are indeed units.

c. The preceding parts show that we have a multiplicative norm on $\mathbb{Z}[\sqrt{-n}]$ such that the elements of norm 1 are precisely the units. By Exercise 13, every nonzero nonunit has a factorization into irreducibles.

    Note that the hypothesis that n is square free was not used in this exercise. Because $a + b\sqrt{-m^2n} = a + (bm)\sqrt{-n}$, we see that the square-free assumption is really no loss of generality. The assumption that $n > 0$ was used in both Part(**a**) and Part(**b**). The square-free assumption is used in the following exercise, however.

17. a. *Property 1:* If $a^2 - nb^2 = 0$, then $a^2 = nb^2$. If $b = 0$, then $a = 0$. If $b \neq 0$, then $n = (a/b)^2$, contradicting the hypothesis that n is square free. Thus $a = 0$ and $b = 0$.

Property 2: Let $\alpha = a + b\sqrt{n}$ and $b = c + d\sqrt{n}$. Then

$$
\begin{aligned}
N(\alpha\beta) &= N((ac + bdn) + (ad + bc)\sqrt{n}) \\
&= (ac + bdn)^2 - n(ad + bc)^2 \\
&= a^2c^2 + 2abcdn + b^2d^2n^2 - a^2d^2n - 2abcdn - b^2c^2n \\
&= a^2c^2 + b^2d^2n^2 - a^2d^2n - b^2c^2n \\
&= (a^2 - nb^2)(c^2 - nd^2) \\
&= (a^2 - nb^2)(c^2 - nd^2) = N(\alpha)N(\beta).
\end{aligned}
$$

b. As an integral domain with a multiplicative norm, the norm of every unit is ± 1 by Theorem 47.7. Now suppose that $\alpha = a + b\sqrt{n}$ has norm ± 1, so that $a^2 - nb^2 = \pm 1$. Then

$$
\frac{1}{\alpha} = \frac{1}{a + b\sqrt{n}} \cdot \frac{a - b\sqrt{n}}{a - b\sqrt{n}} = \frac{a - b\sqrt{n}}{a^2 - nb^2} = \pm(a - b\sqrt{n})
$$

and $(a + (-b)\sqrt{n}) \in \mathbb{Z}[\sqrt{n}]$, so α is a unit.

c. The preceding parts show that we have a multiplicative norm on $\mathbb{Z}[\sqrt{n}]$ such that the elements of norm 1 are precisely the units. By Exercise 13, every nonzero nonunit has a factorization into irreducibles.

18. Given α and β in $\mathbb{Z}[\sqrt{-2}]$, we proceed to construct $\sigma = q_1 + q_2\sqrt{-2}$ and $\rho = \alpha - \beta\sigma$ as described in the proof of Theorem 47.4. Viewing $\alpha = a + b\sqrt{-2} = a + (b\sqrt{2})i$ in $\mathbb{C}$, we have $|a + (b\sqrt{2})i|^2 = a^2 + 2b^2 = N(\alpha)$. Working in $\mathbb{C}$, with $|\;|^2$, we compute

$$\frac{\alpha}{\beta} = \frac{a + b\sqrt{-2}}{c + d\sqrt{-2}} \cdot \frac{c - d\sqrt{-2}}{c - d\sqrt{-2}} = r + s\sqrt{-2}$$

for $r, s \in \mathbb{Q}$. Again, we choose q_1 and q_2 to be integers in $\mathbb{Z}$ as close as possible to r and s respectively.

By construction of σ, we see that $|r - q_1| \leq 1/2$ and $|s - q_2| \leq 1/2$. Thus

$$
\begin{aligned}
N(\frac{\alpha}{\beta} - \sigma) &= N((r + s\sqrt{-2}) - (q_1 + q_2\sqrt{-2})) \\
&= N((r - q_1) + (s - q_2)\sqrt{-2}) \\
&\leq \left(\frac{1}{2}\right)^2 + 2\left(\frac{1}{2}\right)^2 \leq \frac{1}{4} + 2 \cdot \frac{1}{4} = \frac{3}{4}.
\end{aligned}
$$

Thus we obtain

$$N(\rho) = N(\alpha - \beta\sigma) = N(\beta(\frac{\alpha}{\beta} - \sigma)) = N(\beta)N(\frac{\alpha}{\beta} - \sigma) \leq N(\beta)\frac{3}{4},$$

so we do indeed have $N(\rho) < N(\beta)$ as claimed.

48. Automorphisms of Fields

1. The conjugates of $\sqrt{2}$ over $\mathbb{Q}$ are $\sqrt{2}$ and $-\sqrt{2}$.

2. Now $\sqrt{2}$ is the only conjugate of $\sqrt{2}$ over $\mathbb{R}$ because $\sqrt{2} \in \mathbb{R}$.

3. The conjugates of $3 + \sqrt{2}$ over $\mathbb{Q}$ are $3 + \sqrt{2}$ and $3 - \sqrt{2}$; they are the zeros of $(x - 3)^2 - 2 = x^2 - 6x + 7$.

4. The conjugates of $\sqrt{2} - \sqrt{3}$ over $\mathbb{Q}$ are

$$\sqrt{2} - \sqrt{3}, \sqrt{2} + \sqrt{3}, -\sqrt{2} - \sqrt{3}, \text{ and } -\sqrt{2} + \sqrt{3}$$

as indicated by Example 48.17.

5. The conjugates of $\sqrt{2} + i$ over $\mathbb{Q}$ are $\sqrt{2} + i, \sqrt{2} - i, -\sqrt{2} + i$, and $-\sqrt{2} - i$. This is clear because $\mathbb{Q}(\sqrt{2} + i) = (\mathbb{Q}(\sqrt{2}))(i)$.

6. The conjugates of $\sqrt{2} + i$ over $\mathbb{R}$ are $\sqrt{2} + i$ and $\sqrt{2} - i$ because $\sqrt{2} \in \mathbb{R}$.

7. The conjugates of $\sqrt{1 + \sqrt{2}}$ over $\mathbb{Q}$ are

$$\sqrt{1 + \sqrt{2}}, \sqrt{1 - \sqrt{2}}, -\sqrt{1 + \sqrt{2}}, \text{ and } -\sqrt{1 - \sqrt{2}}.$$

This clear because $\mathbb{Q}(\sqrt{1 + \sqrt{2}}) = (\mathbb{Q}(\sqrt{2}))(\sqrt{1 + \sqrt{2}})$.

8. The conjugates of $\sqrt{1+\sqrt{2}}$ over $\mathbb{Q}(\sqrt{2})$ are just $\sqrt{1+\sqrt{2}}$ and $-\sqrt{1+\sqrt{2}}$ because $\sqrt{2} \in \mathbb{Q}(\sqrt{2})$.

9. $\tau_2(\sqrt{3}) = \sqrt{3}$.   **10.** $\tau_2(\sqrt{2} + \sqrt{5}) = -\sqrt{2} + \sqrt{5}$.

11. $(\tau_3\tau_2)(\sqrt{2} + 3\sqrt{5}) = \tau_3[\tau_2(\sqrt{2} + 3\sqrt{5})] = \tau_3(-\sqrt{2} + 3\sqrt{5}) = -\sqrt{2} + 3\sqrt{5}$.

12. $(\tau_5\tau_3)\left(\frac{\sqrt{2}-3\sqrt{5}}{2\sqrt{3}-\sqrt{2}}\right) = \tau_5\left[\tau_3\left(\frac{\sqrt{2}-3\sqrt{5}}{2\sqrt{3}-\sqrt{2}}\right)\right] = \tau_5\left(\frac{\sqrt{2}-3\sqrt{5}}{-2\sqrt{3}-\sqrt{2}}\right) = \frac{\sqrt{2}+3\sqrt{5}}{-2\sqrt{3}-\sqrt{2}}$

13. Note that $\sqrt{45} = 3\sqrt{5}$. We have

$$\begin{aligned}
(\tau_5{}^2\tau_3\tau_2)(\sqrt{2} + \sqrt{45}) &= \tau_5{}^2\tau_3[\tau_2(\sqrt{2} + \sqrt{45})] \\
&= \tau_5{}^2[\tau_3[(-\sqrt{2} + \sqrt{45})]] \\
&= \tau_5{}^2(-\sqrt{2} + \sqrt{45}) \\
&= \tau_5(-\sqrt{2} - \sqrt{45}) = -\sqrt{2} + \sqrt{45}.
\end{aligned}$$

14. Note that $\sqrt{30} = \sqrt{2}\sqrt{3}\sqrt{5}$. Now $\tau_3[\tau_5(\sqrt{2} - \sqrt{3} + (\tau_2\tau_5)(\sqrt{30}))]$

$$\begin{aligned}
&= \tau_3[\tau_5(\sqrt{2} - \sqrt{3} + \tau_2(\tau_5(\sqrt{30})))] \\
&= \tau_3[\tau_5(\sqrt{2} - \sqrt{3} + \tau_2(-\sqrt{30}))] \\
&= \tau_3[\tau_5(\sqrt{2} - \sqrt{3} + \sqrt{30})] \\
&= \tau_3(\sqrt{2} - \sqrt{3} - \sqrt{30}) = \sqrt{2} + \sqrt{3} + \sqrt{30}.
\end{aligned}$$

15. a. Because $\sigma_1(\sqrt{2}) = -\sqrt{2}, \sigma_3(\sqrt{3}) = -\sqrt{3}$, and $\sigma_1(\sqrt{6}) = -\sqrt{6}$, the only elements of $\mathbb{Q}(\sqrt{2}, \sqrt{3})$ left fixed by both σ_1 and σ_3 are those in $\mathbb{Q}$, so $\mathbb{Q}$ is the fixed field.

b. We see that $\sigma_3 = \sigma_1\sigma_2$ leaves $\sqrt{6}$ fixed, because $\sqrt{6} = \sqrt{2}\sqrt{3}$ and σ_3 acts on this product by changing the sign of both factors. Thus the fixed field is $\mathbb{Q}(\sqrt{6})$.

c. Because $\sigma_3(\sqrt{2}) = -\sqrt{2}, \sigma_2(\sqrt{3}) = -\sqrt{3}$, and $\sigma_2(\sqrt{6}) = -\sqrt{6}$, the only elements of $\mathbb{Q}(\sqrt{2}, \sqrt{3})$ left fixed by both σ_3 and σ_2 are those in $\mathbb{Q}$, so $\mathbb{Q}$ is the fixed field.

16. Because τ_3 moves only $\sqrt{3}$, the subfield of $\mathbb{Q}(\sqrt{2}, \sqrt{3}, \sqrt{5})$ left fixed by τ_3 is $\mathbb{Q}(\sqrt{2}, \sqrt{5})$.

17. Because $\tau_3{}^2 = \iota$, the identity, the entire field $\mathbb{Q}(\sqrt{2}, \sqrt{3}, \sqrt{5})$ is left fixed by $\tau_3{}^2$.

18. Because $\tau_2(\sqrt{2}) = -\sqrt{2}$ and $\tau_3(\sqrt{3}) = -\sqrt{3}$, we see that the subfield of $\mathbb{Q}(\sqrt{2}, \sqrt{3}, \sqrt{5})$ left fixed by $\{\tau_2, \tau_3\}$ is $\mathbb{Q}(\sqrt{5})$.

19. Because $\tau_2(\sqrt{2}) = -\sqrt{2}$ and $\tau_5(\sqrt{5}) = -\sqrt{5}$, we see that $\tau_5\tau_2$ leaves both $\sqrt{3}$ and $\sqrt{10}$ fixed, so the subfield of $\mathbb{Q}(\sqrt{2}, \sqrt{3}, \sqrt{5})$ left fixed by $\tau_5\tau_2$ is $\mathbb{Q}(\sqrt{3}, \sqrt{10})$.

20. We see that $\tau_5\tau_3\tau_2$ leaves $\sqrt{15}, \sqrt{6}$, and $\sqrt{10}$ fixed. Because $\sqrt{15} = \sqrt{6}\sqrt{10}/2$, we see that we can describe the fixed field of $\tau_5\tau_3\tau_2$ as $\mathbb{Q}(\sqrt{6}, \sqrt{10})$.

21. Because every product of one, two or three distinct factors formed from $\sqrt{2}, \sqrt{3}$, and $\sqrt{5}$ is moved by one of τ_2, τ_3, or τ_5, we see that the subfield of $\mathbb{Q}(\sqrt{2}, \sqrt{3}, \sqrt{5})$ left fixed by $\{\tau_2, \tau_3, \tau_5\}$ is $\mathbb{Q}$.

22. a. Because $\tau_2(\sqrt{2}) = -\sqrt{2}$, we see that $\tau_2{}^2(\sqrt{2}) = \tau_2(\tau_2(\sqrt{2})) = \tau_2(-\sqrt{2}) = \sqrt{2}$. Because τ_2 moves neither $\sqrt{3}$ nor $\sqrt{5}$, we see that $\tau_2{}^2 = \iota$, the identity map. Thus τ_2 is of order 2 in $G(E/\mathbb{Q})$. Clearly the same argument shows that both τ_3 and τ_5 are of order 2 in $G(E/\mathbb{Q})$ also.

b. $H = \{\iota, \tau_2, \tau_3, \tau_5, \tau_2\tau_3, \tau_2\tau_5, \tau_3\tau_5, \tau_2\tau_3\tau_5\}$.

	ι	τ_2	τ_3	τ_5	$\tau_2\tau_3$	$\tau_2\tau_5$	$\tau_3\tau_5$	$\tau_2\tau_3\tau_5$
ι	ι	τ_2	τ_3	τ_5	$\tau_2\tau_3$	$\tau_2\tau_5$	$\tau_3\tau_5$	$\tau_2\tau_3\tau_5$
τ_2	τ_2	ι	$\tau_2\tau_3$	$\tau_2\tau_5$	τ_3	τ_5	$\tau_2\tau_3\tau_5$	$\tau_3\tau_5$
τ_3	τ_3	$\tau_2\tau_3$	ι	$\tau_3\tau_5$	τ_2	$\tau_2\tau_3\tau_5$	τ_5	$\tau_2\tau_5$
τ_5	τ_5	$\tau_2\tau_5$	$\tau_3\tau_5$	ι	$\tau_2\tau_3\tau_5$	τ_2	τ_3	$\tau_2\tau_3$
$\tau_2\tau_3$	$\tau_2\tau_3$	τ_3	τ_2	$\tau_2\tau_3\tau_5$	ι	$\tau_3\tau_5$	$\tau_2\tau_5$	τ_5
$\tau_2\tau_5$	$\tau_2\tau_5$	τ_5	$\tau_2\tau_3\tau_5$	τ_2	$\tau_3\tau_5$	ι	$\tau_2\tau_3$	τ_3
$\tau_3\tau_5$	$\tau_3\tau_5$	$\tau_2\tau_3\tau_5$	τ_5	τ_3	$\tau_2\tau_5$	$\tau_2\tau_3$	ι	τ_2
$\tau_2\tau_3\tau_5$	$\tau_2\tau_3\tau_5$	$\tau_3\tau_5$	$\tau_2\tau_5$	$\tau_2\tau_3$	τ_5	τ_3	τ_2	ι

c. An automorphism in $G(E/\mathbb{Q})$ is completely determined by its values on $\sqrt{2}, \sqrt{3}$, and $\sqrt{5}$. Each of these is either left alone or mapped into its negative. Thus there are two possibilities for the value of $\sigma \in G(E/\mathbb{Q})$ on $\sqrt{2}$, two possibilities for $\sigma(\sqrt{3})$, and two possibilities for $\sigma(\sqrt{5})$, giving a total of $2 \cdot 2 \cdot 2 = 8$ automorphisms in all. Because $|H| = 8$, we see that $H = G(E/\mathbb{Q})$.

23. The definition is incorrect. Insert "irreducible" before "polynomial".

Two elements, α and β, of an algebraic extension E of a field F are **conjuate over** F if and only if they are both zeros of the same irreducible polynomial $f(x)$ in $F[x]$.

24. The definition is correct.

25. a. We have $\beta = 3 - \sqrt{2}$ is a conjugate of $\alpha = 3 + \sqrt{2}$. They are both zeros of the polynomial $(x - 3)^2 - 2 = x^2 - 6x + 7$ which is irreducible over $\mathbb{Q}$.

b. Because

$$
\begin{aligned}
\psi_{\alpha,\beta}(\sqrt{2}) &= \psi_{\alpha,\beta}(-3 + (3 + \sqrt{2})) \\
&= \psi_{\alpha,\beta}(-3) + \psi_{\alpha,\beta}(3 + \sqrt{2}) \\
&= -3 + (3 - \sqrt{2}) = -\sqrt{2} = \psi_{\sqrt{2}, -\sqrt{2}}(\sqrt{2}),
\end{aligned}
$$

we see that $\psi_{\alpha,\beta}$ and $\psi_{\sqrt{2},-\sqrt{2}}$ are the same map.

26. We have

$$\sigma_2(0) = 0^2 = 0, \quad \sigma_2(1) = 1^2 = 1, \quad \sigma_2(\alpha) = \alpha^2 = \alpha + 1,$$

and

$$\sigma_2(\alpha + 1) = (\alpha + 1)^2 = \alpha^2 + 2 \cdot \alpha + 1 = \alpha^2 + 1 = \alpha + 1 + 1 = \alpha.$$

Thus $\mathbb{Z}_2(\alpha)_{\{\sigma_2\}} = \{0, 1\} = \mathbb{Z}_2$.

27. Using the table for this field in this manual, we find that

$$\sigma_3(0) = 0^3 = 0, \quad \sigma_3(1) = 1^3 = 1, \quad \sigma_3(2) = 2^3 = 2,$$

$$\sigma_3(\alpha) = \alpha^3 = 2\alpha, \quad \sigma_3(2\alpha) = (2\alpha)^3 = \alpha,$$

$$\sigma_3(1 + \alpha) = (1 + \alpha)^3 = 1 + 2\alpha,$$

$$\sigma_3(1 + 2\alpha) = (1 + 2\alpha)^3 = 1 + \alpha,$$

$$\sigma_3(2 + \alpha) = (2 + \alpha)^3 = 2 + 2\alpha,$$

and

$$\sigma_3(2 + 2\alpha) = (2 + 2\alpha)^3 = 2 + \alpha.$$

Thus $\mathbb{Z}_3(\alpha)_{\{\sigma_3\}} = \mathbb{Z}_3$.

28. The map $\sigma_2 : \mathbb{Z}_2(x) \to \mathbb{Z}_2(x)$, where x is an indeterminate, is not an automorphism because the image is $\mathbb{Z}_2(x^2)$. Thus σ is not onto $\mathbb{Z}_2(x)$, but rather maps $\mathbb{Z}_2(x)$ one to one onto a proper subfield of itself.

29. F F T T F T T T T T

30. If α and β are conjugate, then they have the same irreducible polynomial $p(x)$ over F, so both $F(\alpha)$ and $F(\beta)$ are isomorphic to $F[x]/\langle p(x) \rangle$.

31. If $\psi_{\alpha, \beta}$ is an isomorphism of $F(\alpha)$ onto $F(\beta)$, then for every polynomial $f(x) \in F[x]$, we have $f(\alpha) = 0$ if and only if $f(\beta) = 0$, so the monic irreducible polyomial for α over F is the same as the one for β over F.

32. By Corollary 48.5, such an isomorphism must map α onto one of its conjugates over F. Because $\deg(\alpha, F) = n$ there are at most n conjugates of α in $\overline{F}$, for a polynomial of degree n has at most n zeros in a field. On the other hand, Corollary 48.5 asserts that there is exactly one such isomorphism for each conjugate of α over F, so the number of such isomorphisms is equal to the number of conjugates of α over F, which is $\leq n$.

33. We proceed by induction on n. For $n = 1$, Corollary 48.5 shows that σ is completely determined by $\sigma(\alpha_1)$, which must be a conjugate of α_1 over F. Suppose that the statement is true for $n < k$, and let $n = k$. Suppose that σ is known on $\alpha_1, \alpha_2, \cdots, \alpha_{k-1}, \alpha_k$. Let $r = \deg(\alpha_k, F(\alpha_1, \cdots, \alpha_{k-1}))$. Then each element β in $F(\alpha_1, \cdots, \alpha_{k-1}, \alpha_k)$ can be written uniquely in the form

$$\beta = \gamma_0 + \gamma_1 \alpha_k + \gamma_2 \alpha_k^2 + \cdots + \gamma_{r-1} \alpha_k^{r-1}$$

where $\gamma_i \in F(\alpha_1, \cdots, \alpha_{k-1})$ for $i = 0, 1, \cdots, r-1$ according to Theorem 30.23. By our induction assumption, we know $\sigma(\gamma_i)$ for $i = 0, 1, \cdots, r-1$, and we are assuming that we also know $\sigma(\alpha_k)$. The expression for β and the fact that σ is an automorphism shows that we know $\sigma(\beta)$. This completes our proof by induction.

34. By Corollary 48.5, σ maps each zero of $\mathrm{irr}(\alpha, F)$ onto a zero of this same polyomial. Because σ is an automorphism, it is a one-to-one map of E onto E. By counting, it must map the set of zeros in E of this polynomial onto itself, so it is a permutation of this set.

35. Because $S \subseteq H$, it is clear that $E_H \subseteq E_S$. Let $\alpha \in E_S$, so that $\sigma_i(\alpha) = \alpha$ for all $i \in I$. Then $\sigma_i^{-1}(\alpha) = \alpha$ also. It follows at once that $\sigma_i^n(\alpha) = \alpha$ for all $i \in I$ and all $n \in \mathbb{Z}$. Theorem 7.6 shows that every element of H is a product of a finite number of such powers of the σ_i. Because products of automorphisms are computed by function composition, it follows that α is left fixed by each element in H. Therefore $\alpha \in E_H$, so $E_S \subseteq E_H$ and therefore $E_S = E_H$.

36. a. Suppose that $\zeta^i = \zeta^j$ for $i < j \leq p - 1$. Then $\zeta^{j-i} = 1$ and ζ would be a zero of $x^{j-i} - 1$ which is of degree less than $p - 1$, contradicting the fact that $\Phi_p(x)$ is irreducible. Thus these powers ζ^i for $1 \leq i \leq p - 1$ are distinct. Because $\zeta^i \neq 1$ for $1 \leq i \leq p - 1$ but $(\zeta^i)^p = (\zeta^p)^i = 1^i = 1$, we see that ζ^i is a zero of $x^p - 1$ that is different from 1 for $1 \leq i \leq p - 1$, so these distinct powers of ζ must account for all $p - 1$ zeros of $\Phi_p(x)$.

b. Let $\sigma, \tau \in G(\mathbb{Q}(\zeta)/\mathbb{Q})$. Suppose that $\sigma(\zeta) = \zeta^i$ and that $\tau(\zeta) = \zeta^j$. Then

$$
\begin{aligned}
(\sigma\tau)(\zeta) &= \sigma(\tau(\zeta)) = \sigma(\zeta^j) = [\sigma(\zeta)]^j = (\zeta^i)^j \\
&= \zeta^{ij} = \zeta^{ji} = (\zeta^j)^i = [\tau(\zeta)]^i \\
&= \tau(\zeta^i) = \tau(\sigma(\zeta)) = (\tau\sigma)(\zeta).
\end{aligned}
$$

Because $(\sigma\tau)(\zeta) = (\tau\sigma)(\zeta)$, Corollary 48.5 shows that $\sigma\tau = \tau\sigma$. Thus $G(\mathbb{Q}(\zeta)/\mathbb{Q})$ is abelian.

c. We know that $B = \{1, \zeta, \zeta^2, \cdots, \zeta^{p-2}\}$ is a basis for $\mathbb{Q}(\zeta)$ over $\mathbb{Q}$. Let $\beta \in \mathbb{Q}(\zeta)$. We can write

$$\frac{\beta}{\zeta} = a_0 + a_1\zeta + a_2\zeta^2 + \cdots + a_{p-2}\zeta^{p-2}$$

for $a_i \in \mathbb{Q}$ for $0 \le i \le p-2$. Multiplying by ζ, we see that

$$\beta = a_0\zeta + a_1\zeta^2 + a_2\zeta^3 + \cdots + a_{p-2}\zeta^{p-1} \tag{1}$$

so these powers of ζ do span $\mathbb{Q}(\zeta)$. They are linearly independent because a linear combination of them equal to zero yields a linear combination of the elements in B equal to zero upon division by ζ. Thus the set $\{\zeta, \zeta^2, \cdots, \zeta^{p-1}\}$ is a basis for $\mathbb{Q}(\zeta)$ over $\mathbb{Q}$.

By Theorem 48.3, and Part(**a**), there exist automorphisms σ_i for $i = 1, 2, \cdots, p-1$ in $G(\mathbb{Q}(\zeta)/\mathbb{Q})$ such that $\sigma_i(\zeta) = \zeta^i$. Thus if β in Eq.(1) is left fixed by all such σ_i, we must have $a_0 = a_1 = \cdots = a_{p-2}$ so $\beta = a_0(\zeta + \zeta^2 + \cdots + \zeta^{p-1}) = -a_0$ because ζ is a zero of $\Phi_p(x)$. Thus the elements of $\mathbb{Q}(\zeta)$ left fixed lie in $\mathbb{Q}$, so $\mathbb{Q}$ is the fixed field of $G(\mathbb{Q}(\zeta)/\mathbb{Q})$.

37. Yes, if α and β are transcendentals over F, then $\phi : F(\alpha) \to F(\beta)$, where $\phi(a) = a$ for each $a \in F$ and $\phi\left(\frac{f(\alpha)}{g(\alpha)}\right) = \frac{f(\beta)}{g(\beta)}$ for $f(x), g(x) \in F[x]$ and $g(x) \ne 0$, is an isomorphism. Both $F(\alpha)$ and $F(\beta)$ are isomorphic to $F(x)$ as we saw in Case II under the heading **Simple Extensions** in Section 29.

38. In the notation of Exercise 37, taking $\alpha = x$, we must find all β transcendental over F such that $F(x) = F(\beta)$. This means that not only must β be a quotient of polynomials in x that does not lie in the field F, but also, we must be able to solve and express x as a quotient of polynomials in β. This is only possible if β is a quotient $\frac{ax+b}{cx+d}$ of linear polynomials in $F[x]$, and for this quotient not to be in F, we must require that $ad - bc \ne 0$, so $\frac{a}{c} \ne \frac{b}{d}$.

39. a. Let σ be an automorphism of E and let $\alpha \in E$. Then $\sigma(\alpha^2) = \sigma(\alpha\alpha) = \sigma(\alpha)\sigma(\alpha) = \sigma(\alpha)^2$, so σ indeed carries squares into squares.

b. Because the positive numbers in $\mathbb{R}$ are precisely the squares in $\mathbb{R}$, this follows at once from Part(**a**).

c. From $a < b$, we deduce that $b - a > 0$. By Part(**b**), we see that $\sigma(b - a) = \sigma(b) - \sigma(a) > 0$, so $\sigma(a) < \sigma(b)$.

d. Let σ be an automorphism of $\mathbb{R}$. Let $a \in \mathbb{R}$, and find sequences $\{r_i\}$ and $\{s_i\}$ of rational numbers, both converging to a, and satisfying

$$r_i < r_{i+1} < a < s_{i+1} < s_i$$

for all $i \subset \mathbb{Z}^+$. By Part(**c**), we see that

$$\sigma(r_i) < \sigma(r_{i+1}) < \sigma(a) < \sigma(s_{i+1}) < \sigma(s_i). \tag{2}$$

The automorphism σ of $\mathbb{R}$ must leave the prime field $\mathbb{Q}$ fixed, because $\sigma(1) = 1$. Thus the inequality (2) becomes

$$r_i < r_{i+1} < \sigma(a) < s_{i+1} < s_i$$

for all $i \in \mathbb{Z}^+$. Because the sequences $\{r_i\}$ and $\{s_i\}$ converge to a, we see that $\sigma(a) = a$, so σ is the identity automorphism.

49. The Isomorphism Extension Theorem

1. (See the answer in the text.)

2. Extensions are τ_1 given by

$$\tau_1(\sqrt{2}) = \sqrt{2}, \quad \tau_1(\sqrt{3}) = -\sqrt{3}, \quad \tau_1(\sqrt{5}) = \sqrt{5},$$

and τ_2 given by

$$\tau_2(\sqrt{2}) = \sqrt{2}, \quad \tau_2(\sqrt{3}) = \sqrt{3}, \quad \tau_2(\sqrt{5}) = -\sqrt{5}.$$

3. (See the answer in the text.)

4. The extensions are the identity map of $\mathbb{Q}(\sqrt[3]{2})$ onto itself, and τ_1 given by $\tau_1(\alpha_1) = \alpha_2$, that is, the map $\psi_{\alpha_1, \alpha_2}$, and $\tau_2 = \psi_{\alpha_1, \alpha_3}$.

5. (See the answer in the text.)

6. The extensions are

$$\tau_1 \text{ given by } \tau_1(i) = i, \quad \tau_1(\sqrt{3}) = -\sqrt{3}, \quad \tau_1(\alpha_1) = \alpha_1,$$

$$\tau_2 \text{ given by } \tau_2(i) = i, \quad \tau_2(\sqrt{3}) = -\sqrt{3}, \quad \tau_2(\alpha_1) = \alpha_2,$$

$$\tau_3 \text{ given by } \tau_3(i) = i, \quad \tau_3(\sqrt{3}) = -\sqrt{3}, \quad \tau_3(\alpha_1) = \alpha_3,$$

$$\tau_4 \text{ given by } \tau_4(i) = -i, \quad \tau_4(\sqrt{3}) = -\sqrt{3}, \quad \tau_4(\alpha_1) = \alpha_1,$$

$$\tau_5 \text{ given by } \tau_5(i) = -i, \quad \tau_5(\sqrt{3}) = -\sqrt{3}, \quad \tau_5(\alpha_1) = \alpha_2,$$

$$\tau_6 \text{ given by } \tau_6(i) = -i, \quad \tau_6(\sqrt{3}) = -\sqrt{3}, \quad \tau_6(\alpha_1) = \alpha_3.$$

7. (See the answer in the text.)

8. F T F T F T T T F

9. Now $\sigma : K \to K$ is an isomorphism, so $\sigma^{-1} : \sigma[K] \to K$ is an isomorphism. Because K is algebraically closed and is algebraic over $\sigma[K]$, Theorem 49.3 shows that σ^{-1} has an extension to an isomorphism τ mapping K onto a subfield of K. But σ^{-1} is already onto K, and because τ must be a one-to-one map, we see that it cannot be defined on any elements of K not already in $\sigma[K]$. Thus $\sigma[K] = K$, so σ is an automorphism of K.

10. Let E be an algebraic extension of F and let τ be an isomorphism of E onto a subfield of $\overline{F}$ that leaves F fixed. Because E is an algebraic extension of F, the field $\overline{F}$ is an algebraic extension of E and is an algebraic closure of E. By Theorem 49.3, τ can be extended to an isomorphism σ of $\overline{F}$ onto a subfield of $\overline{F}$. By Exercise 9, such an isomorphism σ is an automorphism of $\overline{F}$.

11. By Theorem 49.3, the identity map of F onto F has an extension to an isomorphism τ mapping E onto a subfield of $\overline{F}$. By Theorem 49.3, τ can be extended to an isomorphism σ mapping $\overline{E}$ onto a subfield of $\overline{F}$. Then σ^{-1} is an isomorphism mapping $\sigma[\overline{E}]$ onto $\overline{F}$. By Theorem 49.3, σ^{-1} can be extended to an isomorphism of $\overline{F}$ onto a subfield of $\overline{E}$. Because σ^{-1} is already onto $\overline{E}$ and its extension must be one to one, we see that the domain of σ^{-1} must already be $\overline{F}$. Thus $\sigma[\overline{E}] = \overline{F}$ and σ is an isomorphism of $\overline{E}$ onto $\overline{F}$.

12. We should note that $\overline{\mathbb{Q}(x)}$ is an algebraic closure of $\overline{\mathbb{Q}}(x)$. We know that π is transcendental over $\mathbb{Q}$. Therefore, $\sqrt{\pi}$ must be transcendental over $\mathbb{Q}$, for if it were algebraic, then $\pi = (\sqrt{\pi})^2$ would be algebraic over $\mathbb{Q}$, because algebraic numbers form a closed set under field operations. Therefore the map $\tau : \mathbb{Q}(\sqrt{\pi}) \to \mathbb{Q}(x)$ where $\tau(a) = a$ for $a \in \mathbb{Q}$ and $\tau(\sqrt{\pi}) = x$ is an isomorphism. Theorem 49.3 shows that τ can be extended to an isomorphism σ mapping $\overline{\mathbb{Q}(\sqrt{\pi})}$ onto a subfield of $\overline{\mathbb{Q}(x)}$. Then σ^{-1} is an isomorphism mapping $\sigma[\overline{\mathbb{Q}(\sqrt{\pi})}]$ onto a subfield of $\overline{\mathbb{Q}(\sqrt{\pi})}$ which can be extended to an isomorphism of $\overline{\mathbb{Q}(x)}$ onto a subfield of $\overline{\mathbb{Q}(\sqrt{\pi})}$. But because σ^{-1} is already onto $\overline{\mathbb{Q}(\sqrt{\pi})}$, we see that σ must actually be onto $\overline{\mathbb{Q}(x)}$, so σ provides the required isomorphism of $\overline{\mathbb{Q}(\sqrt{\pi})}$ with $\overline{\mathbb{Q}(x)}$.

13. Let E be a finite extension of F. Then by Theorem 31.11, $E = F(\alpha_1, \alpha_2, \cdots, \alpha_n)$ where each α_i is algebraic over F. Now suppose that $L = F(\alpha_1, \alpha_2, \cdots, \alpha_{k+1})$ and $K = F(\alpha_1, \alpha_2, \cdots, \alpha_k)$. Every isomorphism of L onto a subfield of $\overline{F}$ and leaving F fixed can be viewed as an extension of an isomorphism of K onto a subfield of $\overline{F}$. The extension of such an isomorphism τ of K to an isomorphism σ of L onto a subfield of $\overline{F}$ is completely determined by $\sigma(\alpha_{k+1})$. Let $p(x)$ be the irreducible polynomial for α_{k+1} over K, and let $q(x)$ be the polynomial in $\tau[K][x]$ obtained by applying τ to each of the coefficients of $p(x)$. Because $p(\alpha_{k+1}) = 0$, we must have $q(\sigma(\alpha_{k+1})) = 0$, so the number of choices for $\sigma(\alpha_{k+1})$ is at most $\deg(q(x)) = \deg(p(x)) = [L : K]$. Thus $\{L : K\} \le [L : K]$, that is

$$\{F(\alpha_1, \cdots, \alpha_{k+1}) : F(\alpha_1, \cdots, \alpha_k)\} \le [F(\alpha_1, \cdots, \alpha_{k+1}) : F(\alpha_1, \cdots, \alpha_k)]. \tag{1}$$

We have such an inequality (1) for each $k = 1, 2, \cdots, n-1$. Using the multiplicative properties of the index and of the degree (Corollaries 49.10 and 31.6), we obtain upon multiplication of these $n-1$ inequalities the desired result, $\{E : F\} \le [E : F]$.

50. Splitting Fields

1. The splitting field is $\mathbb{Q}(\sqrt{3})$ and the degree over $\mathbb{Q}$ is 2.

2. Now $x^4 - 1 = (x - 1)(x + 1)(x^2 + 1)$. The splitting field is $\mathbb{Q}(i)$ and the degree over $\mathbb{Q}$ is 2.

3. The splitting field is $\mathbb{Q}(\sqrt{2}, \sqrt{3})$ and the degree over $\mathbb{Q}$ is 4.

4. The splitting field has degree 6 over $\mathbb{Q}$. Replace $\sqrt[3]{2}$ by $\sqrt[3]{3}$ in Example 50.9.

5. Now $x^3 - 1 = (x - 1)(x^2 + x + 1)$. The splitting field has degree 2 over $\mathbb{Q}$.

6. The splitting field has degree $2 \cdot 6 = 12$ over $\mathbb{Q}$. See Example 50.9 for the splitting field of $x^3 - 2$.

7. We have $|G(\mathbb{Q}(\sqrt[3]{2})/\mathbb{Q}| = 1$, because $\sqrt[3]{2} \in \mathbb{R}$ and the other conjugates of $\sqrt[3]{2}$ do not lie in $\mathbb{R}$ (see Example 50.9). They yield isomorphisms into $\mathbb{C}$ rather than automorphisms of $\mathbb{Q}(\sqrt[3]{2})$.

8. We have $|G(\mathbb{Q}(\sqrt[3]{2}, i\sqrt{3})/\mathbb{Q}| = 6$, because $\mathbb{Q}(\sqrt[3]{2}, i\sqrt{3})$ is the splitting field of $x^3 - 2$ and is of degree 6, as shown in Example 50.9.

9. $|G(\mathbb{Q}(\sqrt[3]{2}, i\sqrt{3})/\mathbb{Q}(\sqrt[3]{2})| = 2$, because $\mathbb{Q}(\sqrt[3]{2}, i\sqrt{3})$ is the splitting field of $x^2 + 3$ over $\mathbb{Q}(\sqrt[3]{2})$.

10. Theorem 33.3 shows that the only field of order 8 in $\overline{\mathbb{Z}_2}$ is the splitting field of $x^8 - x$ over $\mathbb{Z}_2$. Because a field of order 8 can be obtained by adjoining to $\mathbb{Z}_2$ a root of any cubic polynomial that is irreducible in $\mathbb{Z}_2[x]$, it must be that all roots of every irreducible cubic lie in this unique subfield of order 8 in $\overline{\mathbb{Z}_2}$.

11. The definition is incorrect. Insert "irreducible" before "polynomial".

Let $F \le E \le \overline{F}$ where $\overline{F}$ is an algebraic closure of a field F. The field E is a **splitting field over** F if and only if E contains all the zeros in $\overline{F}$ of every irreducible polynomial in $F[x]$ that has a zero in E.

12. The definition is incorrect. Replace "lower degree" by "degree one".

A polynomial $f(x)$ in $F[x]$ **splits in an extension field** E of F if and only if it factors in $E[x]$ into a product of polynomials of degree one.

13. We have $1 \le [E : F] \le n!$. The example $E = F = \mathbb{Q}$ and $f(x) = x^2 - 1$ shows that the lower bound 1 cannot be improved unless we are told that $f(x)$ is irreducible over F. Example 50.9 shows that the upper bound $n!$ cannot be improved.

14. T F T T T F F T T

15. Let $F = \mathbb{Q}$ and $E = \mathbb{Q}(\sqrt{2})$. Then $f(x) = x^4 - 5x^2 + 6 = (x^2 - 2)(x^2 - 3)$ has a zero in E, but does not split in E.

16. a. This multiplicative relation is not necessarily true. Example 50.9 and Exercise 7 show that $6 = |G(\mathbb{Q}(\sqrt[3]{2}, i\sqrt{3})/\mathbb{Q})| \ne |G(\mathbb{Q}(\sqrt[3]{2}, i\sqrt{3})/\mathbb{Q}(\sqrt[3]{2}))| \cdot |G(\mathbb{Q}\sqrt[3]{2})/\mathbb{Q})| = 2 \cdot 1 = 2$.

b. Yes, because each field is a splitting field of the one immediately under it. If E is a splitting field over F then $|G(E/F)| = \{E : F\}$, and the index is multiplicative by Corollary 49.10.

17. Let E be the splitting field of a set S of polynomials in $F[x]$. If $E = F$, then E is the splitting field of x over F. If $E \ne F$, then find a polynomial $f_1(x)$ in S that does not split in F, and form its splitting field, which is a subfield E_1 of E where $[E_1 : F] > 1$. If $E = E_1$, then E is the splitting field of $f_1(x)$ over F. If $E \ne E_1$, find a polynomial $f_2(x)$ in S that does not split in E_1, and form its splitting field $E_2 \le E$ where $[E_2 : E_1] > 1$. If $E = E_2$, then E is the splitting field of $f_1(x)f_2(x)$ over F. If $E \ne E_2$, then continue the construction in the obvious way. Because by hypothesis E is a *finite* extension of F, this process must eventually terminate with some $E_r = E$, which is then the splitting field of the product $g(x) = f_1(x)f_2(x) \cdots f_r(x)$ over F.

18. Find $\alpha \in E$ that is not in F. Now α is algebraic over F, and must be of degree 2 because $[E : F] = 2$ and $[F(\alpha) : F] = \deg(\alpha, F)$. Thus $\mathrm{irr}(\alpha, F) = x^2 + bx + c$ for some $b, c \in F$. Because $\alpha \in E$, this polynomial factors in $E[x]$ into a product $(x - \alpha)(x - \beta)$, so the other root β of $\mathrm{irr}(\alpha, F)$ lies in E also. Thus E is the splitting field of $\mathrm{irr}(\alpha, F)$.

19. Let E be a splitting field over F. Let α be in E but not in F. By Corollary 50.6, the polynomial $\mathrm{irr}(\alpha, F)$ splits in E since it has a zero α in E. Thus E contains all conjugates of α over F.

Conversely, suppose that E contains all conjugates of $\alpha \in E$ over F, where $F \le E \le \overline{F}$. Because an automorphism σ of $\overline{F}$ leaving F fixed carries every element of $\overline{F}$ into one of its conjugates over F, we see that $\sigma(\alpha) \in E$. Thus σ induces a one-to-one map of E into E. Because the same is true of σ^{-1}, we see that σ maps E onto E, and thus induces an automorphism of E leaving F fixed. Theorem 50.3 shows that under these conditions, E is a splitting field of F.

20. Because $\mathbb{Q}(\sqrt[3]{2})$ lies in $\mathbb{R}$ and the other two conjugates of $\sqrt[3]{2}$ do not lie in $\mathbb{R}$, we see that no map of $\sqrt[3]{2}$ into any conjugate other than $\sqrt[3]{2}$ itself can give rise to an automorphism of $\mathbb{Q}(\sqrt[3]{2})$; the other two maps give rise to isomorphisms of $\mathbb{Q}(\sqrt[3]{2})$ onto a subfield of $\overline{\mathbb{Q}}$. Because any automorphism of $\mathbb{Q}(\sqrt[3]{2})$ must leave the prime field $\mathbb{Q}$ fixed, we see that the identity is the only automorphism of $\mathbb{Q}(\sqrt[3]{2})$. [For an alternate argument, see Exercise 39 of Section 48.]

21. The conjugates of $\sqrt[3]{2}$ over $\mathbb{Q}(i\sqrt{3}\,)$ are

$$\sqrt[3]{2}, \quad \sqrt[3]{2}\frac{-1+i\sqrt{3}}{2}, \quad \text{and} \quad \sqrt[3]{2}\frac{-1-i\sqrt{3}}{2}.$$

Maps of $\sqrt[3]{2}$ into each of them give rise to the only three automorphisms in $G(\mathbb{Q}(\sqrt[3]{2}, i\sqrt{3}\,)/\mathbb{Q}(i\sqrt{3}\,))$. Let σ be the automorphism such that $\sigma(\sqrt[3]{2}\,) = \sqrt[3]{2}\frac{-1+i\sqrt{3}}{2}$. Then σ must be a generator of this group of order 3, because σ is not the identity map, and every group of order 3 is cyclic. Thus the automorphism group is isomorphic to $\mathbb{Z}_3$.

22. a. Each automorphism of E leaving F fixed is a one-to-one map that carries each zero of $f(x)$ into one of its conjugates, which must be a zero of an irreducible factor of $f(x)$ and hence is also a zero of $f(x)$. Thus each automorphism gives rise to a one-to-one map of the set of zeros of $f(x)$ onto itself, that is, it acts as a permutation on the zeros of $f(x)$.

b. Because E is the splitting field of $f(x)$ over F, we know that $E = F(\alpha_1, \alpha_2, \cdots, \alpha_n)$ where $\alpha_1, \alpha_2, \cdots, \alpha_n$ are the zeros of $f(x)$. As Exercise 33 of Section 48 shows, an automorphism σ of E leaving F fixed is completely determined by the values $\sigma(\alpha_1), \sigma(\alpha_2), \cdots, \sigma(\alpha_n)$ that is, by the permutation of the zeros of $f(x)$ given by σ.

c. We associate with each $\sigma \in G(E/F)$ its permutation of the zeros of $f(x)$ in E. Part(**b**) shows that different elements of $G(E/F)$ produce different permutations of the zeros of $f(x)$. Because multiplication $\sigma\tau$ in $G(E/F)$ is function composition and because multiplication of the permutations of zeros is again composition of these same functions, with domain restricted to the zeros of $f(x)$, we see that $G(E/F)$ is isomorphic to a subgroup of the group of all permutations of the zeros of $f(x)$.

23. a. We have $|G(E/\mathbb{Q})| = 2 \cdot 3 = 6$, because $\{\mathbb{Q}(i\sqrt{3}\,) : \mathbb{Q}\} = 2$ since $\text{irr}(i\sqrt{3}, \mathbb{Q}) = x^2 + 3$ and $\{\mathbb{Q}(\sqrt[3]{2}\,) : \mathbb{Q}(i\sqrt{3}\,)\} = 3$ because $\text{irr}(\sqrt[3]{2}, \mathbb{Q}(i\sqrt{3}\,)) = x^3 - 2$. The index is multiplicative by Corollary 49.10.

b. Because E is the splitting field of $x^3 - 2$ over $\mathbb{Q}$, Exercise 22 shows that $G(E/\mathbb{Q})$ is isomorphic to a subgroup of the group of all permutations of the three zeros of $x^3 - 2$ in E. Because the group of all permutations of three objects has order 6 and $|G(E/\mathbb{Q})| = 6$ by Part(**a**), we see that $G(E/\mathbb{Q})$ is isomorphic to the full symmetric group on three letters, that is, to S_3.

24. We have $x^p = (x-1)(x^{p-1} + \cdots + x + 1)$, and Corollary 23.17 shows that the second of these factors, the cyclotomic polynomial $\Phi_p(x)$, is irreducible over the field $\mathbb{Q}$. Let ζ be a zero of $\Phi_p(x)$ in its splitting field over $\mathbb{Q}$. Exercise 36a of Section 48 shows that then $\zeta, \zeta^2, \zeta^3, \cdots, \zeta^{p-1}$ are distinct and are all zeros of $\Phi_p(x)$. Thus all zeros of $\Phi_p(x)$ lie in the simple extension $\mathbb{Q}(\zeta)$, so $\mathbb{Q}(\zeta)$ is the splitting field of $x^p - 1$ and of course has degree $p - 1$ over $\mathbb{Q}$ because $\Phi_p(x) = \text{irr}(\zeta, \mathbb{Q})$ has degree $p - 1$.

25. By Corollary 49.5, there exists an isomorphism $\phi : \overline{F} \to \overline{F'}$ leaving each element of F fixed. Because the coefficients of $f(x) \in F[x]$ are all left fixed by ϕ, we see that ϕ carries each zero of $f(x)$ in $\overline{F}$ into a zero of $f(x)$ in $\overline{F'}$. Because the zeros of $f(x)$ in $\overline{F}$ generate its splitting field E in $\overline{F}$, we see that $\phi[E]$ is contained in the splitting field E' of $f(x)$ in $\overline{F'}$. But the same argument can be made for ϕ^{-1}; we must have $\phi^{-1}[E'] \subseteq E$. Thus ϕ maps E onto E', so these two splitting fields of $f(x)$ are isomorphic.

51. Separable Extensions

1. Because $\sqrt[3]{2}\sqrt{2} = 2^{1/3}2^{1/2} = 2^{5/6}$, we have $\sqrt[6]{2} = 2/(\sqrt[3]{2}\sqrt{2})$ so $\mathbb{Q}(\sqrt[6]{2}\,) \subseteq \mathbb{Q}(\sqrt[3]{2}, \sqrt{2}\,)$. Because $\sqrt[3]{2} = (\sqrt[6]{2}\,)^2$ and $\sqrt{2} = (\sqrt[6]{2}\,)^3$, we have $\mathbb{Q}(\sqrt[3]{2}, \sqrt{2}\,) \subseteq \mathbb{Q}(\sqrt[6]{2}\,)$, so $\mathbb{Q}(\sqrt[6]{2}\,) = \mathbb{Q}(\sqrt[3]{2}, \sqrt{2}\,)$. We can take $\alpha = \sqrt[6]{2}$.

2. Because $(\sqrt[4]{2})^3(\sqrt[6]{2}) = 2^{3/4}2^{1/6} = 2^{9/12}2^{2/12} = 2^{11/12}$, we see that $\sqrt[12]{2} = 2/[(\sqrt[4]{2})^3(\sqrt[6]{2})]$ so $\mathbb{Q}(\sqrt[12]{2}) \subseteq \mathbb{Q}(\sqrt[4]{2}, \sqrt[6]{2})$. Because $\sqrt[4]{2} = (\sqrt[12]{2})^3$ and $\sqrt[6]{2} = (\sqrt[12]{2})^2$, we have $\mathbb{Q}(\sqrt[4]{2}, \sqrt[6]{2}) \subseteq \mathbb{Q}(\sqrt[12]{2})$, so $\mathbb{Q}(\sqrt[12]{2}) = \mathbb{Q}(\sqrt[4]{2}, \sqrt[6]{2})$. We can take $\alpha = \sqrt[12]{2}$.

3. We try $\alpha = \sqrt{2} + \sqrt{3}$. Squaring and cubing, we find that $\alpha^2 = 5 + 2\sqrt{2}\sqrt{3}$ and $\alpha^3 = 11\sqrt{2} + 9\sqrt{3}$. Because

$$\sqrt{2} = \frac{\alpha^3 - 9\alpha}{2} \text{ and } \sqrt{3} = \frac{11\alpha - \alpha^3}{2},$$

we see that $\mathbb{Q}(\sqrt{2} + \sqrt{3}) = \mathbb{Q}(\sqrt{2}, \sqrt{3})$.

4. Of course $\mathbb{Q}(i\sqrt[3]{2}) \subseteq \mathbb{Q}(i, \sqrt[3]{2})$. Because $i = -(i\sqrt[3]{2})^3/2$ and $\sqrt[3]{2} = -2/(i\sqrt[3]{2})^2$, we see that $\mathbb{Q}(i, \sqrt[3]{2}) \subseteq \mathbb{Q}(i\sqrt[3]{2})$. Thus $\mathbb{Q}(i, \sqrt[3]{2}) = \mathbb{Q}(i\sqrt[3]{2})$, so we can take $\alpha = i\sqrt[3]{2}$.

5. The definition is incorrect. Replace $F[x]$ by $\overline{F}[x]$ at the end.

   Let $\overline{F}$ be an algebraic closure of a field F. The **multiplicity of a zero** $\alpha \in \overline{F}$ of a polynomial $f(x) \in F[x]$ is $\nu \in \mathbb{Z}^+$ if and only if $(x - \alpha)^\nu$ is the highest power of $x - \alpha$ that is a factor of $f(x)$ in $\overline{F}[x]$.

6. The definition is correct.

7. (See the answer in the text.)

8. F T T F F T T T T T

9. We are given that α is separable over F, so by definition, $F(\alpha)$ is a separable extension over F. Because β is separable over F, it follows that β is separable over $F(\alpha)$ because $q(x) = \text{irr}(\beta, F(\alpha))$ divides $\text{irr}(\beta, F)$ so β is a zero of $q(x)$ of multiplicity 1. Therefore $F(\alpha, \beta)$ is a separable extension of F by Theorem 51.9. Corollary 51.10 then asserts that each element of $F(\alpha, \beta)$ is separable over F. In particular, $\alpha \pm \beta, \alpha\beta$, and α/β if $\beta \neq 0$ are all separable over F.

10. We know that $[\mathbb{Z}_p(y) : \mathbb{Z}_p(y^p)]$ is at most p. If we can show that $\{1, y, y^2, \cdots, y^{p-1}\}$ is an independent set over $\mathbb{Z}_p(y^p)$, then by Theorem 30.19, this set could be enlarged to a basis for $\mathbb{Z}_p(y)$ over $\mathbb{Z}_p(y^p)$. But because a basis can have at most p elements, it would already be a basis, and $[\mathbb{Z}_p(y) : \mathbb{Z}_p(y^p)] = p$, showing that $\text{irr}(y, \mathbb{Z}_p(y^p))$ would have degree p and must therefore be $x^p - y^p$. Thus our problem is reduced to showing that $S = \{1, y, y^2, \cdots, y^{p-1}\}$ is an independent set over $\mathbb{Z}_p(y^p)$.

   Suppose that

$$\frac{r_0(y^p)}{s_0(y^p)} \cdot 1 + \frac{r_1(y^p)}{s_1(y^p)} \cdot y + \frac{r_2(y^p)}{s_2(y^p)} \cdot y^2 + \cdots + \frac{r_{p-1}(y^p)}{s_{p-1}(y^p)} \cdot y^{p-1} = 0$$

where $r_i(y^p), s_i(y^p) \in \mathbb{Z}_p[y^p]$ for $i = 0, 1, 2, \cdots, p-1$. We want to show that all these coefficients in $\mathbb{Z}_p(y^p)$ must be zero. Clearing denominators, we see that it is no loss of generality to assume that all $s_i(y^p) = 1$ for $i = 0, 1, 2, \cdots, p-1$. Now the powers of y appearing in $r_i(y^p)(y^i)$ are all congruent to i modulo p, and consequently no terms in this expression can be combined with any terms of $r_j(y^p)(y^j)$ for $j \neq i$. Because y is an indeterminant, we then see that this linear combination of elements in S can be zero only if all the coefficients $r_i(y^p)$ are zero, so S is an independent set over $\mathbb{Z}_p(y^p)$, and we are done.

11. Let E be an algebraic extension of a perfect field F and let K be a finite extension of E. To show that E is perfect, we must show that K is a separable extension of E. Let α be an element of K. Because $[K : E]$ is finite, α is algebraic over E. Because E is algebraic over F, then α is algebraic over F by

Exercise 31 of Section 31. Because F is perfect, α is a zero of $\operatorname{irr}(\alpha, F)$ of multiplicity 1. Because $\operatorname{irr}(\alpha, E)$ divides $\operatorname{irr}(\alpha, F)$, we see that α is a zero of $\operatorname{irr}(\alpha, E)$ of multiplicity 1, so α is separable over E by the italicized remark preceding Theorem 51.9. Thus each $\alpha \in K$ is separable over E, so K is separable over E by Corollary 51.10.

12. Because K is algebraic over E and E is algebraic over F, we have K algebraic over F by Exercise 31 of Section 31. Let $\beta \in K$ and let $\beta_0, \beta_1, \cdots, \beta_n$ be the coefficients in E of $\operatorname{irr}(\beta, E)$. Because β is a zero of $\operatorname{irr}(\beta, E)$ of algebraic multiplicity 1, we see that $F(\beta_0, \beta_1, \cdots, \beta_n, \beta)$ is a separable extension of $F(\beta_0, \beta_1, \cdots, \beta_n)$, which in turn is a separable extension of F by Corollary 51.10. Thus we are back to a tower of finite extensions, and deduce from Theorem 51.9 that $F(\beta_0, \beta_1, \cdots, \beta_n, \beta)$ is a separable extension of F. In particular, β is separable over F. This shows that every element of K is separable over F, so by definition, K is separable over F.

13. Exercise 9 shows that the set S of all elements in E that are separable over F is closed under addition, multiplication, and division by nonzero elements. Of course 0 and 1 are separable over F, so Exercise 9 further shows that S contains additive inverses and reciprocals of nonzero elements. Therefore S is a subfield of E.

14. a. We know that the nonzero elements of E form a cyclic group E^* of order $p^n - 1$ under multiplication, so all elements of E are zeros of $x^{p^n} - x$. (See Section 33.) Thus for $\alpha \in E$, we have

$$
\begin{aligned}
\sigma_p^{n}(\alpha) &= \sigma_p^{n-1}(\sigma_p(\alpha)) = \sigma_p^{n-1}(\alpha^p) = \sigma_p^{n-2}(\sigma_p(\alpha^p)) \\
&= \sigma_p^{n-2}(\sigma_p(\alpha))^p = \sigma_p^{n-2}((\alpha^p)^p) = \sigma_p^{n-2}(\alpha^{p^2}) \\
&= \cdots = \alpha^{p^n} = \alpha
\end{aligned}
$$

so σ_p^{n} is the identity automorphism. If α is a generator of the group E^*, then $\alpha^{p^i} \neq \alpha$ for $i < n$, so we see that n is indeed the order of σ_p.

b. Section 33 shows that E is an extension of $\mathbb{Z}_p$ of order n, and is the splitting field of any irreducible polynomial of degree n in $\mathbb{Z}[x]$. Because E is a separable extension of the finite perfect field $\mathbb{Z}_p$, we see that $|G(E/F)| = \{E : F\} = [E : F] = n$. Since $\sigma_p \in G(E/F)$ has order n, we see $G(E/F)$ is cyclic of order n.

15. a. Let $f(x) = \sum_{i=0}^{\infty} a_i x^i$ and $g(x) = \sum_{i=0}^{\infty} b_i x^i$. Then

$$
\begin{aligned}
D(f(x) + g(x)) &= D\left(\sum_{i=0}^{\infty}(a_i + b_i)x^i\right) \\
&\quad - \sum_{i=1}^{\infty}(i \cdot 1)(a_i + b_i)x^{i-1} \\
&= \sum_{i=1}^{\infty}(i \cdot 1)a_i x^{i-1} + \sum_{i=1}^{\infty}(i \cdot 1)b_i x^{i-1} \\
&= D(f(x)) + D(g(x)).
\end{aligned}
$$

thus D is a homomorphism of $\langle F[x], + \rangle$.

b. If F has characteristic zero, then $\operatorname{Ker}(D) = F$.

c. If F has characteristic p, then $\operatorname{Ker}(D) = F[x^p]$.

16. a. Let $f(x) = \sum_{i=0}^{\infty} a_i x^i$. Then

$$D(af(x)) = D\left(\sum_{i=0}^{\infty} aa_i x^i\right) = \sum_{i=1}^{\infty} (i \cdot 1)aa_i x^{i-1}$$

$$= a\sum_{i=1}^{\infty} (i \cdot 1)a_i x^{i-1} = aD(f(x)).$$

b. We use induction on $n = \deg(f(x)g(x))$. If $n = 0$, then $f(x), g(x), f(x)g(x) \in F$ and $D(f(x)) = D(g(x)) = D(f(x)g(x)) = 0$ by Part(**b**) and Part(**c**) of Exercise 15. Suppose the formula is true for $n < k$, and let us prove it for $n = k < 0$. Write $f(x) = h(x) + a_r x^r$ where $a_r x^r$ is the term of highest degree in $f(x)$. Similarly write $g(x)$ as $g(x) = q(x) + b_s x^s$. Then

$$f(x)g(x) = h(x)q(x) + h(x)b_s x^s + a_r x^r q(x) + a_r b_s x^{r+s}.$$

Of these four terms, all are of degree less than $k = r + s$ except for the last term, so by Part(**a**) of Exercise 15 and our induction hypothesis, we have

$$\begin{aligned}
D(f(x)g(x)) = {}& h(x)q'(x) + h'(x)q(x) \\
& + h(x)(s \cdot 1)b_s x^{s-1} + h'(x)b_s x^s \\
& + a_r x^r q'(x) + (r \cdot 1)a_r x^{r-1}q(x) \\
& + [(r+s) \cdot 1]a_r b_s x^{r+s-1}.
\end{aligned}$$

We notice that we have

$$h(x)q'(x) + h(x)(s \cdot 1)b_s x^{s-1} = h(x)g'(x)$$

and

$$h'(x)q(x) + (r \cdot 1)a_r x^{r-1}q(x) = f'(x)q(x),$$

so we can continue with

$$\begin{aligned}
D(f(x)g(x)) = {}& h(x)g'(x) + f'(x)q(x) + h'(x)b_s x^s + a_r x^r q'(x) \\
& + (s \cdot 1)a_r b_s x^{r+s-1} + (r \cdot 1)a_r b_s x^{r+s-1} \\
= {}& h(x)g'(x) + f'(x)q(x) \\
& + a_r x^r [q'(x) + (s \cdot 1)b_s x^{s-1}] \\
& + [h'(x) + (r \cdot 1)a_r x^{r-1}]b_s x^s \\
= {}& h(x)g'(x) + f'(x)q(x) + a_r x^r g'(x) + f'(x)b_s x^s \\
= {}& [h(x) + a_r x^r]g'(x) + f'(x)[q(x) + b_s x^s] \\
= {}& f(x)g'(x) + f'(x)g(x).
\end{aligned}$$

c. We proceed by induction on m. If $m = 1$, the relation becomes $D((f(x))^1) = ((1 \cdot 1)f(x)^0 f'(x)$ which reduces to $D(f(x)) = f'(x)$. Thus the relation holds for $m = 1$. Suppose it is true for $m < k$ where $k > 1$. We show it holds for $m = k$. Using Part(**b**), we obtain

$$\begin{aligned}
D(f(x)^k) = {}& D(f(x)(f(x)^{k-1})) \\
= {}& f(x)[(k-1)f(x)^{k-2}f'(x)] + f'(x)f(x)^{k-1} \\
= {}& [f(x)(k-1)f(x)^{k-2} + f(x)^{k-1}]f'(x) \\
= {}& [(k-1)f(x)^{k-1} + f(x)^{k-1}]f'(x) \\
= {}& kf(x)^{k-1}f'(x)
\end{aligned}$$

which completes our induction proof.

17. In $\overline{F}[x]$, let $f(x) = (x - \alpha)^\nu g(x)$ where $g(\alpha) \neq 0$ and $\nu \geq 1$ because $f(\alpha) = 0$. Then by Exercise 16, we have

$$f'(x) = (x - \alpha)^\nu g'(x) + \nu(x - \alpha)^{\nu - 1} g(x).$$

Remembering that $\nu \geq 1$ and that $g(\alpha) \neq 0$, we see that $f'(\alpha) = 0$ if and only if $\nu > 1$, that is, if and only if α is a zero of $f(x)$ of multiplicity > 1.

18. Let $f(x)$ be an irreducible polynomial in $F[x]$ where F is a field of characteristic 0. Suppose that α is a zero of $f(x)$ in $\overline{F}$. Because $f(x) \in F[x]$ has minimal degree among all nonzero polynomials having α as a zero, we see that $f'(\alpha) \neq 0$, for the degree of $f'(x)$ is always one less than the degree of $f(x)$ in the characteristic 0 case. By Exercise 17, α is a zero of $f(x)$ of multiplicity 1. By Theorem 51.2, all zeros of $f(x)$ have this same multiplicity 1, so $f(x)$ is separable.

19. Let α be a zero of irreducible $q(x)$ in the algebraic closure $\overline{F}$. The argument in Exercise 18 shows that $q'(\alpha) \neq 0$ unless $q'(x)$ should be the zero polynomial. Now $q'(x) = 0$ if and only if each exponent of each term of $q(x)$ is divisible by p. If this is not the case, then $q'(\alpha) \neq 0$ so α has multiplicity 1 by Exercise 17, and so do other zeros of $q(x)$ by Theorem 51.2, so $q(x)$ is a separable polynomial. This proves the "only if" part of the exercise.

Suppose now that every exponent in $q(x)$ is divisible by p. Let $g(x)$ be the polynomial obtained from $q(x)$ by dividing each exponent by p. Then $\alpha \in \overline{F}$ is a zero of $q(x)$ if and only if α^p is a zero of $g(x)$. Let $g(x)$ factor into $(x - \alpha^p)h(x)$ in $\overline{F}[x]$. Then $q(x) = (x^p - \alpha^p)h(x^p) = (x - \alpha)^p h(x^p)$ in $\overline{F}[x]$, showing that α is a zero of $q(x)$ of algebraic multiplicity at least p, so $q(x)$ is not separable.

20. The polynomials $f(x)$ and $f'(x)$ have a common nonconstant factor in $\overline{F}[x]$ if and only if they have a common zero in $\overline{F}$, because a zero of the common nonconstant factor must be a zero of each polynomial, and a common zero α give rise to a common factor $x - \alpha$. Thus by Exercise 17, $f(x)$ and $f'(x)$ having a common factor in $\overline{F}[x]$ is equivalent to $f(x)$ having a zero of multiplicity greater than 1. Therefore there is no nonconstant factor of $f(x)$ and $f'(x)$ in $\overline{F}[x]$ if and only if $f(x)$ has no zero in $\overline{F}$ of multiplicity greater than 1.

21. If $f(x)$ and $f'(x)$ have no nonconstant factor in $\overline{F}[x]$, then they certainly have no nonconstant factor in $F[x]$. Suppose now that they have no nonconstant factor in $F[x]$ so that a gcd of $f(x)$ and $f'(x)$ in $F[x]$ is 1. By Theorem 46.9,

$$1 = h(x)f(x) + g(x)f'(x)$$

for some polynomials $h(x), g(x) \in F[x]$. Viewing this equation in $\overline{F}[x]$, we see that every common factor of $f(x)$ and $f'(x)$ must divide 1, so the only such common factors are elements of $\overline{F}$, and 1 is a gcd of $f(x)$ and $f'(x)$ in $\overline{F}[x]$ also. Thus $f(x)$ and $f'(x)$ have no common nonconstant factor in $F[x]$ if and only if they have no common nonconstant factor in $\overline{F}[x]$. By Exercise 20, this is equivalent to $f(x)$ having no zero in $\overline{F}$ of multiplicity greater than 1.

22. Compute a gcd of $f(x)$ and $f'(x)$ using the Euclidean algorithm. Then $f(x)$ has a zero of multiplicity > 1 in $\overline{F}$ if and only if this gcd is of degree > 0.

52. Totally Inseparable Extensions

1. The separable closure is $\mathbb{Z}_3(y^3, z^9)$ because $(y^3)^4 = u$ and $(z^9)^2 = v$ and 3 does not divide 4 or 2. The field $\mathbb{Z}_3(y, z)$ is clearly totally inseparable over (y^3, z^9).

2. Clearly the separable closure contains y^3. Therefore it must contain $(y^2 z^{18})^3/(y^3)^2 = z^{54}$, and hence must contain z^{27}. Because it must also contain $y^2 z^{18}$, we see that the separable closure is $\mathbb{Z}_3(y^3, y^2 z^{18}, z^{27})$. Clearly $\mathbb{Z}_3(y, z)$ is totally inseparable over this field.

3. The totally inseparable closure is $\mathbb{Z}_3(y^4, z^2)$.

4. The totally inseparable closure must contain y^4, and therefore $(y^2 z^{18})^2/y^4 = z^{36}$, so it must also contain z^4. Of course it must also contain $y^2 z^{18}$ and therefore $y^2 z^{18}/(z^4)^4 = y^2 z^2$. We see the totally inseparable closure is $\mathbb{Z}_3(y^4, y^2 z^2, z^4)$. Note that $(y^2 z^2)^{27} = (y^{12})^4 (y^2 z^{18})^3$, and that $(z^4)^{27} = (y^2 z^{18})^6/y^{12}$.

5. F T F F F F T F T T

6. If E is a separable extension of F, then there are no elements of E totally inseparable over F, so the totally inseparable closure of F in E is just F, which is a subfield of E.

Suppose that E does contain some elements totally inseparable over F, and let K be the union of F with the set of all such totally inseparable elements. We need only show that for $\alpha, \beta \in K$, the elements $\alpha \pm \beta, \alpha\beta$, and $1/\alpha$ if $\alpha \neq 0$, are either in F or are totally inseparable over F. Suppose that α is not in F, but is totally inseparable over F, so that $\alpha^{p^r} \in F$. Then for any element b in F, we have $(\alpha \pm b)^{p^r} = \alpha^{p^r} \pm b^{p^r}$ and this sum or difference is in F. Also $(b\alpha)^{p^r} = b^{p^r}\alpha^{p^r}$ is in F, and $(1/\alpha)^{p^r} = 1/\alpha^{p^r}$ is in F if $\alpha \neq 0$. This shows that for α in K but not in F and for b in F, the elements $\alpha \pm b, b\alpha$, and $1/\alpha, \alpha \neq 0$, are in K. The other case we have to worry about is where α and β are both totally inseparable over F, that is, they are both in K, but neither one is in F. Then $\alpha^{p^r} \in F$ and $\beta^{p^s} \in F$ for some $r, s \in \mathbb{Z}^+$. Suppose that $s \geq r$. Then $(\alpha \pm \beta)^{p^s} = \alpha^{p^s} \pm \beta^{p^s} = (\alpha^{p^r})^{p^{s-r}} \pm \beta^{p^s}$ is in F, and $(\alpha\beta)^{p^s} = (\alpha^{p^r})^{p^{s-r}}\beta^{p^s}$ is in F. Thus $\alpha \pm \beta$ and $\alpha\beta$ are either already in F or are totally inseparable over F. This shows that K is closed under the field operations of addition, subtraction, multiplication, and contains multiplicative inverses of nonzero elements.

7. Suppose that F is perfect. If $x^p - a$ has no zero in F for some $a \in F$, then $F(\sqrt[p]{a})$ is a proper extension of F and is totally inseparable over F, contradicting the hypothesis that F is perfect. Thus $x^p - a$ has a zero in F for every $a \in F$, that is, $F^p = F$.

Conversely, suppose that $F^p = F$ and let $f(x)$ be an irreducible polynomial in $F[x]$. We must show that $f(x)$ is a separable polynomial. Let E be the separable closure of F in the splitting field K of $f(x)$ in $\overline{F}$. Let $[E : F] = n$. Now the map $\sigma_p : E \to E^p$ is an isomorphism. Because $F^p = F$ and σ_p is one to one, no $\alpha \in E$ that is not in F is carried into F. Because σ_p is an isomorphism, the extension E of degree n over F is carried into an extension E^p of $F^p = F$ of degree n. Because $E^p \leq E$, we see that $n = [E : F] = [E : E^p][E^p : F] = [E : E^p]n$, so $[E : E^p] = 1$ and $E = E^p$. But then E has no totally inseparable extension, for an element α of such an extension must satisfy $\alpha^{p^r} = \beta \in E$, where $\alpha \notin E$. But because $E^p = E$, we see that $E^{p^r} = E$ and the polynomial $x^{p^r} = \beta$ has a zero γ in E, so that $x^{p^r} - \beta = x^{p^r} - \gamma^{p^r} = (x - \gamma)^{p^r}$, showing that γ is the only zero of this polynomial. Thus no such $\alpha \notin E$ exists and we must have $E = K$, so the splitting field of $f(x)$ is a separable extension of F, and therefore $f(x)$ is a separable polynomial.

8. The solution of Exercise 7 showed that if $F^p = F$, then $E^p = E$. Conversely, suppose that $E^p = E$. Let $n = [E : F]$. Because σ_p is an isomorphism, it must be that $[E^p : F^p] = n$. Of course $F^p \leq F$. Then we have

$$n = [E^p : F^p] = [E : F^p] = [E : F][F : F^p] = n[F : F^p],$$

so $[F : F^p] = 1$ and $F = F^p$.

53. Galois Theory

1. We have $\{K : \mathbb{Q}\} = [K : \mathbb{Q}] = 8$.

2. We have $|G(K/\mathbb{Q})| = [K : \mathbb{Q}] = 8$ because K is a normal extension of $\mathbb{Q}$.

3. We have $|\lambda(\mathbb{Q})| = |G(K/\mathbb{Q})| = 8$.

4. We have $|\lambda(\mathbb{Q}(\sqrt{2}, \sqrt{3}))| = [K : \mathbb{Q}(\sqrt{2}, \sqrt{3})] = 2$.

5. We have $|\lambda(\mathbb{Q}(\sqrt{6}))| = [K : \mathbb{Q}(\sqrt{6})] = 4$.

6. We have $|\lambda(\mathbb{Q}(\sqrt{30}))| = [K : \mathbb{Q}(\sqrt{30})] = 4$.

7. We have $|\lambda(\mathbb{Q}(\sqrt{2} + \sqrt{6}))| = [K : \mathbb{Q}(\sqrt{2} + \sqrt{6})] = 2$, because $\deg(\sqrt{2} + \sqrt{6}, \mathbb{Q}) = 4$.

8. We have $|\lambda(K)| = [K : K] = 1$.

9. Now $x^4 - 1 = (x^2 - 1)(x^2 + 1) = (x - 1)(x + 1)(x^2 + 1)$ so the splitting field of $x^4 + 1$ over $\mathbb{Q}$ is the same as the splitting field of $x^2 + 1$ over $\mathbb{Q}$. This splitting field is $\mathbb{Q}(i)$. It is of degree 2 over $\mathbb{Q}$, and its Galois group is cyclic of order 2 with generator σ where $\sigma(i) = -i$.

10. Because $729 = 9^3$, Theorem 53.7 shows that the Galois group of $GF(729)$ over $GF(9)$ is cyclic of order 3, generated by σ_9 where $\sigma_9(\alpha) = \alpha^9$ for $\alpha \in GF(729)$.

11. See the answer in the text. The answer to Exercise 23 of Section 50 in this manual explains why the group is isomorphic to S_3. We might explain the statement in the text answer that the notation was chosen to reflect the notation in Example 8.7, where S_3 consisted of permutations of $\{1, 2, 3\}$. Here, we are permuting $\{\alpha_1, \alpha_2, \alpha_3\}$, and we defined our permutations so they have the effect on the subscripts 1, 2, and 3 that S_3 has on the numbers 1, 2, and 3. It is worth indicating how this can be verified. Note that $\zeta = \frac{-1+i\sqrt{3}}{2} = -\frac{1}{2} + i\frac{\sqrt{3}}{2} = \cos\frac{2\pi}{3} + i\sin\frac{2\pi}{3}$ is a cube root of unity, and $\frac{-1-i\sqrt{3}}{2} = \zeta^2$. Thus our three zeros of $x^3 - 2$ can be written as

$$\alpha_1, \quad \alpha_2 = \alpha_1\zeta, \quad \text{and} \quad \alpha_3 = \alpha_1\zeta^2.$$

According to the definitions of the six automorphism in the text, we see each ρ_i maps ζ into ζ, but each μ_i maps ζ into ζ^2.

We illustrate with two computations how our choice of notation mirrors the effect of our notation for S_3 in Example 8.7. In that example, ρ_1 maps 1 to 2, 2 to 3, and 3 to 1. Here we have $\rho_1(\alpha_1) = \alpha_2$ by definition. Computing we find

$$\rho_1(\alpha_2) = \rho_1(\alpha_1\zeta) = \rho_1(\alpha_1)\rho_1(\zeta) = \alpha_2\zeta = (\alpha_1\zeta)\zeta = \alpha_3,$$

and

$$\rho_1(\alpha_3) = \rho_1(\alpha_1\zeta^2) = \rho_1(\alpha_1)(\rho_1(\zeta))^2 = \alpha_2\zeta^2 = (\alpha_1\zeta)\zeta^2 = \alpha_1$$

because $\zeta^3 = 1$. For our second illustration, we use μ_2, which we expect to leave subscript 2 alone and swap subscripts 1 and 3. Remember that $\mu_2(\zeta) = \zeta^2$, and $\mu_2(\alpha_1) = \alpha_3$. Computing,

$$\mu_2(\alpha_2) = \mu_2(\alpha_1\zeta) = \mu_2(\alpha_1)\mu_2(\zeta) = \alpha_3\zeta^2 = (\alpha_1\zeta^2)\zeta^2 = \alpha_1\zeta = \alpha_2$$

and

$$\mu_2(\alpha_3) = \mu_2(\alpha_1\zeta^2) = \mu_2(\alpha_1)\mu_2(\zeta^2) = \alpha_3\zeta^4 = (\alpha_1\zeta^2)\zeta^4 = \alpha_1$$

because $\zeta^6 = (\zeta^3)^2 = 1^2 = 1$.

12. Now $x^4 - 5x^2 + 6 = (x^2 - 2)(x^2 - 3)$ so the spitting field is $\mathbb{Q}(\sqrt{2}, \sqrt{3})$. Its Galois group over $\mathbb{Q}$ is isomorphic to the Klein 4-group $\mathbb{Z}_2 \times \mathbb{Z}_2$. See Example 53.3 for a description of the action of each element on $\sqrt{2}$ and on $\sqrt{3}$.

13. Now $x^3 - 1 = (x-1)(x^2 + x + 1)$. Because a primitive cube root of unity is $\frac{-1 + i\sqrt{3}}{2}$, we see that its splitting field over $\mathbb{Q}$ is $\mathbb{Q}(i\sqrt{3})$. The Galois group is cyclic of order 2 and is generated by σ where $\sigma(i\sqrt{3}) = -i\sqrt{3}$.

14. Let $F = \mathbb{Q}, K_1 = \mathbb{Q}(\sqrt{2})$ and $K_2 = \mathbb{Q}(i)$. The fields are not isomorphic because the additive inverse of unity is a square in K_2 but is not a square in K_1. However, the Galois groups over $\mathbb{Q}$ are isomorphic, for they are both cyclic of order 2.

15. F F T T T F F T F T

16. Because $F(K/E) \leq G(K/F)$ and $G(K/F)$ is abelian, we see that $G(K/E)$ is abelian, for a subgroup of an abelian group is abelian. Because $G(E/F) \simeq G(K/F)/G(K/E)$ and $G(K/F)$ is abelian, we see that $G(E/F)$ is abelian, for a factor group of an abelian group, where multiplication is done by choosing representatives, must again be abelian.

17. To show that $N_{K/F}(\alpha) \in F$, we need only show that it is left fixed by each $\tau \in G(K/F)$. From the given formla and the fact that τ is an automorphism, we have

$$\tau(N_{K/F}(\alpha)) = \prod_{\sigma \in G(K/F)} (\tau\sigma)(\alpha).$$

But as σ runs through the elements of $G(K/F), \tau\sigma$ again runs through all elements, because $G(K/F)$ is a group. Thus only the order of the factors in the product is changed, and because multiplication in K is commutative, the product is unchanged. Thus $\tau(N_{K/F}(\alpha)) = N_{K/F}(\alpha)$ for all $\tau \in G(K/F)$, so $N_{K/F}(\alpha) \in F$. Precisely the same argument shows that $Tr_{K/F}(\alpha) \in F$, only this time it is the order of the summands in the sum that gets changed when computing $\tau(Tr_{K/F}(\alpha))$.

18. a. $N_{K/\mathbb{Q}}(\sqrt{2}) = \sqrt{2}\sqrt{2}(-\sqrt{2})(-\sqrt{2}) = 4$, because two of the elements of the Galois group leave $\sqrt{2}$ fixed, and two carry it into $-\sqrt{2}$. The computations shown for Part(**b**) - Part(**h**) are based similarly on action of the Galois group on $\sqrt{2}$ and $\sqrt{3}$.

b. $N_{K/\mathbb{Q}}(\sqrt{2} + \sqrt{3}) = (\sqrt{2} + \sqrt{3})(\sqrt{2} - \sqrt{3})(-\sqrt{2} + \sqrt{3})(-\sqrt{2} - \sqrt{3}) = (-1)^2 = 1$.

c. $N_{K/\mathbb{Q}}(\sqrt{6}) = (\sqrt{6})(-\sqrt{6})(-\sqrt{6})(\sqrt{6}) = 36$

d. $N_{K/\mathbb{Q}}(2) = (2)(2)(2)(2) = 16$

e. $Tr_{K/\mathbb{Q}}(\sqrt{2}) = \sqrt{2} + \sqrt{2} + (-\sqrt{2}) + (-\sqrt{2}) = 0$

f. $Tr_{K/\mathbb{Q}}(\sqrt{2} + \sqrt{3}) = (\sqrt{2} + \sqrt{3}) + (-\sqrt{2} + \sqrt{3}) + (\sqrt{2} - \sqrt{3}) + (-\sqrt{2} - \sqrt{3}) = 0$.

g. $Tr_{K/\mathbb{Q}}(\sqrt{6}) = \sqrt{6} + (-\sqrt{6}) + (-\sqrt{6}) + \sqrt{6} = 0$

h. $Tr_{K/\mathbb{Q}}(2) = 2 + 2 + 2 + 2 = 8$

19. Let $f(x) = \text{irr}(\alpha, F)$. Because $K = F(\alpha)$ is normal over F, it is a splitting field of $f(x)$. Let the factorization of $f(x)$ in $K[x]$ be

$$f(x) = (x - \alpha_1)(x - \alpha_2) \cdots (x - \alpha_n) \tag{1}$$

where $\alpha = \alpha_1$. Now $|G(K/F)| = n$, and α is carried onto each α_i for $i = 1, 2, \cdots, n$ by precisely one element of $G(K/G)$. Thus $N_{K/F}(\alpha) = \alpha_1\alpha_2 \cdots \alpha_n$. If we multiply the linear factors in (1) together, we see that the constant term a_0 is $(-1)^n \alpha_1 \alpha_2 \cdots \alpha_n = (-1)^n N_{K/F}(\alpha)$. Similarly, we see that $Tr_{K/F}(\alpha) = \alpha_1 + \alpha_2 + \cdots + \alpha_n$. If we pick up the coefficient a_{n-1} of x^{n-1} in $f(x)$ by multiplying the linear factors in (1), we find that $a_{n-1} = -\alpha_1 - \alpha_1 - \cdots - \alpha_n = -Tr_{K/F}(\alpha)$.

20. Let $\alpha_1, \alpha_2, \cdots, \alpha_r$ be the distinct zeros of $f(x)$ in $\overline{F}$ and form the splitting field $K = F(\alpha_1, \alpha_2, \cdots, \alpha_r)$ of $f(x)$ in $\overline{F}$. Note that $r \leq n$ because $f(x)$ has at most n distinct zeros. Because all irreducible factors of $f(x)$ are separable, we see that K is normal over F. Now each $\sigma \in G(K/F)$ provides a permutation of $S = \{\alpha_1, \alpha_2, \cdots, \alpha_r\}$ and distinct elements of $G(K/F)$ correspond to distinct permutations of S because an automorphism of K leaving F fixed is uniquely determined by its values on the elements of S. Because permutation multiplication and multiplication in $G(K/F)$ are both function composition, we see that $G(K/F)$ is isomorphic to a subgroup of the group of all permutations of S, which is isomorphic to a subgroup of S_r. By the Theorem of Lagrange, it follows that $|G(K/F)|$ divides $r!$ which in turn divides $n!$ because $r \leq n$.

21. Let K be the splitting field of $f(x)$ over F. Because each $\sigma \in G(K/F)$ carries a zero of $f(x)$ into a zero of $f(x)$ and is a one-to-one map, it induces a permutation of the set S of distinct zeros in $\overline{F}$ of $f(x)$. The action of $\sigma \in G(K/G)$ on all elements of K is completely determined by its action on the elements of the set S. Because permutation multiplication and multiplication in $G(K/F)$ are both function composition, we see that $G(K/F)$ is isomorphic in a natural way to a subgroup of the group of all permutations of S.

22. a. Exercise 17 of Section 51 shows that $x^n - 1$ has no zeros of multiplicity greater than 1 as long as $n \cdot 1$ is not equal to zero in F. Thus the splitting field of $x^n - 1$ over F is a normal extension. If ζ is a primitive nth root of unity, then $1, \zeta, \zeta^2, \cdots, \zeta^{n-1}$ are distinct elements, and are all zeros of $x^n - 1$. Thus the splitting field of $x^n - 1$ over F is $F(\zeta)$.

b. The action of $\sigma \in G(F(\zeta)/F)$ is completely determined by $\sigma(\zeta)$ which must be one of the conjugates ζ^s of ζ over F. Let $\sigma, \tau \in G(F(\zeta)/F)$, and suppose $\sigma(\zeta) = \zeta^s$ and $\tau(\zeta) = \zeta^t$. Then

$$\begin{aligned} (\sigma\tau)(\zeta) &= \sigma(\tau(\zeta)) = \sigma(\zeta^t) = [\sigma(\zeta)]^t = (\zeta^s)^t = \zeta^{st} = \zeta^{ts} \\ &= (\zeta^t)^s = [\tau(\zeta)]^s = \tau(\zeta^s) = \tau(\sigma(\zeta)) = (\tau\sigma)(\zeta) \end{aligned}$$

so $\sigma\tau = \tau\sigma$ and $G(F(\zeta)/F)$ is abelian.

23. a. Because K is cyclic over F, we know that $G(K/F)$ is a cyclic group. Now $G(K/E)$ is a subgroup of $G(K/F)$, and is thus cyclic as a subgroup of a cyclic group. Therefore K is cyclic over E. Because E is a normal extension of F, we know that $G(E/F) \simeq G(K/F)/G(K/E)$ so $G(E/F)$ is isomorphic to a factor group of a cyclic group, and is thus cyclic. (A factor group of a cyclic group A is generated by a coset containing a generator of A.) Therefore E is cyclic over F.

b. By Galois theory, we know that there is a one-to-one correspondence between subgroups H of $G(K/F)$ and fields $E = K_H$ such that $F \leq E \leq K$. Because $G(K/F)$ is cyclic, it contains precisely one subgroup of each order d that divides $|G(K/F)| = [K : F]$. Such a subgroup corresponds to a field E where $F \leq E \leq K$ and $[K : E] = d$, so that $[E : F] = m = n/d$. Now as d runs through all divisors of n, the quotients $m = n/d$ also run through all divisors of n, so we are done.

24. a. For $\tau \in G(K/F)$, we have a natural extension of τ to an automorphism $\overline{\tau}$ of $K[x]$ where $\overline{\tau}(\alpha_0 + \alpha_1 x + \cdots + \alpha_n x^n) = \tau(\alpha_0) + \tau(\alpha_1)x + \cdots + \tau(\alpha_n)x^n$. Clearly the polynomials left fixed by $\overline{\tau}$ for *all* $\tau \in G(K/F)$ are precisely those in $F(x)$. For $f(x) = \prod_{\sigma \in G(K/F)}(x - \sigma(\alpha))$, we have

$$\overline{\tau}(f(x)) = \prod_{\sigma \in G(K/F)} (x - (\tau\sigma)(\alpha)).$$

Now as σ runs through all elements of $G(K/F)$, we see that $\tau\sigma$ also runs through all elements because $G(K/F)$ is a group. Thus $\overline{\tau}(f(x)) = f(x)$ for each $\tau \in G(K/F)$, so $f(x) \in F[x]$.

b. Because $\sigma(\alpha)$ is a conjugate of α over F for all $\sigma \in G(K/F)$, we see that $f(x)$ has precisely the conjugates of α as zeros. Because $f(\alpha) = 0$, we know by Theorem 29.13 that $p(x) = \mathrm{irr}(\alpha, F)$ divides $f(x)$. Let $f(x) = p(x)q_1(x)$. If $q_1(x) \neq 0$, then it has as zero some conjugate of α whose irreducible polynomial over F is again $p(x)$, so $p(x)$ divides $q_1(x)$ and we have $f(x) = p(x)^2 q_2(x)$. We continue this process until we finally obtain $f(x) = p(x)^r c$ for some $c \in F$. Because $p(x)$ and $f(x)$ are both monic, we must have $f(x) = p(x)^r$.

Now $f(x) = p(x)$ if and only if $\deg(\alpha, F) = |G(K/F)| = [K : F]$. Because $\deg(\alpha, F) = [F(\alpha) : F]$, we see that this occurs if and only if $[F(\alpha) : F] = [K : F]$ so that $[K : F(\alpha)] = 1$ and $K = F(\alpha)$.

25. In the one-to-one correspondence between subgroups of $G(K/F)$ and fields E where $F \leq E \leq K$, the diagram of subgroups of $G(K/F)$ is the inverted diagram of such subfields E of K. Now $E \vee L$ is the smallest subfield containing both E and L, and thus must correspond to the largest subgroup contained in both $G(K/E)$ and $G(K/L)$. Thus

$$G(K/(E \vee L)) = G(K/E) \cap G(K/L).$$

26. Continuing to work with the one-to-one corespondence and diagrams mentioned in the solution to Exercsie 25, we note that $E \cap L$ is the largest subfield of K contained in both E and L. Thus its group must be the smallest subgroup of $G(K/F)$ containing both $G(K/E)$ and $G(K/L)$. Therefore

$$G(K/E \cap L)) = G(K/E) \vee G(K/L),$$

which is defined as the intersection of all subgroups of $G(K/F)$ that contain both $G(K/E)$ and $G(K/L)$, and is called the **join** of the subgroups $G(K/E)$ and $G(K/L)$.

54. Illustrations of Galois Theory

1. Recall that if $x^4 + 1$ has a factorization into polynomials of lower degree in $\mathbb{Q}[x]$, then it has such a factorization in $\mathbb{Z}[x]$; see Theorem 23.11. The polynomial does not have a linear factor, for neither 1 nor -1 are zeros of the polynomial. Suppose that

$$x^4 + 1 = (x^2 + ax + b)(x^2 + cx + d)$$

for $a, b, c, d \in \mathbb{Z}$. Equating coefficients of x^3, x^2, x, and 1 in that order, we find that

$$a + c = 0, \quad ac + b + d = 0, \quad ad + bc = 0, \quad \text{and} \quad bd = 1.$$

If $b = d = 1$, then $ac + 2 = 0$ so $ac = -2$ and $a^2 = 2$, which is impossible for an integer a. The other possibility, $b = d = -1$, leads to $a^2 = -2$ which is also impossible. Thus $x^4 + 1$ is irreducible in $\mathbb{Q}[x]$.

2. The fields corresponding to the subgroups $G(K/\mathbb{Q}), H_2, H_4, H_7$, and $\{\rho_0\}$ are either derived in the text or are obvious. We turn to the other subgroups, H_1, H_3, H_5, and H_8. Both H_1 and H_3 have order 4 and must have fixed fields of degree 2 over $\mathbb{Q}$. Recalling that the fixed field of H_2 is $\mathbb{Q}(i)$, we see that the other two obvious extensions of degree 2, namely $\mathbb{Q}(\sqrt{2})$ and $\mathbb{Q}(i\sqrt{2})$, must be fixed fields of H_1 and H_3. We find that $\delta_1(\sqrt{2}) = \delta_1((\sqrt[4]{2})^2) = (i\sqrt[4]{2})^2 = -\sqrt{2} \neq \sqrt{2}$. Thus H_3, which contains δ_1, must have $\mathbb{Q}(i\sqrt{2})$ as its fixed field, and H_1 must have $\mathbb{Q}(\sqrt{2})$ as its fixed field.

For the fixed field of $H_5 = \{\rho_0, \mu_2\}$ we need to find some elements left fixed by μ_2. Because $\mu_2(\alpha) = -\alpha$ and $\mu_2(i) = -i$, the product $i\alpha$ is an obvious choice. Now $i\sqrt[4]{2}$ is a zero of $x^4 - 2$ which is irreducible, so $\mathbb{Q}(i\sqrt[4]{2})$ is of degree 4 over $\mathbb{Q}$ and left fixed by H_5.

Because ρ_2 leaves i fixed and maps α into $-\alpha$, it leaves i and $\alpha^2 = \sqrt{2}$ fixed. Thus the fixed field of $H_6 = \{\rho_0, \rho_2\}$ is $\mathbb{Q}(i, \sqrt{2})$.

To find an element left fixed by $H_8 = \{\rho_0, \delta_2\}$, we form

$$\beta = \rho_0(\alpha) + \delta_2(\alpha) = \alpha - i\alpha = \sqrt[4]{2}(1 - i).$$

(Note that because H_8 is a group, this sum of elements of H_8 applied to any one element in the field is sure to be left fixed by H_8.) Now $\beta^4 = 2(-4) = -8$, so β is a zero of $x^4 + 8$. This polynomial does not have $\pm 1, \pm 2, \pm 4,$ or ± 8 as a zero, so it has no linear factors. If

$$x^4 + 8 = (x^2 + ax + b)(x^2 + cx + d),$$

then $a + c = 0, ac + b + d = 0, ad + bc = 0,$ and $bd = 8$. From $a + c = 0$ and $ad + bc = 0$, we find that $ad - ba = 0$ so $a(d - b) = 0$ and either $a = 0$ or $b = d$. Because $bd = 8$, we do not have $b = d$, so $a = 0$. But then $b + d = 0$ so $b = -d$, which again cannot satisfy $bd = 8$. Thus $x^4 + 8$ is irreducible, and $\mathbb{Q}(\sqrt[4]{2}(1 - i))$ has degree 4 over $\mathbb{Q}$ and is left fixed by H_8, so it is the fixed field of H_8.

3. The choices for primitive elements given in the text answers, the corresponding polynomials and the irreducibility of the polynomials are obvious or proved in the text or the preceding solution, except for the first and fourth answers given.

For the case $\mathbb{Q}(\sqrt{2}, i)$, let $\beta = \sqrt[4]{2}$ and $\gamma = i$. The proof of Theorem 51.15 shows that for $a \in \mathbb{Q}, \beta + a\gamma$ is a primitive element if

$$(\beta_i - \beta)/(\gamma - \gamma_j) \neq a$$

where β_i can be any conjugate of β and γ_j is any conjugate other than γ of γ. Now $\gamma - \gamma_j$ is always $i - (-i) = 2i$, and because $\beta = \alpha = \sqrt[4]{2}$ in Table 54.5, it is clear from the table that $[(\text{conjugate of } \alpha) - \alpha]/(2i)$ is never a nonzero element of $\mathbb{Q}$. Thus we can take $a = 1$, and we find that $\sqrt[4]{2} + i$ is a primitive element. Let $\delta = \sqrt[4]{2} + i$. Then $\delta - i = \sqrt[4]{2}$ so

$$(\delta - i)^4 = \delta^4 - 4\delta^3 i - 6\delta^2 + 4\delta i + 1 = 2$$

so

$$\delta^4 - 6\delta^2 - 1 = (4\delta^3 - 4\delta)i$$

Squaring both sides, we obtain

$$\delta^8 - 12\delta^6 + 34\delta^4 + 12\delta^2 + 1 = -16\delta^6 + 32\delta^4 - 16\delta^2$$

so $\delta^8 + 4\delta^6 + 2\delta^4 + 28\delta^2 + 1 = 0$. Thus δ is a zero of $x^8 + 4x^6 + 2x^4 + 28x^2 + 1 = 0$. Because we know that $\mathbb{Q}(\delta)$ is of degree 8 over $\mathbb{Q}$, this must be irr$(\mathbb{Q}, \delta)$.

For $\mathbb{Q}(\sqrt{2}, i)$, we have $[(\text{conjugate of } \sqrt{2}) - \sqrt{2}]/2i$ is never a nonzero element of $\mathbb{Q}$, so $\sqrt{2} + i$ is a primitive element. If $\delta = \sqrt{2} + i$, then $\delta - i = \sqrt{2}$ and $\delta^2 - 2\delta i - 1 = 2$. Then $\delta^2 - 3 = 2\delta i$ so $\delta^4 - 6\delta^2 + 9 = -4\delta^2$ and $\delta^4 - 2\delta^2 + 9 = 0$. Thus δ is a zero of $x^4 - 2x^2 + 9$, and because $\mathbb{Q}(\delta)$ is of degree 4 over $\mathbb{Q}$, we see that this polynomial is irreducible.

4. a. If ζ is a primitive 5th root of unity, then $1, \zeta, \zeta^2, \zeta^3$, and ζ^4 are five distinct elements of $\mathbb{Q}(\zeta)$, and $(\zeta^k)^5 = (\zeta^5)^k = 1^k = 1$ shows that these five elements are five zeros of $x^5 = 1$. Thus $x^5 - 1$ splits in $\mathbb{Q}(\zeta)$.

b. We know that $x^5 - 1 = (x - 1)\Phi_5(x)$ where $\Phi_5(x) = x^4 + x^3 + x^2 + 1$ is the irreducible (Corollary 23.17) cyclotomic polynomial having ζ as a root. Every automorphism of $K = \mathbb{Q}(\zeta)$ over $\mathbb{Q}$ must map ζ into one of the four roots $\zeta, \zeta^2, \zeta^3, \zeta^4$ of this polynomial.

c. Let $\sigma_j \in G(K/\mathbb{Q})$ be the automorphism such that $\sigma(\zeta) = \zeta^j$ for $j = 1, 2, 3, 4$. Then $(\sigma_j \sigma_k)(\zeta) = \sigma_j(\zeta^k) = (\zeta^j)^k = \zeta^{jk} = \sigma_m(\zeta)$ where m is the product of j and k in $\mathbb{Z}_5$. Thus $G(K/\mathbb{Q})$ is isomorphic to the group $\{1, 2, 3, 4\}$ of nonzero elements of $\mathbb{Z}_5$ under multiplication. It is cyclic of order 4, generated by σ_2.

d.

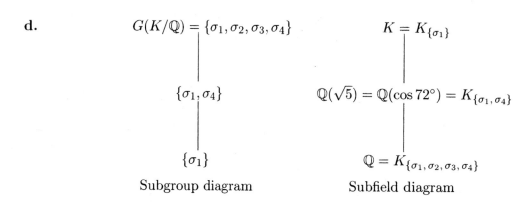

$$G(K/\mathbb{Q}) = \{\sigma_1, \sigma_2, \sigma_3, \sigma_4\} \qquad\qquad K = K_{\{\sigma_1\}}$$

$$\{\sigma_1, \sigma_4\} \qquad\qquad \mathbb{Q}(\sqrt{5}) = \mathbb{Q}(\cos 72°) = K_{\{\sigma_1, \sigma_4\}}$$

$$\{\sigma_1\} \qquad\qquad \mathbb{Q} = K_{\{\sigma_1, \sigma_2, \sigma_3, \sigma_4\}}$$

Subgroup diagram   Subfield diagram

To find $K_{\{\sigma_1, \sigma_4\}}$, note that $\zeta = \cos 72° + i \sin 72°$ and that $\zeta^4 = \cos(-72°) + i \sin(-72°) = \cos(72°) - i \sin(72°)$. Therefore $\alpha = \sigma_1(\zeta) + \sigma_4(\zeta) = \zeta + \zeta^4 = 2 \cos 72°$ is left fixed by σ_1 and σ_4. Alternatively, doing a bit of computation, we find that

$$\alpha^2 = (\zeta + \zeta^4)^2 = \zeta^2 + 2 + \zeta^3,$$
$$\alpha = \zeta + \zeta^4.$$

Now ζ is a zero of $\Phi_5(x) = x^4 + x^3 + x^2 + x + 1$, so we see that $\alpha^2 + \alpha - 1 = 0$, so α is a zero of $x^2 + x - 1$ which has zeros $(-1 \pm \sqrt{5})/2$. Thus we can also describe $\mathbb{Q}(\alpha)$ as $\mathbb{Q}(\sqrt{5})$.

5. The splitting field of $x^5 - 2$ over $\mathbb{Q}(\zeta)$ is $\mathbb{Q}(\zeta, \sqrt[5]{2})$, because $\sqrt[5]{2}, \zeta\sqrt[5]{2}, \zeta^2\sqrt[5]{2}, \zeta^3\sqrt[5]{2}$, and $\zeta^4\sqrt[5]{2}$ are the five zeros of $x^5 - 2$. The Galois group $\{\sigma_0, \sigma_1, \sigma_2, \sigma_3, \sigma_4\}$ is described by the table.

	σ_0	σ_1	σ_2	σ_3	σ_4
$\sqrt[5]{2} \rightarrow$	$\sqrt[5]{2}$	$\zeta\sqrt[5]{2}$	$\zeta^2\sqrt[5]{2}$	$\zeta^3\sqrt[5]{2}$	$\zeta^4\sqrt[5]{2}$

We have $(\sigma_j \sigma_k)(\sqrt[5]{2}) = \sigma_j(\zeta^k(\sqrt[5]{2})) = \zeta^k \zeta^j(\sqrt[5]{2}) = \zeta^{j+k}(\sqrt[5]{2})$ from which we see that the Galois group is isomorphic to $\langle \mathbb{Z}_5, + \rangle$, so it is cyclic of order 5.

6. a. If ζ is a primitive 7th root of unity, then $1, \zeta, \zeta^2, \zeta^3, \zeta^4, \zeta^5$, and ζ^6 are seven distinct elements of $\mathbb{Q}(\zeta)$, and $(\zeta^k)^7 = (\zeta^7)^k = 1^k = 1$ shows that these seven elements are seven zeros of $x^7 = 1$. Thus $x^7 - 1$ splits in $\mathbb{Q}(\zeta)$.

b. We know that $x^7 - 1 = (x - 1)\Phi_7(x)$ where $\Phi_7(x) = x^6 + x^5 + x^4 + x^3 + x^2 + 1$ is the irreducible (Corollary 23.17) cyclotomic polynomial having ζ as a root. Every automorphism of $K = \mathbb{Q}(\zeta)$ over $\mathbb{Q}$ must map ζ into one of the six roots $\zeta, \zeta^2, \zeta^3, \zeta^4, \zeta^5, \zeta^6$ of this polynomial.

c. Let $\sigma_j \in G(K/\mathbb{Q})$ be the automorphism such that $\sigma(\zeta) = \zeta^j$ for $j = 1, 2, 3, 4, 5, 6$. Then $(\sigma_j\sigma_k)(\zeta) = \sigma_j(\zeta^k) = (\zeta^j)^k = \zeta^{jk} = \sigma_m(\zeta)$ where m is the product jk in $\mathbb{Z}_7$. Thus $G(K/\mathbb{Q})$ is isomorphic to the group $\{1, 2, 3, 4, 5, 6\}$ of nonzero elements of $\mathbb{Z}_7$ under multiplication. It is cyclic of order 6, generated by σ_3.

d.

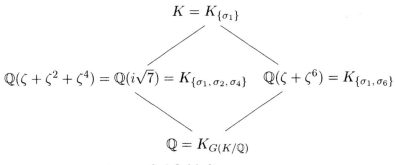

$$G(K/\mathbb{Q}) = \{\sigma_1, \sigma_2, \sigma_3, \sigma_4, \sigma_5, \sigma_6\}$$

$$\{\sigma_1, \sigma_2, \sigma_4\} \qquad \{\sigma_1, \sigma_6\}$$

$$\{\sigma_1\}$$

Subgroup diagram

$$K = K_{\{\sigma_1\}}$$

$$\mathbb{Q}(\zeta + \zeta^2 + \zeta^4) = \mathbb{Q}(i\sqrt{7}) = K_{\{\sigma_1, \sigma_2, \sigma_4\}} \quad \mathbb{Q}(\zeta + \zeta^6) = K_{\{\sigma_1, \sigma_6\}}$$

$$\mathbb{Q} = K_{G(K/\mathbb{Q})}$$

Subfield diagram

Clearly $\alpha = \zeta + \zeta^2 + \zeta^4$ is left fixed by $\{\sigma_1, \sigma_2, \sigma_4\}$. Computing, we find that

$$\begin{aligned} \alpha^2 &= \zeta^2 + \zeta^4 + \zeta + 2\zeta^3 + 2\zeta^6 + 2\zeta^5, \\ \alpha &= \zeta + \zeta^2 + \zeta^4. \end{aligned}$$

Thus we find that $\alpha^2 + \alpha = 2(\zeta^6 + \zeta^5 + \zeta^4 + \zeta^3 + \zeta^2 + \zeta)$. Because ζ is a zero of

$$\Phi_7(x) = x^6 + x^5 + x^4 + x^3 + x^2 + x + 1,$$

we see at once that $\alpha^2 + \alpha + 2 = 0$. The zeros of $x^2 + x + 2$ are $(-1 \pm i\sqrt{7})/2$, and we see that $\mathbb{Q}(\alpha) = \mathbb{Q}(i\sqrt{7})$.

Working in an analogous way for the subgroup $\{\sigma_1, \sigma_6\}$, we form the element $\beta = \zeta + \zeta^6$ which is left fixed by this subgroup. Computing, we find that

$$\begin{aligned} \beta^3 &= (\zeta + \zeta^6)^3 = \zeta^3 + 3\zeta + 3\zeta^6 + \zeta^4, \\ \beta^2 &= (\zeta + \zeta^6)^2 = \zeta^2 + 2 + \zeta^5, \\ \beta &= \zeta + \zeta^6. \end{aligned}$$

Recalling that $\Phi_7(\zeta) = 0$ as above, we find that $\beta^3 + \beta^2 - 2\beta - 1 = 0$. Thus β is a zero of $x^3 + x^2 - 2x + 1$ which is irreducible because it has no zero in $\mathbb{Z}$.

7. Now $x^8 - 1 = (x^4 + 1)(x^2 + 1)(x - 1)(x + 1)$. Example 54.7 shows that the splitting field of $x^4 + 1$ contains i, which is a zero of $x^2 + 1$. Thus the splitting field of $x^8 - 1$ is the same as the splitting field of $x^4 + 1$, whose group was completely described in Example 54.7. This is the "easiest way possible" for us to describe this group.

8. Using the quadratic formula to find α such that $\alpha^4 - 4\alpha^2 - 1 = 0$, we find that $\alpha^2 = (4 \pm \sqrt{20})/2 = 2 \pm \sqrt{5}$ so the possible values for α are $\pm\sqrt{2 + \sqrt{5}}$ and $\pm i\sqrt{\sqrt{5} - 2}$. We see that the splitting field of $x^4 - 4x^2 - 1 = 0$ can be generated by adjoining in succession $\sqrt{5}$, $\sqrt{\sqrt{5} + 2}$, and $\sqrt{2 - \sqrt{5}}$. Thus it has degree $2^3 = 8$ over $\mathbb{Q}$. It can obviously be generated by adjoining $\alpha_1 = \sqrt{\sqrt{5} + 2}$ and $\alpha_2 = i\sqrt{\sqrt{5} - 2}$. Let $K = \mathbb{Q}(\alpha_1, \alpha_2)$. The eight elements of $G(K/\mathbb{Q})$ are given by this table.

	ρ_0	ρ_1	ρ_2	ρ_3	μ_1	μ_2	δ_1	δ_2
$\alpha_1 \rightarrow$	α_1	α_2	$-\alpha_1$	$-\alpha_2$	α_2	$-\alpha_2$	$-\alpha_1$	α_1
$\alpha_2 \rightarrow$	α_2	$-\alpha_1$	$-\alpha_2$	α_1	α_1	$-\alpha_1$	α_2	$-\alpha_2$

The group is isomorphic to D_4 and the notation here is chosen to coincide with the notation used in Example 8.10. The subgroup diagram is identical with that in Fig. 54.6(a). Here is the subfield diagram.

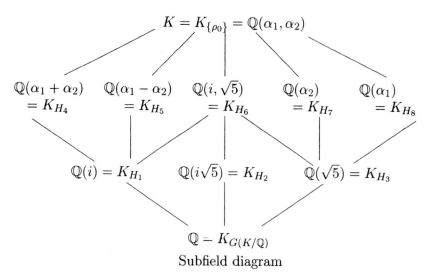

Subfield diagram

We now check most of this diagram. Note that α_1^2 is left fixed by $H_3 = \{\rho_0, \rho_2, \delta_1, \delta_2\}$ and that $\alpha_1^2 = \sqrt{5} + 2$, so the fixed field of H_3 is $\mathbb{Q}(\sqrt{5})$. Note also that $\alpha_1\alpha_2 = i$ is left fixed by $H_1 = \{\rho_0, \rho_2, \mu_1, \mu_2\}$ so the fixed field of H_1 is $\mathbb{Q}(i)$. Then $H_6 = H_1 \cap H_3$ leaves both i and $\sqrt{5}$ fixed. Also K_{H_2} must be the only remaining extension of $\mathbb{Q}$ of degree 2, and $\mathbb{Q}(i\sqrt{5})$ fits the bill.

The remaining fields are trivial to check because they are described in terms of α_1 and α_2. For example, to see that $\mathbb{Q}(\alpha_1)$ is the fixed field of H_8, we need only note that $\{\rho_0, \delta_2\} = H_8$ is the set of elements leaving α_1 fixed in the action table shown earlier.

9. We develop some formulas to use in this exercise and the next one. For simple notation, we denote symmetric expressions in the indeterminates y_1, y_2, y_3 by the notation $S(\text{formula})$ where the formula indicates the nature of *one* summand of the expression. Thus we write

$$\begin{aligned} s_1 &= y_1 + y_2 + y_3 = S(y_i), \\ s_2 &= y_1 y_2 + y_1 y_3 + y_2 y_3 = S(y_i y_j), \\ s_3 &= y_1 y_2 y_3 = S(y_i y_j y_k). \end{aligned}$$

Note that the subscripts i, j, k do not run independently through values from 1 to 3; we always have $i \neq j \neq k$. The formula simply indicates the nature of one, typical term in the symmetric expression.

a. We now express some other symmetric expression in terms of s_1, s_2, and s_3. Theorem 54.2 asserts that this is possible. If you write them out for subscript values from 1 to 3, you can see why they hold. Equation (1) is the answer to Part(**a**).

$$y_1^2 + y_2^2 + y_3^2 = S(y_i^2) = [S(y_i)]^2 - 2S(y_iy_j) = s_1^2 - 2s_2. \tag{1}$$

b. We have

$$S(y_i^2 y_j) = S(y_iy_j)S(y_i) - 3S(y_iy_jy_k) = s_1s_2 - 3s_3. \tag{2}$$

We can now get the answer to Part(**b**), using formula (2),

$$\frac{y_1}{y_2} + \frac{y_2}{y_1} + \frac{y_1}{y_3} + \frac{y_3}{y_1} + \frac{y_2}{y_3} + \frac{y_3}{y_2} = S\left(\frac{y_i}{y_j}\right) =$$

$$\frac{y_1^2 y_3 + y_2^2 y_3 + y_1^2 y_2 + y_3^2 y_2 + y_2^2 y_1 + y_3^2 y_1}{y_1 y_2 y_3} = \frac{S(y_i^2 y_j)}{y_1 y_2 y_3}$$

$$= \frac{s_1 s_2 - 3 s_3}{s_3}.$$

10. We have $x^3 - 4x^2 + 6x - 2 = (x - \alpha_1)(x - \alpha_2)(x - \alpha_3)$. Therefore the elementary symmetric expressions in α_1, α_2, and α_3 are given by

$$
\begin{aligned}
s_1 &= \alpha_1 + \alpha_2 + \alpha_3 = 4, \\
s_2 &= \alpha_1\alpha_2 + \alpha_1\alpha_3 + \alpha_2\alpha_3 = 6, \\
s_3 &= \alpha_1\alpha_2\alpha_3 = 2.
\end{aligned}
$$

We feel free to make use of the formulas (1) and (2) of the solution to the preceding exercise, and using the notation there, we also have the relation

$$S(y_i^2 y_j^2) = [S(y_iy_j)]^2 - 2S(y_i^2 y_j y_k) = s_2^2 - 2s_1s_3. \tag{3}$$

a. x - 4

b. Let $x^3 + b_2x^2 + b_1x + b_0 = (x - \alpha_1^2)(x - \alpha_2^2)(x - \alpha_3^2)$. Now $b_2 = -\alpha_1^2 - \alpha_2^2 - \alpha_3^2 = -(s_1^2 - 2s_2)$ by formula (1). Evaluating with the values of s_1, s_2, and s_3, we find that $b_2 = -(16 - 12) = -4$. Also $b_1 = \alpha_1^2\alpha_2^2 + \alpha_1^2\alpha_3^2 + \alpha_2^2\alpha_3^2 = s_2^2 - 2s_1s_3$ by formula (3). Evaluating, we find that $b_1 = 36 - 16 = 20$. Finally, $b_0 = -\alpha_1^2\alpha_2^2\alpha_3^2 = -s_3^2 = -4$. Thus the answer is

$$x^3 - 4x^2 + 20x - 4.$$

11. By Cayley's Theorem, every finite group G is isomorphic to a subgroup of S_n where n is the order of G. Now Theorem 54.2 shows that for each positive integer n, there exists a normal extension K of a field E such that $G(K/E) \simeq S_n$. If H is a subgroup of $G(K/E)$ isomorphic to G, then H is the Galois group of K over K_H, where $L = K_H$ is the fixed field of H. Thus H is isomorphic to G and is the Galois group $G(K/L)$ of K over L.

12. a. If $\Delta(f) = 0$, then $\alpha_i = \alpha_j$ for some $i \neq j$. Thus $\mathrm{irr}(\alpha_i, F) = \mathrm{irr}(\alpha_j, F)$. Because the irreducible factors of $f(x)$ are all separable and do not have zeros of multiplicity greater than 1, we see that $f(x)$ must have $\mathrm{irr}(\alpha_i, F)^2$ as a factor.

b. Clearly $[\Delta(f)]^2$ is a symmetric expression in the α_i, and hence left fixed by any permutation of the α_i, and thus is invariant under $G(K/F)$. Therefore $[\Delta(f)]^2$ is in F.

c. Consider the effect of a transposition (α_i, α_j) on $\Delta(f)$; it is no loss of generality to suppose $i < j$. The factor $\alpha_i - \alpha_j$ is carried into $\alpha_j - \alpha_i$, so it changes sign. For $k > j, \alpha_k - \alpha_j$ and $\alpha_k - \alpha_i$ for $k > j$ are carried into each other, so they do not contribute a sign change. The same is true of $\alpha_i - \alpha_k$ and $\alpha_j - \alpha_k$ for $k < i$. For $i < k < j$, the terms $\alpha_k - \alpha_i$ and $\alpha_j - \alpha_k$ are carried into $\alpha_k - \alpha_j = -(\alpha_j - \alpha_k)$ and into $\alpha_i - \alpha_k = -(\alpha_k - \alpha_i)$, so they contribute two sign changes. Thus the transposition contributes $1 + 2(j - i - 1)$ sign changes, which is an odd number, and carries $\Delta(f)$ into $-\Delta(f)$. Thus a permutation leaves $\Delta(f)$ fixed if and only if it can be expressed as a product of an even number of transpositions, that is, if and only if it is in $\overline{A}_n$. Hence $G(K/F) \subseteq \overline{A}_n$ if and only if it leaves $\Delta(f)$ fixed, that is, if and only if $\Delta(f) \in F$.

13. Let α and β be algebraic integers and let K be the splitting field of $\mathrm{irr}(\alpha, \mathbb{Q}) \cdot \mathrm{irr}(\beta, \mathbb{Q})$. Now

$$g(x) = \prod_{\sigma \in G(K/\mathbb{Q})} (x - \sigma(\alpha))$$

is a power of $\mathrm{irr}(\alpha, \mathbb{Q})$, and thus has integer coefficients and leading coefficient 1 because α is an algebraic integer. The same is true of

$$h(x) = \prod_{\mu \in G(K/\mathbb{Q})} (x - \mu(\beta)).$$

Now

$$
\begin{aligned}
k(x) &= \prod_{\sigma, \mu \in G(K/\mathbb{Q})} [x - (\sigma(\alpha) + \mu(\beta))] \\
&= \prod_{\sigma \in G(K/\mathbb{Q})} \left[\prod_{\mu \in G(K/\mathbb{Q})} [(x - \sigma(\alpha)) - \mu(\beta)] \right] \\
&= \prod_{\sigma \in G(K/\mathbb{Q})} h(x - \sigma(\alpha)).
\end{aligned}
$$

Because $h(x)$ has integer coefficients, $h(x - \sigma(\alpha))$ is a polynomial in $x - \sigma(\alpha)$ with integer coefficients. We can view $k(x)$ as a symmetric expression in α and its conjugates over the field $\mathbb{Q}$ involving only integers in $\mathbb{Q}$. By Theorem 54.2, the symmetric expression in α and its conjugates can be expressed as polynomials in the elementary symmetric functions of α and its conjugates, that is, in terms of the coefficients of $g(x)$ or their negatives. Thus $k(x)$ has integer coefficients. Now one zero of $k(x)$ is $\alpha + \beta$, corresponding to the factor where σ and μ are both the identity permutation. Thus $\mathrm{irr}(\alpha + \beta, \mathbb{Q})$ is a factor of the monic polynomial $k(x)$. Because a factorization in $\mathbb{Q}[x]$ can always be implemented in $\mathbb{Z}[x]$ by Theorem 23.11, we see that $\mathrm{irr}(\alpha + \beta, \mathbb{Q})$ is monic with integer coefficients, and hence $\alpha + \beta$ is an algebraic integer. If α is a zero of $f(x)$, then $-\alpha$ is a zero of $f(-x)$ which again has integer coefficients and is monic, so $-\alpha$ is again an algebraic integer.

One can argue that $\alpha\beta$ is an algebraic integer by the same technique that we used for $\alpha + \beta$, considering

$$\prod_{\sigma, \mu \in G(K/\mathbb{Q})} [x - \sigma(\alpha)\mu(\beta)] = \prod_{\sigma \in G(K/\mathbb{Q})} \left[\prod_{\mu \in G(K/\mathbb{Q})} [x - \sigma(\alpha)\mu(\beta)] \right].$$

Thus the algebraic integers are closed under addition and multiplication, and include additive inverses, and of course 0 which is a zero of x. Hence they form a subring of $\mathbb{C}$.

55. Cyclotomic Extensions

1. Let $\zeta = \cos \frac{\pi}{4} + i \sin \frac{\pi}{4} = \frac{1}{\sqrt{2}} + \frac{1}{\sqrt{2}}i$. Using the relations

$$\zeta^8 = 1, \quad \zeta + \zeta^7 = \sqrt{2}, \quad \text{and} \quad \zeta^3 + \zeta^5 = -\sqrt{2}$$

we have

$$
\begin{aligned}
\Phi(x) &= (x - \zeta)(x - \zeta^7)(x - \zeta^3)(x - \zeta^5) \\
&= [x^2 - (\zeta + \zeta^7)x + 1][x^2 - (\zeta^3 + \zeta^5)x + 1] \\
&= (x^2 - \sqrt{2}x + 1)(x^2 + \sqrt{2}x + 1) \\
&= x^4 + 1.
\end{aligned}
$$

2. The group is $G = \{1, 3, 7, 9, 11, 13, 17, 19\}$ under multiplication modulo 20. We find that $3^2 = 9, 3^3 = 7$, and $3^4 = 1$, so 3 and 7 have order 4, and 9 has order 2. Taking their additive inverses, we see that 17 and 13 have order 4, and 11 has order 2. Of course $19 = -1$ has order 2. Because G is abelian, it is isomorphic to $\mathbb{Z}_4 \times \mathbb{Z}_2$ by Theorem 11.12.

3. **a.** We have $60 = 2^2 \cdot 3 \cdot 5$, so $\varphi(60) = 2 \cdot 2 \cdot 4 = 16$.

   b. Now $1000 = 2^3 \cdot 5^3$, so $\varphi(1000) = 2^2 \cdot 5^2 \cdot 4 = 400$.

   c. Factoring, $8100 = 2^2 \cdot 3^4 \cdot 5^2$, so $\varphi(8100) = 2 \cdot 3^3 \cdot 5 \cdot 2 \cdot 4 = 2160$

4. We just use Theorem 55.8 with the Fermat primes 3, 5, and 17, repeatedly multiplying by 2 the numbers $3, 5, 15 = 3 \cdot 5, 17, 51 = 3 \cdot 17$, and $85 = 5 \cdot 17$ until we have generated the first 30 possibilites, which we list in five columns. (The next Fermat prime, 257, is not needed for the first 30 values of n).

3	12	30	60	102
4	15	32	64	120
5	16	34	68	128
6	17	40	80	136
8	20	48	85	160
10	24	51	96	170

5. Now 360 and 180 are divisible by 3^2, so the 360-gon and the 180-gon are not constructible. However, $360/3 = 120 = 8 \cdot 3 \cdot 5$ so the regular 120-gon is constructible, and therefore an angle of $3°$ is constructible.

6. **a.** We have $[K : \mathbb{Q}] = \varphi(12) = 4$ because the integers ≤ 12 and relatively prime to 12 are 1, 5, 7, and 11.

   b. The group $G(K/\mathbb{Q})$ is isomorphic to $\{1, 5, 7, 11\}$ under multiplication modulo 12, and $5^2, 7^2$, and 11^2 are all congruent to 1 modulo 12. Thus for all $\sigma \in G(K/\mathbb{Q})$, we must have $\sigma^2 = \iota$, the identity automorphism. The group is isomorphic to $\mathbb{Z}_2 \times \mathbb{Z}_2$ by Theorem 11.12.

7. We have $\Phi_3(x) = \frac{x^3-1}{x-1} = x^2 + x + 1$ for every field of characteristic $\neq 3$, because ζ and ζ^2 are both primitive cube roots of unity.

In $\mathbb{Z}_3[x]$, $x_8 - 1 = (x^4 + 1)(x^2 + 1)(x - 1)(x + 1)$ so the four primitive 8th roots of unity must be zeros of $x^4 + 1$, and $\Phi_8(x) = x^4 + 1 = (x^2 + x + 2)(x^2 + 2x + 2)$.

8. In $\mathbb{Z}_3[x]$, we have $x^6 - 1 = (x^2 - 1)^3 = (x - 1)^3(x + 1)^3$, so the polynomial already splits in $\mathbb{Z}_3$, and the splitting field has 3 elements.

9. T T F T T F T T F

10. The nth roots of unity form a cyclic group of order n under multiplication, for if ζ is a primitive nth root of unity, then
$$\zeta^j \zeta^k = \zeta^{j+k} = \zeta^{j+k \pmod n}.$$

Each element ζ^j of this group generates a subgroup of some order d dividing n, and is thus a primitive dth root of unity. Also, if d divides n, then $\zeta^{n/d}$ is a primitive dth root of unity. Thus the collection of all primitive dth roots of unity for d dividing n contains all the nth roots of unity. Because the primitive dth roots of unity are the zeros of $\Phi_d(x)$ in a field of characteristic not dividing d, we see that $x^n - 1 = \prod_{d|n} \Phi_d(x)$.

11. We have $\Phi_1(x) = x-1$, $\Phi_2(x) = x+1$, and $\Phi_3(x) = \frac{x^3+1}{x-1} = x^2+x+1$. Also, $x^4-1 = (x^2+1)(x-1)(x+1)$ so $\Phi_4(x) = x^2 + 1$, and $\Phi_5(x) = x^4 + x^3 + x^2 + x + 1$ (see Corollary 23.17). A primitive 6th root of unity is a primitive cube root of -1, and hence a zero of $x^3 + 1 = (x + 1)(x^2 - x + 1)$. Because there are two primitive 6th roots of unity, we see that $\Phi_6(x) = x^2 - x + 1$.

12. By Exercises 10 and 11,
$$\begin{aligned} x^{12} - 1 &= \Phi_1(x)\Phi_2(x)\Phi_3(x)\Phi_4(x)\Phi_6(x)\Phi_{12}(x) \\ &= (x - 1)(x + 1)(x^2 + x + 1)(x^2 + 1)(x^2 - x + 1)\Phi_{12}(x) \\ &= (x^4 - 1)(x^4 + x^2 + 1)\Phi_{12}(x) \\ &= (x^8 + x^6 - x^2 - 1)\Phi_{12}(x). \end{aligned}$$

Polynomial long division yields $\frac{x^{12}-1}{x^8+x^6-x^2-1} = x^4 - x^2 + 1$, so $\Phi_{12}(x) = x^4 - x^2 + 1$.

13. Let ζ be a primitive nth root of unity where n is odd. The positive powers of ζ which equal 1 are then $n, 2n, 3n, \cdots$. Because n is odd, $(-\zeta)^n = -1$. Consequently $(-\zeta)^{2n} = 1$, so the multiplicative order r of $-\zeta$ is either $2n$ or $< n$. If $r < n$, then $1 = [(-\zeta)^r]^2 = [(-\zeta)^2]^r = (\zeta^2)^r = \zeta^{2r}$, and we would then have to have $n = 2r$, contradicting the fact that n is odd. Therefore $r = 2n$, and $-\zeta$ is a primitive $2n$th root of unity.

We have shown that if ζ is a primitive nth root of unity for n odd, then $-\zeta$ is a primitive $2n$th root of unity. Now formula (1) in the text shows that for n odd, $\varphi(n) = \varphi(2n)$, so if $\zeta_1, \zeta_2, \cdots, \zeta_{\varphi(n)}$ are all the primitive nth roots of unity for n odd, then their negatives account for all the primitive $2n$th roots of unity. Now ζ is a zero of $\Phi_n(x)$ if and only $-\zeta$ is zero of $\Phi_n(-x)$, which, by Definition 55.2, shows that $\Phi_{2n}(x)$ must be either $\Phi_n(-x)$ or $-\Phi_n(-x)$, depending on whether the degree $\varphi(n)$ of $\Phi_n(x)$ is even or odd. But formula (1) in the text shows that if n is odd, then $\varphi(n)$ is even. Thus we have $\Phi_{2n}(x) = \Phi_n(-x)$.

14. Let ζ be a primitive mnth root of unity. Then ζ^m is a primitive nth root of unity and ζ^n is a primitive mth root of unity, so the splitting field of $x^{mn} - 1$ contains the splitting field of $(x^m - 1)(x^n - 1)$.

The splitting field of $(x^m - 1)(x^n - 1)$ contains ζ^n and ζ^m and thus contains $\zeta^m \zeta^n$. Now $\zeta^m \zeta^n = \zeta^{m+n}$ has order the least positive integer r such that mn divides $r(m+n)$. No prime dividing m divides $m + n$ because m and n are relatively prime. Similarly, no prime dividing n divides $m + n$. Consequently, mn must divide r, so $\zeta^m \zeta^n$ is a primitive mnth root of unity. Thus the splitting field of $(x^m - 1)(x^n - 1)$ contains a primitive mnth root of unity, and thus contains the splitting field of $x^{mn} - 1$. This completes the demonstration that the splitting fields of $x^{mn} - 1$ and of $(x^n - 1)(x^m - 1)$ are the same.

15. Let ζ be a primitive mnth root of unity, so that $K = \mathbb{Q}(\zeta)$ is the splitting field of $x^{mn} - 1$. Now form $F = \mathbb{Q}(\zeta^n)$ and $E = \mathbb{Q}(\zeta^m)$ as shown in the diagram.

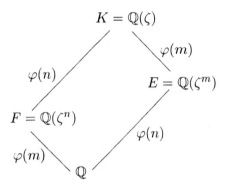

Now ζ^m is a primitive nth root of unity and ζ^n is a primitive mth root of unity. Thus $[F : \mathbb{Q}] = \varphi(m)$ and $[E : \mathbb{Q}] = \varphi(n)$ as labeled on the lower part of the diagram. Formula (1) for $\varphi(n)$ in the text shows that because m and n are relatively prime, $\varphi(mn) = \varphi(m)\varphi(n)$. Because $[K : \mathbb{Q}] = \varphi(mn)$, we see that $[K : F] = \varphi(n)$ and also that $[K : E] = \varphi(m)$ as labeled on the upper part of the diagram. Thus $G(K/F)$ is a subgroup of $G(K/\mathbb{Q})$ of order $\varphi(n)$ and $G(K/E)$ is a subgroup of order $\varphi(m)$.

We check the conditions of Exercise 50 of Section 11 to show that $G(K/\mathbb{Q}) \simeq G(K/F) \times G(K/E)$. Because $G(K/\mathbb{Q})$ is abelian (see Theorem 55.4), condition (b) holds. For condition (c), suppose that $\sigma \in G(K/\mathbb{Q})$ is in both $G(K/F)$ and $G(K/E)$. Then $\sigma(\zeta^m) = \zeta^m$ and $\sigma(\zeta^n) = \zeta^n$. Suppose that $\sigma(\zeta) = \zeta^r$. Then $\sigma(\zeta^m) = \zeta^{rm} = \zeta^m$ so $r \equiv 1 \pmod{n}$. Also $\sigma(\zeta^n) = \zeta^{rn} = \zeta^n$ so $r \equiv 1 \pmod{m}$. Because n and m are relatively prime, we see that $r \equiv 1 \pmod{mn}$, so $r = 1$ and σ is the identity automorphism. Thus $G(K/F) \cap G(K/E)$ consists of just the identity automorphism. To demonstrate condition (a) that $G(K/F) \vee G(K/E) = G(K/\mathbb{Q})$, form the $\varphi(m)\varphi(n)$ elements $\sigma\mu$ where $\sigma \in F(K/E)$ and $\mu \in G(K/F)$. We claim that these products are all distinct, so that they must comprise all of $G(K/\mathbb{Q})$. Suppose that $\sigma\mu = \sigma_1\mu_1$ for $\sigma, \sigma_1 \in G(K/E)$ and $\mu, \mu_1 \in G(K/F)$. Then $\sigma_1^{-1}\sigma = \mu_1\mu^{-1}$ is in both $G(K/E)$ and $G(K/F)$, and thus must be the identity automorphism. Therefore $\sigma = \sigma_1$ and $\mu = \mu_1$. Exercise 51 of Section 11 now shows that $G(K/\mathbb{Q}) \simeq F(K/F) \times G(K/E)$.

56. Insolvability of the Quintic

1. No, The splitting field E cannot be obtained by adjoining a square root of an element of $\mathbb{Z}_2$ to $\mathbb{Z}_2$ because all elements in $\mathbb{Z}_2$ are already squares. However, K is an extension by radicals, for $x^3 - 1 = (x - 1)(x^2 + x + 1)$ so K is also the splitting field of $x^3 - 1$. Thus $K = \mathbb{Z}_2(\zeta)$ where ζ is a primitive cube root of unity, and $\zeta^3 = 1$ is in $\mathbb{Z}_2$.

2. Yes, because if α is a zero of $f(x) = ax^8 + bx^6 + cx^4 + dx^2 + e$ then α^2 is a zero of $g(x) = ax^4 + bx^3 + cx^2 + dx + e$. Because $g(x)$ is a quartic, $F(\alpha^2)$ is an extension of F by radicals, and thus $F(\alpha)$ is an extension of F by radicals.

3. T T T F T F T F F T

4. We have

$$
\begin{aligned}
f(x) &= ax^2 + bx + c = a(x^2 + \frac{b}{a}x) + c \\
&= a(x + \frac{b}{2 \cdot a})^2 + c - \frac{b^2}{4 \cdot a} \text{ if } 2 \cdot a \neq 0.
\end{aligned}
$$

Thus if $\alpha \in \overline{F}$ satisfies

$$
a(\alpha + \frac{b}{2 \cdot a})^2 = \frac{b^2 - 4 \cdot ac}{4 \cdot a}
$$

so that

$$
\alpha + \frac{b}{2 \cdot a} = \pm\sqrt{\frac{b^2 - 4 \cdot ac}{4 \cdot a^2}} \text{ and } \alpha = \frac{-b \pm \sqrt{b^2 - 4 \cdot ac}}{2 \cdot a},
$$

then α is a zero of $ax^2 + bx + c$.

5. Let α be a zero of $ax^4 + bx^2 + c$. Then α^2 is a zero of $ax^2 + bx + c$ which is solvable by radicals by Exercise 4. If $\alpha_1 = \alpha, \alpha_2, \alpha_3$, and α_4 are the zeros of $ax^4 + bx^2 + c$, then the tower of field starting with F and adjoining in sequence $\alpha_1^2, \alpha_2^2, \alpha_3^2, \alpha_4^2, \alpha_1, \alpha_2, \alpha_3, \alpha_4$ is an extension where each successive field of the tower is either equal to the preceding field or is obtained from it by adjoining a square root of an element of the preceding field. Thus the splitting field is an extension by radicals, so the quartic is solvable.

6. We can achieve any refinement of a subnormal series by inserting, one at a time, a finite number of groups. Let $H_i < H_{i+1}$ be two adjacent terms of the series, so that H_{i+1}/H_i is abelian, and suppose that an additional subgroup K is inserted between them so that $H_i < K < H_{i+1}$. Then K/H_i is abelian because it can be regarded as a subgroup of H_{i+1}/H_i. By Theorem 34.7, H_{i+1}/K is isomorphic to $(H_{i+1}/H_i)/(K/H_i)$, which is the factor group of an abelian group, and hence is abelian. For an alternate argument, note that because H_{i+1}/H_i is abelian, H_i must contain the commutator subgroup of H_{i+1}, so K also contains this commutator subgroup and H_{i+1}/K is abelian.

7. Let $H_i < H_{i+1}$ be two adjacent groups in a subnormal series with solvable quotient groups, so that H_{i+1}/H_i is a solvable group. By definition, there exists a subnormal series

$$
H_i/H_i < K_1/H_i < K_2/H_i < \cdots < K_r/H_i < H_{i+1}/H_i
$$

with abelian quotient groups. We claim that the refinement of the original series at this ith level to

$$
K_0 = H_i < K_1 < K_2 < \cdots < K_r < H_{i+1} = K_{r+1}
$$

has abelian quotient groups at this level, for by Theorem 34.7, $K_j/K_{j+1} \simeq (K_j/H_i)/(K_{j-1}/H_i)$, which is abelian by our construction. Making such a refinement at each level of the given subnormal series, we obtain a subnormal series with abelian quotient groups.

8. a. The generalization of this to an n-cycle and a transposition in S_n is proved in the solution to Exercise 39 of Section 9.

b. Let K be the splitting field of the irreducible polynomial $f(x)$ of degree 5 over $\mathbb{Q}$. Because each element of $G(K/\mathbb{Q})$ corresponds to a permutation of the five zeros of $f(x)$ in K, and multiplication is function composition for both automorphisms and permutations, we can view $G(K/\mathbb{Q})$ as a subgroup of S_5. Now $|G(K/\mathbb{Q})|$ is divisible by 5, for a zero α of $f(x)$ generates $\mathbb{Q}(\alpha) < K$ of degree 5 over $\mathbb{Q}$, and degrees of towers are multiplicative. By Sylow theory, a group of order divisible by 5 contains an element of order 5, which we can view as a cycle of length 5 in S_5.

The automorphism σ of $\mathbb{C}$ where $\sigma(a + bi) = a - bi$ induces an automorphism of K, which must carry one complex root $a + bi$ of $f(x)$ into the other one $a - bi$ and leave the real roots of $f(x)$ fixed. Thus this automorphism of K is of order 2, so we can view it as a transposition in S_5.

c. We find that $f'(x) = 10x^4 - 20x^3 = x^3(10x - 20)$, so $f'(x) > 0$ where $x > 2$ or $x < 0$, and $f'(x) < 0$ for $0 < x < 2$. Because $f(-1) = -2, f(0) = 5$, and $f(2) = -11$, we see that $f(x)$ has one real zero between -1 and 0, one beween 0 and 2, and one greater than 2. These are all the real zeros because $f(x)$ increases for $x > 2$ and for $x < -1$. Thus $f(x)$ has exactly three real zeros and exactly two complex zeros so the group of the polynomial is isomorphic to S_5 and the polynomial is not solvable by radicals.

APPENDIX: Matrix Algebra

1. $\begin{bmatrix} 2 & 1 \\ 2 & 7 \end{bmatrix}$
2. $\begin{bmatrix} 4+i & -3+i & 1 \\ 7-i & 1+2i & 2-i \end{bmatrix}$
3. $\begin{bmatrix} -3+2i & -1-4i \\ 2 & -i \\ 0 & -i \end{bmatrix}$

4. $\begin{bmatrix} 3 & 1 \\ 5 & 15 \end{bmatrix}$
5. $\begin{bmatrix} 5 & 16 & -3 \\ 0 & -18 & 24 \end{bmatrix}$
6. Undefined

7. $\begin{bmatrix} 1 & -i \\ 4-6i & -2-2i \end{bmatrix}$
8. $\begin{bmatrix} 1 & -1 \\ 1 & 0 \end{bmatrix}^4 = \begin{bmatrix} 0 & -1 \\ 1 & -1 \end{bmatrix}^2 = \begin{bmatrix} -1 & 1 \\ -1 & 0 \end{bmatrix}$

9. $\begin{bmatrix} 1 & -i \\ i & 1 \end{bmatrix}^4 = \begin{bmatrix} 2 & -2i \\ 2i & 2 \end{bmatrix}^2 = \begin{bmatrix} 8 & -8i \\ 8i & 8 \end{bmatrix}$

10. $\begin{bmatrix} 1 & 0 \\ 1 & 1 \end{bmatrix}\begin{bmatrix} 0 & 1 \\ 2 & 1 \end{bmatrix} = \begin{bmatrix} 0 & 1 \\ 2 & 2 \end{bmatrix}$, $\begin{bmatrix} 0 & 1 \\ 2 & 1 \end{bmatrix}\begin{bmatrix} 1 & 0 \\ 1 & 1 \end{bmatrix} = \begin{bmatrix} 1 & 1 \\ 3 & 1 \end{bmatrix}$

11. $\begin{bmatrix} 0 & -1 \\ 1 & 0 \end{bmatrix}$
12. $\begin{bmatrix} 1/2 & 0 & 0 \\ 0 & 1/4 & 0 \\ 0 & 0 & -1 \end{bmatrix}$
13. $(3)(2)(-8) = -48$

14. Given that A^{-1} and B^{-1} exist, the associative property for matrix multiplication yields $(B^{-1}A^{-1})(AB) = B^{-1}(A^{-1}A)B = B^{-1}I_nB = B^{-1}B = I_n$ so AB is invertible. Similarly, we compute $(A^{-1}B^{-1})(BA) = A^{-1}(B^{-1}B)A = A^{-1}I_nA = A^{-1}A = I_n$ so BA is invertible also.